I0057531

Robert W. Talbot (Ed.)

Atmospheric Mercury

MDPI

This book is a reprint of the Special Issue that appeared in the online, open access journal, *Atmosphere* (ISSN 2073-4433) from 2013–2014, available at:

http://www.mdpi.com/journal/atmosphere/special_issues/atmospheric-mercury

Guest Editor
Robert W. Talbot
Institute for Climate and Atmospheric Science
Department of Earth & Atmospheric Sciences
University of Houston
USA

Editorial Office
MDPI AG
St. Alban-Anlage 66
Basel, Switzerland

Publisher
Shu-Kun Lin

Managing Editor
Lucy Lu

1. Edition 2016

MDPI • Basel • Beijing • Wuhan • Barcelona • Belgrade

ISBN 978-3-03842-290-7 (Hbk)
ISBN 978-3-03842-291-4 (PDF)

Articles in this volume are Open Access and distributed under the Creative Commons Attribution license (CC BY), which allows users to download, copy and build upon published articles even for commercial purposes, as long as the author and publisher are properly credited, which ensures maximum dissemination and a wider impact of our publications. The book taken as a whole is © 2016 MDPI, Basel, Switzerland, distributed under the terms and conditions of the Creative Commons by Attribution (CC BY-NC-ND) license (http://creativecommons.org/licenses/by-nc-nd/4.0/).

Table of Contents

List of Contributors.. VII

About the Guest Editor... XIII

Preface to "Atmospheric Mercury" XV

Matthew T. Parsons, Daniel McLennan, Monique Lapalme, Curtis Mooney, Corinna Watt and Rachel Mintz
Total Gaseous Mercury Concentration Measurements at Fort McMurray, Alberta, Canada
Reprinted from: *Atmosphere* **2013**, 4(4), 472–493
http://www.mdpi.com/2073-4433/4/4/472..1

Xin Lan, Robert Talbot, Patrick Laine, Barry Lefer, James Flynn and Azucena Torres
Seasonal and Diurnal Variations of Total Gaseous Mercury in Urban Houston, TX, USA
Reprinted from: *Atmosphere* **2014**, 5(2), 399–419
http://www.mdpi.com/2073-4433/5/2/399..25

Xiaohong Xu, Umme Akhtar, Kyle Clark and Xiaobin Wang
Temporal Variability of Atmospheric Total Gaseous Mercury in Windsor, ON, Canada
Reprinted from: *Atmosphere* **2014**, 5(3), 536–556
http://www.mdpi.com/2073-4433/5/3/536..48

Amanda S. Cole, Alexandra Steffen, Chris S. Eckley, Julie Narayan, Martin Pilote, Rob Tordon, Jennifer A. Graydon, Vincent L. St. Louis, Xiaohong Xu and Brian A. Branfireun
A Survey of Mercury in Air and Precipitation across Canada: Patterns and Trends
Reprinted from: *Atmosphere* **2014**, 5(3), 635–668
http://www.mdpi.com/2073-4433/5/3/635..73

Gang S. Lee, Pyung R. Kim, Young J. Han, Thomas M. Holsen and Seung H. Lee
Tracing Sources of Total Gaseous Mercury to Yongheung Island off the Coast
of Korea
Reprinted from: *Atmosphere* **2014**, *5*(2), 273–291
http://www.mdpi.com/2073-4433/5/2/273..109

Xinrong Ren, Winston T. Luke, Paul Kelley, Mark Cohen, Fong Ngan,
Richard Artz, Jake Walker, Steve Brooks, Christopher Moore,
Phil Swartzendruber, Dieter Bauer, James Remeika, Anthony Hynes, Jack Dibb,
John Rolison, Nishanth Krishnamurthy, William M. Landing,
Arsineh Hecobian, Jeffery Shook and L. Greg Huey
Mercury Speciation at a Coastal Site in the Northern Gulf of Mexico: Results from
the Grand Bay Intensive Studies in Summer 2010 and Spring 2011
Reprinted from: *Atmosphere* **2014**, *5*(2), 230–251
http://www.mdpi.com/2073-4433/5/2/230..130

Cheryl Tatum Ernest, Deanna Donohoue, Dieter Bauer, Arnout Ter Schure and
Anthony J. Hynes
Programmable Thermal Dissociation of Reactive Gaseous Mercury, a Potential
Approach to Chemical Speciation: Results from a Field Study
Reprinted from: *Atmosphere* **2014**, *5*(3), 575–596
http://www.mdpi.com/2073-4433/5/3/575..155

Jesse O. Bash, Annmarie G. Carlton, William T. Hutzell and O. Russell Bullock Jr.
Regional Air Quality Model Application of the Aqueous-Phase Photo Reduction
of Atmospheric Oxidized Mercury by Dicarboxylic Acids
Reprinted from: *Atmosphere* **2014**, *5*(1), 1–15
http://www.mdpi.com/2073-4433/5/1/1..179

Yanxu Zhang and Lyatt Jaeglé
Decreases in Mercury Wet Deposition over the United States during 2004–2010:
Roles of Domestic and Global Background Emission Reductions
Reprinted from: *Atmosphere* **2013**, *4*(2), 113–131
http://www.mdpi.com/2073-4433/4/2/113..196

Steve Brooks, Xinrong Ren, Mark Cohen, Winston T. Luke, Paul Kelley, Richard Artz, Anthony Hynes, William Landing and Borja Martos
Airborne Vertical Profiling of Mercury Speciation near Tullahoma, TN, USA
Reprinted from: *Atmosphere* **2014**, *5*(3), 557–574
http://www.mdpi.com/2073-4433/5/3/557...219

Franz Slemr, Andreas Weigelt, Ralf Ebinghaus, Carl Brenninkmeijer, Angela Baker, Tanja Schuck, Armin Rauthe-Schöch, Hella Riede, Emma Leedham, Markus Hermann, Peter van Velthoven, David Oram, Debbie O'Sullivan, Christoph Dyroff, Andreas Zahn and Helmut Ziereis
Mercury Plumes in the Global Upper Troposphere Observed during Flights with the CARIBIC Observatory from May 2005 until June 2013
Reprinted from: *Atmosphere* **2014**, *5*(2), 342–369
http://www.mdpi.com/2073-4433/5/2/342...240

Peter Rafaj, Janusz Cofala, Jeroen Kuenen, Artur Wyrwa and Janusz Zyśk
Benefits of European Climate Policies for Mercury Air Pollution
Reprinted from: *Atmosphere* **2014**, *5*(1), 45–59
http://www.mdpi.com/2073-4433/5/1/45...272

List of Contributors

Umme Akhtar Current Affiliation: Department of Chemical Engineering, University of Toronto, Toronto, ON M5S 3E5, Canada; Department of Civil and Environmental Engineering, University of Windsor, 401 Sunset Ave, Windsor, ON N9B 3P4, Canada.

Richard Artz Air Resources Laboratory, National Oceanic and Atmospheric Administration, 5830 University Research Court, College Park, MD 20740, USA.

Angela Baker Atmospheric Chemistry Division, Max-Planck-Institut für Chemie (MPI), Hahn-Meitner-Weg 1, D-55128 Mainz, Germany.

Jesse O. Bash National Exposure Research Laboratory, US Environmental Protection Agency, Research Triangle Park, NC 27711, USA.

Dieter Bauer State Key Laboratory of Information Engineering in Survey, Mapping and Remote Sensing, Wuhan University, Wuhan 430079, China; Rosenstiel School of Marine and Atmospheric Science, University of Miami, 4600 Rickenbacker Causeway, Miami, FL 33149, USA.

Brian A. Branfireun Department of Biology and Centre for Environment and Sustainability, University of Western Ontario, 1151 Richmond Street, London, ON N6A 5B7, Canada.

Carl Brenninkmeijer Atmospheric Chemistry Division, Max-Planck-Institut für Chemie (MPI), Hahn-Meitner-Weg 1, D-55128 Mainz, Germany.

Steve Brooks Air Resources Laboratory, National Oceanic and Atmospheric Administration, 5830 University Research Court, College Park, MD 20740, USA; Department of Mechanical, Aerospace and Biomedical Engineering, University of Tennessee Space Institute, 411 BH Goethert Parkway, Tullahoma, TN 37388, USA.

O. Russell Bullock Jr. National Exposure Research Laboratory, US Environmental Protection Agency, Research Triangle Park, NC 27711, USA.

Annmarie G. Carlton Department of Environmental Sciences, Rutgers University, New Brunswick, NJ 08903, USA.

Kyle Clark Department of Civil and Environmental Engineering, University of Windsor, 401 Sunset Ave, Windsor, ON N9B 3P4, Canada.

Janusz Cofala International Institute for Applied Systems Analysis (IIASA), A-2361 Laxenburg, Austria.

Mark Cohen Air Resources Laboratory, National Oceanic and Atmospheric Administration, 5830 University Research Court, College Park, MD 20740, USA.

Amanda S. Cole Air Quality Processes Research, Science & Technology Branch, Environment Canada, 4905 Dufferin St., Toronto, ON M3H 5T4, Canada.

Jack Dibb Earth Systems Research Center, Institute for the Study of Earth, Oceans and Space, University of New Hampshire, 8 College Road, Durham, NH 03824, USA.

Deanna Donohoue State Key Laboratory of Information Engineering in Survey, Mapping and Remote Sensing, Wuhan University, Wuhan 430079, China; Interdisciplinary Graduate School of Engineering Sciences, Kyushu University, Fukuoka 8168580, Japan.

Christoph Dyroff Institute of Meteorology and Climate Research, Karlsruhe Institute of Technology (KIT), Hermann-von-Helmholtz-Platz 1, D-76344 Eggenstein-Leopoldshafen, Germany.

Ralf Ebinghaus Institute of Coastal Research, Helmholtz-Zentrum Geesthacht (HZG), Max-Planck-Straße 1, D-21502 Geesthacht, Germany.

Chris S. Eckley Environmental Protection Agency, Region 10, 1200 6th Ave, Seattle, WA 98101, USA.

Cheryl Tatum Ernest State Key Laboratory of Information Engineering in Survey, Mapping and Remote Sensing, Wuhan University, Wuhan 430079, China; Key Laboratory of Research and Development of Chinese Medicine of Zhejiang Province, Zhejiang Academy of Traditional Chinese Medicine, 132 Tianmushan Road, Hangzhou 310007, China.

James Flynn Institute for Climate and Atmospheric Sciences, Department of Earth & Atmospheric Sciences, University of Houston, Houston, TX 77204, USA.

Jennifer A. Graydon Department of Biological Sciences, University of Alberta, Edmonton, AB T6G 2E9, Canada.

Young J. Han Department of Environmental Science, Kangwon National University, 192-1, Hyoja-2-dong, Chuncheon, 200-701 Gangwon-do, Korea.

Arsineh Hecobian School of Earth and Atmospheric Science, Georgia Institute of Technology, 311 Ferst Drive, Atlanta, GA 30332, USA; Current Affiliation: Department of Atmospheric Science, Colorado State University, 200 West Lake Street, Fort Collins, CO 80523, USA.

Markus Hermann Leibniz-Institut für Troposphärenforschung (TROPOS), Permoserstrasse 15, D-04318 Leipzig, Germany.

Thomas M. Holsen Department of Civil and Environmental Engineering, Clarkson University, Potsdam, NY 13699, USA.

L. Greg Huey School of Earth and Atmospheric Science, Georgia Institute of Technology, 311 Ferst Drive, Atlanta, GA 30332, USA.

William T. Hutzell National Exposure Research Laboratory, US Environmental Protection Agency, Research Triangle Park, NC 27711, USA.

Anthony J. Hynes State Key Laboratory of Information Engineering in Survey, Mapping and Remote Sensing, Wuhan University, Wuhan 430079, China; Rosenstiel School of Marine and Atmospheric Science, University of Miami, 4600 Rickenbacker Causeway, Miami, FL 33149, USA.

Lyatt Jaeglé Department of Atmospheric Sciences, University of Washington, Seattle, WA 98195, USA.

Paul Kelley Cooperative Institute for Climate and Satellites, and Air Resources Laboratory, National Oceanic and Atmospheric Administration, University of Maryland, 5830 University Research Court, College Park, MD 20740, USA.

Pyung R. Kim Department of Environmental Science, Kangwon National University, 192-1, Hyoja-2-dong, Chuncheon, 200-701 Gangwon-do, Korea.

Nishanth Krishnamurthy Department of Earth, Ocean, and Atmospheric Science, Florida State University, 117 North Woodward Avenue, Tallahassee, FL 32306, USA.

Jeroen Kuenen Climate, Air and Sustainability, TNO, Utrecht 3584 CB, The Netherlands.

Patrick Laine Portnoy Environmental Incorporation, Houston, TX 77043, USA.

Xin Lan Institute for Climate and Atmospheric Sciences, Department of Earth & Atmospheric Sciences, University of Houston, Houston, TX 77204, USA.

William M. Landing Department of Earth, Ocean, and Atmospheric Science, Florida State University, 117 North Woodward Avenue, Tallahassee, FL 32306, USA.

Monique Lapalme Meteorological Service of Canada, Environment Canada, 9250 49 St NW, Edmonton, AB T6B 1K5, Canada.

Gang S. Lee Department of Environmental Science, Kangwon National University, 192-1, Hyoja-2-dong, Chuncheon, 200-701 Gangwon-do, Korea.

Seung H. Lee Department of Environmental & Energy Engineering, Anyang University, 22 Samdeok-ro, Manan-gu, Anyang, 430-714 Gyeonggi-do, Korea.

Emma Leedham Atmospheric Chemistry Division, Max-Planck-Institut für Chemie (MPI), Hahn-Meitner-Weg 1, D-55128 Mainz, Germany.

Barry Lefer Institute for Climate and Atmospheric Sciences, Department of Earth & Atmospheric Sciences, University of Houston, Houston, TX 77204, USA.

Winston T. Luke Air Resources Laboratory, National Oceanic and Atmospheric Administration, 5830 University Research Court, College Park, MD 20740, USA.

Borja Martos Department of Mechanical, Aerospace and Biomedical Engineering, University of Tennessee Space Institute, 411 BH Goethert Parkway, Tullahoma, TN 37388, USA.

Daniel McLennan Meteorological Service of Canada, Environment Canada, 9250 49 St NW, Edmonton, AB T6B 1K5, Canada.

Rachel Mintz Meteorological Service of Canada, Environment Canada, 9250 49 St NW, Edmonton, AB T6B 1K5, Canada.

Curtis Mooney Meteorological Service of Canada, Environment Canada, 9250 49 St NW, Edmonton, AB T6B 1K5, Canada.

Christopher Moore Division of Atmospheric Sciences, Desert Research Institute, 2215 Raggio Parkway, NV 89512, USA.

Julie Narayan Air Quality Processes Research, Science & Technology Branch, Environment Canada, 4905 Dufferin St., Toronto, ON M3H 5T4, Canada.

Fong Ngan Air Resources Laboratory, National Oceanic and Atmospheric Administration, 5830 University Research Court, College Park, MD 20740, USA; Cooperative Institute for Climate and Satellites, University of Maryland, 5825 University Research Court, College Park, MD 20740, USA.

David Oram National Centre for Atmospheric Science, University of East Anglia (UEA), Norwich NR4 7TJ, UK.

Debbie O'Sullivan National Centre for Atmospheric Science, University of East Anglia (UEA), Norwich NR4 7TJ, UK; Current Affiliation: Meteorological Office, Exeter, EX1 3PB, UK.

Matthew T. Parsons Meteorological Service of Canada, Environment Canada, 9250 49 St NW, Edmonton, AB T6B 1K5, Canada.

Martin Pilote Aquatic Contaminants Research Division, Environment Canada, 105 McGill, Montreal, QC H2Y 2E7, Canada.

Peter Rafaj International Institute for Applied Systems Analysis (IIASA), A-2361 Laxenburg, Austria.

Armin Rauthe-Schöch Atmospheric Chemistry Division, Max-Planck-Institut für Chemie (MPI), Hahn-Meitner-Weg 1, D-55128 Mainz, Germany.

James Remeika Rosenstiel School of Marine and Atmospheric Science, University of Miami, 4600 Rickenbacker Causeway, Miami, FL 33149, USA.

Xinrong Ren Air Resources Laboratory, National Oceanic and Atmospheric Administration, and Cooperative Institute for Climate and Satellites, University of Maryland, 5825 University Research Court, College Park, MD 20740, USA; Department of Earth, Ocean, and Atmospheric Science, Florida State University, 117 North Woodward Avenue, Tallahassee, FL 32306, USA.

Hella Riede Atmospheric Chemistry Division, Max-Planck-Institut für Chemie (MPI), Hahn-Meitner-Weg 1, D-55128 Mainz, Germany.

John Rolison Department of Earth, Ocean, and Atmospheric Science, Florida State University, 117 North Woodward Avenue, Tallahassee, FL 32306, USA; Current Affiliation: Department of Chemistry, Otago University, Dunedin 9016, New Zealand.

Tanja Schuck Atmospheric Chemistry Division, Max-Planck-Institut für Chemie (MPI), Hahn-Meitner-Weg 1, D-55128 Mainz, Germany; Current Affiliation: NRW State Agency for Nature, Environment and Consumer Protection, Recklinghausen, Germany.

Arnout Ter Schure Department of Physiology and Pharmacology, Sapienza— University of Rome, P.le Aldo Moro 5, 00185 Rome, Italy.

Jeffery Shook School of Earth and Atmospheric Science, Georgia Institute of Technology, 311 Ferst Drive, Atlanta, GA 30332, USA; Current Affiliation: Talcott Mountain Science Center, 324 Montevideo Road, Avon, CT 06001, USA.

Franz Slemr Atmospheric Chemistry Division, Max-Planck-Institut für Chemie (MPI), Hahn-Meitner-Weg 1, D-55128 Mainz, Germany.

Vincent L. St. Louis Department of Biological Sciences, University of Alberta, Edmonton, AB T6G 2E9, Canada.

Alexandra Steffen Air Quality Processes Research, Science & Technology Branch, Environment Canada, 4905 Dufferin St., Toronto, ON M3H 5T4, Canada.

Phil Swartzendruber Rosenstiel School of Marine and Atmospheric Science, University of Miami, 4600 Rickenbacker Causeway, Miami, FL 33149, USA; Current Affiliation: Puget Sound Clean Air Agency, 1904 Third Avenue, Seattle, WA 98101, USA.

Robert Talbot Institute for Climate and Atmospheric Sciences, Department of Earth & Atmospheric Sciences, University of Houston, Houston, TX 77204, USA.

Rob Tordon Environment Canada, 45 Alderney Drive, Dartmouth, NS B2Y 2N6, Canada.

Azucena Torres Institute for Climate and Atmospheric Sciences, Department of Earth & Atmospheric Sciences, University of Houston, Houston, TX 77204, USA.

Peter van Velthoven Royal Netherlands Meteorological Institute (KNMI), P.O. Box 201, NL-3730 AE De Bilt, The Netherlands.

Jake Walker Grand Bay National Estuarine Research Reserve, 6005 Bayou Heron Road, Moss Point, MS 39562, USA.

Xiaobin Wang Department of Civil and Environmental Engineering, University of Windsor, 401 Sunset Ave, Windsor, ON N9B 3P4, Canada.

Corinna Watt Meteorological Service of Canada, Environment Canada, 9250 49 St NW, Edmonton, AB T6B 1K5, Canada.

Andreas Weigelt Institute of Coastal Research, Helmholtz-Zentrum Geesthacht (HZG), Max-Planck-Straße 1, D-21502 Geesthacht, Germany.

Artur Wyrwa AGH University of Science and Technology, 30-059 Krakow, Poland.

Xiaohong Xu Department of Civil and Environmental Engineering, University of Windsor, 401 Sunset Ave., Windsor, ON N9B 3P4, Canada.

Andreas Zahn Institute of Meteorology and Climate Research, Karlsruhe Institute of Technology (KIT), Hermann-von-Helmholtz-Platz 1, D-76344 Eggenstein-Leopoldshafen, Germany.

Yanxu Zhang Department of Atmospheric Sciences, University of Washington, Seattle, WA 98195, USA.

Helmut Ziereis Institut für Physik der Atmosphäre, Deutsches Zentrum für Luft- und Raumfahrt (DLR), D-82230 Wessling, Germany.

Janusz Zyśk AGH University of Science and Technology, 30-059 Krakow, Poland.

About the Guest Editor

Robert Talbot's interests encompass regional to global scale atmospheric circulations and associated transport of trace constituents. He is Professor of Atmospheric Chemistry and Director of the Institute for Climate and Atmospheric Science (ICAS) at the University of Houston. He has been lead author or co-author of more than 350 papers with more than 20 focused on various aspects of atmospheric mercury. He was the first to discover that in the upper troposphere–lower stratosphere, elemental mercury is depleted to near zero in its mixing ratio. The reason(s) for this are still unresolved. He has studied mercury in continental and marine environments both in New England and along the Gulf Coast of the U.S.

Preface to "Atmospheric Mercury"

This book provides a brief introduction to atmospheric mercury. It is a relatively new topic in atmospheric chemistry. These papers give a glimpse of various aspects of the subject but are in no way a comprehensive look at atmospheric mercury. At first, atmospheric mercury seemed to be a clear-cut subject, but now we realize it is extremely complex and largely misunderstood. Improvements need to be realized on both the measurement and modeling sides of the subject. The papers in this book provide information on urban and rural environments in which mercury cycling is quite different. This subject has now reached global importance owing to the Minamata Convention on Mercury, which is prompting evaluation of how scientific knowledge can contribute to its implementation and effectiveness.

Robert W. Talbot
Guest Editor

Total Gaseous Mercury Concentration Measurements at Fort McMurray, Alberta, Canada

Matthew T. Parsons, Daniel McLennan, Monique Lapalme, Curtis Mooney, Corinna Watt and Rachel Mintz

Abstract: Observations are described from total gaseous mercury (TGM) concentrations measured at the Wood Buffalo Environmental Association (WBEA) Fort McMurray—Patricia McInnes air quality monitoring station—from 21 October 2010 through 31 May 2013, inclusively. Fort McMurray is approximately 380 km north-northeast of Edmonton, Alberta, and approximately 30 km south of major Canadian oil sands developments. The average TGM concentration over the period of this study was 1.45 ± 0.18 ng·m^{-3}. Principal component analysis suggests that observed TGM concentrations are correlated with meteorological conditions including temperature, relative humidity, and solar radiation, and also ozone concentration. There is no significant correlation between ambient concentrations of TGM and anthropogenic pollutants, such as nitrogen oxides (NO$_X$) and sulphur dioxide (SO$_2$). Principal component analysis also shows that the highest TGM concentrations observed are a result of forest fire smoke near the monitoring station. Back trajectory analysis highlights the importance of long-range transport, indicating that unseasonably high TGM concentrations are generally associated with air from the southeast and west, while unseasonably low TGM concentrations are a result of arctic air moving over the monitoring station. In general, TGM concentration appears to be driven by diel and seasonal trends superimposed over a combination of long-range transport and regional surface-air flux of gaseous mercury.

Reprinted from *Atmosphere*. Cite as: Parsons, M.T.; McLennan, D.; Lapalme, M.; Mooney, C.; Watt, C.; Mintz, R. Total Gaseous Mercury Concentration Measurements at Fort McMurray, Alberta, Canada. *Atmosphere* **2013**, *4*, 472–493.

1. Introduction

Total gaseous mercury (TGM) is ubiquitous in the atmosphere, persisting for up to 1.5 years, thus making mercury a pollutant of global concern due to potential long-range transport [1]. Mercury can exist in several forms in the atmosphere. The bulk (95–97%) of atmospheric mercury exists as gaseous elemental mercury (GEM), while the remainder consists of gaseous oxidized mercury (GOM) (also referred to as reactive gaseous mercury), and particulate bound mercury (PBM) [2–4]. TGM consists of both GEM and GOM, with typical ambient TGM concentrations in

1

the range of 1.3–1.7 ng·m^{-3} in the northern hemisphere [1,5,6]; within the province of Alberta in Canada, mean hourly TGM concentrations have been measured in the range of 1.36–1.65 ng·m^{-3} [7,8]. Each form of mercury varies in its removal efficiencies via wet or dry deposition to surfaces where it can undergo further reactions, such as methylation to enhance toxicity and undergo bioaccumulation within the ecosystem [1,9]. A unique feature of mercury is its ability to re-emit following deposition, effectively increasing the atmospheric lifetime and global distribution of mercury. Thus, mercury can be found—to varying degrees—in all ecosystems, even in locations greatly removed from any major sources.

Atmospheric mercury has many anthropogenic sources including combustion of coal and other fuels (especially those with elevated mercury content) and mining activities; all of which are commonplace in and around the Canadian oil sands region of northern Alberta. Industries within the Regional Municipality of Wood Buffalo, which contains the urban center of Fort McMurray and the Canadian oil sands development, were reported as releasing 51 kg of industrial mercury emissions to the air in 2011, which is approximately 2% of industrial mercury air emissions for all of Canada in 2011 [10]. Likewise, industries in the region were also reported as releasing 6% and 8% of industrial NO$_X$ and SO$_2$ emissions, respectively, for all of Canada [10]. In an attempt to better understand the effects of oil, gas, and bitumen extraction in the Canadian oil sands on atmospheric mercury concentrations, Environment Canada has operated—and continues to operate—a continuous ambient TGM analyzer at the Wood Buffalo Environmental Association (WBEA) Fort McMurray—Patricia McInnes air quality monitoring station (Patricia McInnes station herein). The analyzer measures ambient concentration of TGM in the community of Fort McMurray, Alberta, the urban hub of the Canadian oil sands region, approximately 30 km south of the nearest oil sands production facility (see Section 3.1 below for a site description and map of the area). No previous studies on ambient atmospheric mercury have been conducted in the Canadian oil sands region, despite the concentration of mining activity and other potential mercury sources present in this internationally scrutinized region subject to significant industrial and environmental regulations [11].

Several communities and groups are concerned that ecosystem-wide mercury concentrations are on the rise in the region, affecting local water and food sources, and exacerbated by the steady increase in industrial development in the area [12]. For example, Kelly *et al.* [13] measured greater mercury concentrations in water and snow near oil sands developments compared to locations further removed from developments. On the other hand, Wiklund *et al.* [14] note that, since the 1990s, increasing Canadian oil sands industrial development does not coincide with the decreasing trend in mercury concentrations in sediment cores downstream of the oil sands region. Thus, uncertainty remains regarding mercury sources, transport, and fate in the region, and it is important to consider the role of atmospheric mercury in these processes. In an effort to enhance pollutant monitoring over all aspects of the ecosystem in this region, the Canadian and Alberta governments initiated the Joint Canada-Alberta Implementation Plan for Oil Sands Monitoring [15]; the ambient atmospheric mercury measurement programme described here forms one piece of the air monitoring component of this plan. The objective of this study is to better understand the underlying factors driving atmospheric mercury concentrations in the oil sands region, which may, in turn, help to address knowledge gaps regarding the fate of mercury in the region.

2. Results and Discussion

2.1. General Trends and Statistics

Figure 1 shows TGM concentrations measured at Patricia McInnes station from 21 October 2010 through 31 May 2013, inclusively. Summary statistics for these data are also listed in Table 1. The average TGM concentration at Patricia McInnes station over the course of this study was 1.45 ± 0.18 ng·m^{-3}, which is comparable to that measured at other stations in the province of Alberta (1.36–1.65 ng·m^{-3} [7,8]). The gap in data from 21 May 2011 through 20 October 2011 was due to instrument contamination as noted in Section 3.3 below. There is a slight increasing trend in TGM concentrations measured at the Patricia McInnes station with a rate of 0.051 ± 0.003 ng·m^{-3}·y^{-1} over the range of the study period as determined with linear regression. Note that at approximately 2.5 years in length, this period of record is not sufficiently long to definitively regard this as a true long-term trend, and this trend reported may only represent a medium-term fluctuation within a different longer-term trend. As seen in Figure 1, there are several instances of elevated TGM concentrations. Most of these events coincide with forest fire smoke near the Patricia McInnes station as indicated by the shaded areas in Figure 1. Conversely, not all cases where forest fire smoke was present led to increased TGM concentrations. This may be due to the distance between the forest fire and the monitoring station, *i.e.*, the age of the smoke. Data impacted by forest fire smoke is included in analyses described below except where noted. Figure 1 indicates that, aside from periods impacted by forest fire smoke, high TGM concentrations are generally observed over short time scales, whereas low TGM concentrations are generally observed over longer time scales. Table 1 shows the statistics for the data deemed to be measured during forest fire smoke events (N = 447) in relation to the data that was not impacted by forest fire smoke (N = 17,020).

Figure 1 also shows ambient TGM concentrations to have a seasonal dependence. This is further illustrated in Figure 2, which shows the monthly averaged profile of TGM concentrations measured at Patricia McInnes station. Figure 2 shows a maximum in the spring and a minimum in the fall for data not impacted by forest fire smoke (black data), a pattern that has been observed in other similar locations and linked to meteorological cycles [8,16,17]. Forest fire smoke was only present during the months of May through September, inclusively, as shown in Figure 2 (red data). In general, mean monthly TGM is significantly higher in these months ($p < 0.0001$; d.f. ≥ 654) when forest fire smoke was present than when it was not, with the exception of September ($p = 0.16$; d.f. = 423), according to t-tests of monthly mean values.

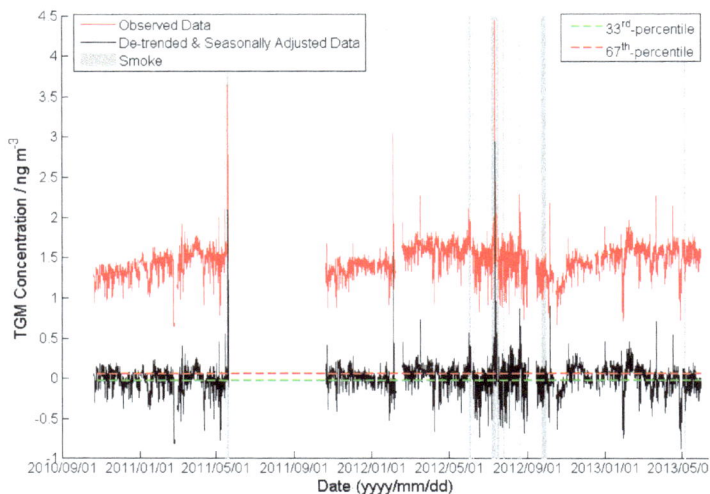

Figure 1. Time series of hourly total gaseous mercury (TGM) concentrations at Patricia McInnes station. Grey shaded areas indicate when forest fire smoke was in the vicinity of the monitoring station. The de-trended/seasonally adjusted TGM time series is shown for reference. As described in the text, TGM concentration data was categorized as "high", "average", or "low", with divisions at the 33rd- and 67th-percentiles in de-trended/seasonally adjusted TGM concentration.

Table 1. Summary statistics for Patricia McInnes station hourly TGM concentration data.

Dataset	Date Range	N	Mean ($ng \cdot m^{-3}$)	Median ($ng \cdot m^{-3}$)	Standard Deviation ($ng \cdot m^{-3}$)	Minimum ($ng \cdot m^{-3}$)	Maximum ($ng \cdot m^{-3}$)
All data	21 October	17467	1.45	1.46	0.18	0.64	4.43
Excluding smoke	2010–31	17020	1.45	1.46	0.17	0.64	3.05
Only smoke	May 2013	447	1.73	1.69	0.34	1.08	4.43

Figure 3 shows the diel profile of mean TGM concentrations for each season (spring: 21 March–20 June; summer: 21 June–20 September; fall: 21 September–20 December; winter: 21 December–20 March). As observed in other studies [16,18], minimum and maximum TGM concentrations are observed in the morning and in the afternoon-evening, respectively. Seasonal diel profiles of TGM concentrations qualitatively resemble those of the mean temperatures and ozone concentrations at Patricia McInnes (Figure 3) and highlight the relationship between these variables under average conditions. This relationship is also identified in the principal component analyses described in Section 2.2.

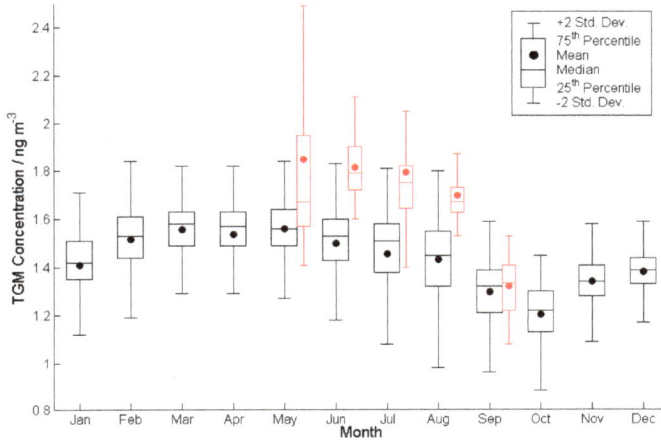

Figure 2. Monthly averaged TGM concentrations measured at Patricia McInnes station. Red data are impacted by forest fire smoke, whereas black data are not.

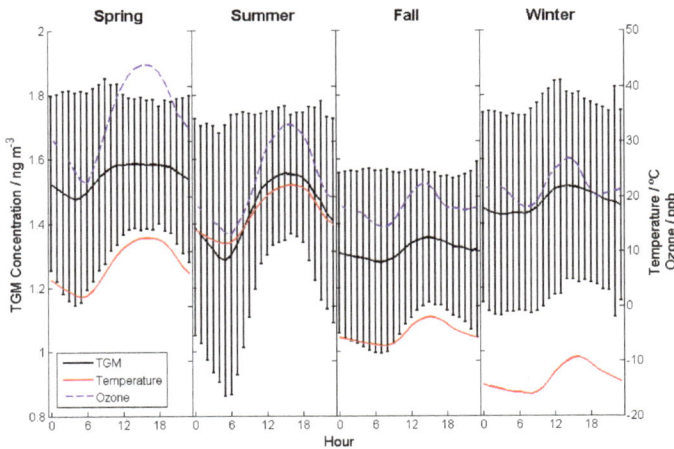

Figure 3. Seasonal diel profiles of mean TGM concentrations at Patricia McInnes station after removing data impacted by forest fire smoke. Seasonal diel profiles of mean temperatures and ozone concentrations at Patricia McInnes are also shown for reference Vertical bars indicate ± 2 standard deviations of TGM concentration data at each hour.

2.2. Principal Component Analysis

In addition to TGM concentration measurements, other air quality and meteorological measurements were considered, as described in Section 3.2. Air quality parameters include nitrous oxide (NO), nitrogen dioxide (NO_2), oxides of nitrogen (NO_X), ozone (O_3), fine particulate matter ($PM_{2.5}$), sulphur dioxide (SO_2), total hydrocarbon (THC), total reduced sulphur (TRS), ammonia (NH_3), and carbon monoxide (CO). Meteorological parameters include temperature (T), relative humidity (RH), wind speed and direction, solar radiation, and snow depth. Principal component analysis (PCA) was performed to identify correlations between parameters and serves as a means for reducing the number of parameters in the dataset into a handful of important factors. PCA was similarly applied in previous work where TGM concentrations were measured alongside a variety of ancillary variables to quantify relationships within the dataset and identify common factors that may influence TGM concentrations [19]. Three data subsets were established in order to better understand the factors affecting TGM and other air quality and meteorological parameters. The data subsets are grouped by rank according to "high", "average", and "low" TGM concentrations (*i.e.*, three bins divided at the 33rd- and 67th-percentiles). In order to avoid biasing data subsets with predominately spring values in the "high" subset or fall values in the "low" subset, for example, the TGM concentration data was de-trended and seasonally adjusted only to *categorize* any given TGM concentration as "high", "average", or "low", but the *original* TGM concentration values were used as input for PCA. De-trending and seasonal adjustment was carried out with additive decomposition with a linear trend component and 4th-order polynomial fit to monthly averaged data (similar to that shown in Figure 2) as a seasonal filter [20,21]. The de-trended and seasonally adjusted time series is shown in Figure 1 for reference. PCA results for each data subset, "high", "average", and "low" TGM concentrations, are shown in Tables 2–4, respectively. For comparison, PCA results for the full, combined dataset are shown in Table 5. These tables show the rotated factor loadings, the amount of variability explained by each factor and the eigenvalues. Only factors with eigenvalues greater than 1 are deemed significant and are interpreted below [22]. Parameters are an important contribution to a given factor when the absolute value of the corresponding rotated factor loading is ≥ 0.5 (bold values in Table 2 through Table 4) [22]. The meaning of each factor was inferred based on the groupings of parameters that load heaviest on that factor (e.g., NO_X is interpreted as anthropogenic combustion processes, and $PM_{2.5}$, CO, and NH_3 together are interpreted as forest fire smoke).

6

Table 2. Principal component analysis (PCA) rotated factor loadings for the highest 1/3rd of de-trended/seasonally adjusted TGM concentrations (*i.e.*, "high" TGM concentrations). Factor loadings with an absolute value greater than or equal to 0.5 are bolded. The shaded row highlights TGM loadings on each factor.

Parameter	Combustion Processes	Diel Trending	Forest Fire Smoke	Snow Depth	Industrial Sulphur
NO	**0.93**	0.04	0.03	0.06	0.01
NO_2	**0.86**	0.28	0.06	0.16	0.14
NO_X	**0.96**	0.17	0.05	0.12	0.08
THC	**0.60**	0.04	0.25	−0.45	0.32
O_3	**−0.55**	**−0.70**	0.01	0.07	−0.10
Temperature	−0.23	**−0.68**	0.19	−0.48	0.01
RH	0.14	**0.86**	0	0.12	0.11
Solar Radiation	−0.03	**−0.78**	−0.03	−0.12	0.09
TGM	−0.05	−0.49	**0.59**	0.28	0.17
$PM_{2.5}$	0.06	−0.05	**0.87**	−0.11	0.17
NH_3	0.02	0.04	**0.74**	0.01	−0.06
CO	0.11	0	**0.79**	−0.10	−0.03
Snow Depth	0.22	0.35	−0.01	**0.81**	0.08
SO_2	0.05	−0.05	−0.06	0.22	**0.83**
TRS	0.25	0.10	0.19	−0.35	**0.70**
Variability (%)	22.7	18.6	16.1	9.5	9.4
Cumulative Variability (%)	22.7	41.3	57.5	67.0	76.4
Eigenvalues	4.62	2.99	1.60	1.13	1.11

Table 3. PCA rotated factor loadings for the middle 1/3rd of de-trended/seasonally adjusted TGM concentrations (*i.e.*, "average" TGM concentrations). Factor loadings with an absolute value greater than or equal to 0.5 are bolded. The shaded row highlights TGM loadings on each factor.

Parameter	Combustion Processes & NH_3	Diel Trending	Industrial Sulphur & PM	Temperature & Snow Depth	CO
NO	**0.89**	0.07	0.14	0.07	0.10
NO_2	**0.70**	0.29	0.36	0.30	0.20
NO_X	**0.87**	0.20	0.28	0.20	0.17
NH_3	**0.64**	−0.11	−0.25	−0.11	−0.35
THC	0.49	0.39	0.42	−0.30	0.07
TGM	−0.06	**−0.76**	0.05	0.28	0.05
O_3	−0.48	**−0.70**	−0.27	−0.07	−0.18
Temperature	−0.19	**−0.60**	0	**−0.63**	0.18
RH	0.09	**0.81**	0.03	0.20	−0.01
Solar Radiation	0.02	**−0.70**	0.05	−0.18	0.01
$PM_{2.5}$	0.29	−0.13	**0.55**	0.02	0.34
SO_2	−0.03	−0.15	**0.71**	0.27	−0.19
TRS	0.19	0.17	**0.69**	−0.16	−0.07
Snow Depth	0.11	0.08	0.04	**0.91**	0.09
CO	0.09	−0.04	−0.10	0.01	**0.86**
Variability (%)	20.8	19.5	12.3	11.5	7.8
Cumulative Variability (%)	20.8	40.2	52.5	64.0	71.8
Eigenvalues	4.80	2.16	1.52	1.23	1.06

Table 4. PCA rotated factor loadings for lowest 1/3rd of de-trended/seasonally adjusted TGM concentrations (*i.e.*, "low" TGM concentrations). Factor loadings with an absolute value greater than or equal to 0.5 are bolded. The shaded row highlights TGM loadings on each factor.

Parameter	Combustion Processes	Diel Trending	Temperature & Snow Depth	Industrial Sulphur & NH$_3$
NO	**0.85**	0.06	0.16	0.08
NO$_2$	**0.74**	0.17	**0.51**	−0.07
NO$_X$	**0.88**	0.12	0.34	0.01
PM$_{2.5}$	**0.59**	−0.12	0.07	−0.31
THC	**0.68**	0.42	−0.16	−0.1
O$_3$	**−0.54**	**−0.71**	−0.06	0
TGM	−0.02	**−0.72**	0.05	0.02
RH	0.03	**0.83**	−0.03	0.02
Solar Radiation	0.05	**−0.64**	−0.34	−0.02
Temperature	−0.14	−0.26	**−0.87**	0.03
Snow Depth	0.07	−0.14	**0.9**	−0.06
SO$_2$	0.16	−0.17	0.28	**−0.61**
TRS	0.49	0.11	−0.07	**−0.51**
NH$_3$	0.21	−0.07	0.08	**0.61**
CO	0.41	−0.10	−0.10	0.20
Variability (%)	24.5	16.7	14.9	7.8
Cumulative Variability (%)	24.5	41.2	56.1	63.9
Eigenvalues	4.61	2.14	1.74	1.10

Focusing on the factors in relation to TGM, Table 2 correlates "high" TGM concentrations with parameters that are associated with forest fire smoke, namely: PM$_{2.5}$, CO, and NH$_3$ [23]. In other words, and not surprisingly, many of the high values in TGM are often a result of forest fire smoke impinging on the station. The next important factor for TGM during "high" TGM concentration observations relates to diel trending that strongly links ozone, temperature, RH, and solar radiation. TGM also shows minimal loading on the factor attributed to snow depth for the "high" TGM concentration data subset, which is not surprising given the strong association between "high" TGM concentration observations and forest fire smoke during the summer when there is no snow cover. PCA results for the "average" TGM concentration data subset (Table 3) show a shift of TGM correlation to the diel trending factor shared with ozone, temperature, RH, and solar radiation. The next important factor for TGM during "average" TGM concentration observations relates to temperature and snow depth. Similarly, PCA results for the "low" TGM concentration data subset (Table 4) also correlate TGM with diel trending ozone, RH, and solar radiation (but not temperature). Note that, in general, temperature is not as strongly associated with diel trending factors in all the PCA result tables, likely due to the wide variability of temperature in this region. Interestingly, for "average" and "low" TGM concentration subsets, TGM is effectively unrelated to pollutants mainly associated with local anthropogenic sources (*i.e.*, NO, NO_2, NO_X, SO_2, TRS). Likewise, "high" TGM concentration data is very minimally, and essentially insignificantly, associated with SO_2 and TRS. Considering the PCA results from the complete dataset in Table 5, the overall ranking of importance for factors driving TGM variability is: diel trending > forest fire smoke > temperature and snow depth > industrial sulphur > combustion processes. Thus, Table 5 further indicates the lack of association between industrial pollutants and TGM. This is despite the fact that the station is within an urban area and near the industrialized center of the Canadian oil sands region.

Table 5. PCA rotated factor loadings for the full, combined dataset. Factor loadings with an absolute value greater than or equal to 0.5 are bolded. The shaded row highlights TGM loadings on each factor.

Parameter	Combustion Processes	Diel Trending	Forest Fire Smoke	Temperature & Snow Depth	Industrial Sulphur
NO	**0.92**	0	0.07	0.06	0.01
NO_2	**0.83**	0.20	0.09	0.29	0.17
NO_X	**0.96**	0.10	0.08	0.19	0.09
THC	**0.62**	0.27	0.15	−0.39	0.28
O_3	**−0.56**	**−0.70**	−0.01	0.01	−0.08
TGM	−0.12	**−0.65**	0.37	0.24	0.14
RH	0.10	**0.84**	0.02	0.12	0.07
Solar Radiation	0.01	**−0.73**	−0.07	−0.25	0.03
Temperature	−0.26	**−0.54**	0.15	**−0.64**	−0.01
$PM_{2.5}$	0.08	−0.07	**0.82**	−0.06	−0.31
CO	0.12	−0.06	**0.71**	−0.03	−0.06
NH_3	0.03	0.04	**0.73**	0	−0.04
Snow Depth	0.17	−0.11	0	**0.89**	0.06
SO_2	0.04	−0.11	−0.06	0.24	**0.80**
TRS	0.27	0.11	0.16	−0.21	**0.73**
Variability (%)	22.5	17.4	13.0	11.4	9.2
Cumulative Variability (%)	22.5	39.9	52.9	64.3	73.5
Eigenvalues	4.53	2.48	1.50	1.42	1.10

2.3. Directional and Back Trajectory Analyses

Although the PCA results suggest a lack of localized industrial source influences on TGM concentrations, the directionality of TGM concentrations was explored and is plotted as a pollutant rose alongside a wind rose in Figure 4. Note that the pollution rose in Figure 4b excludes both data observed in calm wind conditions (where wind speed < 5 km·h^{-1}) and data impacted by forest fire smoke. Although the pollution rose, plotted as bars on the black axis in Figure 4b, at first appears to suggest possible point-sources to the north-northwest (*i.e.*, oil sands development) and southeast (*i.e.*, urban sources within Fort McMurray), a comparison to the wind rose (Figure 4a) indicates similar north-northwest and southeast dominant wind direction. The predominant winds at the station follow river valleys from the north-northwest and the southeast, hence TGM is simply observed from wherever the wind originates

9

with minimal increased weighting toward any direction of a point-source. This is further illustrated with the directionally dependent TGM concentration percentiles, plotted as lines on the red axis in Figure 4b, which do not indicate any significant directional dependence to TGM concentrations. Thus, both the PCA results and Figure 4 suggest a limited role of local point-source influences on ambient TGM concentrations. These results suggest that the TGM observed at Patricia McInnes is predominately due to broad area sources (*i.e.*, surface flux) and long-range transport.

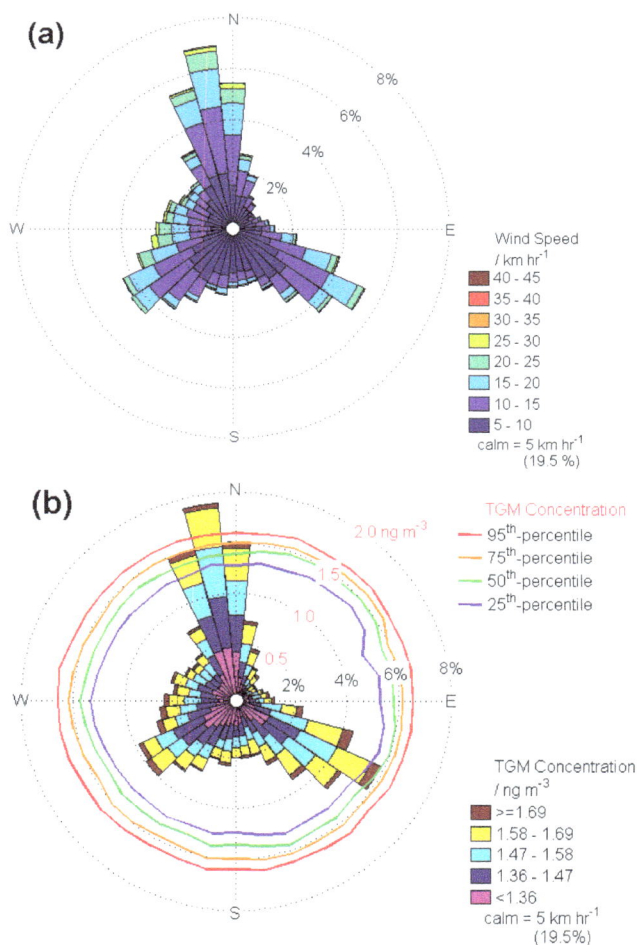

Figure 4. (a) Wind rose and (b) TGM concentration rose for Patricia McInnes station. Data corresponding to calm winds (wind speed < 5 km·h^{-1}) are not shown here. The TGM rose excludes data impacted by forest fire smoke. Frequency of concentration observations are plotted on the black scale and percentiles in TGM concentrations are plotted on the red radial axis in panel (b).

Since the PCA results and Figure 4 suggest that local conditions were not sufficient to explain all observations of TGM concentrations, the role of long-range transport was assessed for the observations at Patricia McInnes. To investigate the role of long-range transport, back trajectories were calculated with the Hybrid Single Particle Lagrangian Integrated Trajectory (HYSPLIT) model [24,25]. Back trajectory analysis has been identified as an important tool for assessing the history of air masses influencing concentration measurements of atmospheric species [26]. The 72-hour HYSPLIT back trajectories from Patricia McInnes station were obtained for all observations of hourly TGM concentrations. Data observed with forest fire smoke were excluded from this analysis. The coordinates of the starting point of each back trajectory were plotted in ArcMap10 as points in Figure 5a where the colour of each point represents the de-trended/seasonally adjusted TGM concentration observed with each back trajectory (as defined in Section 2.2 and Figure 1). De-trended/seasonally adjusted TGM concentrations were used rather than actual TGM concentrations to gain insight into which regions were most frequently associated with *unseasonably* high or low TGM concentrations for any given time of year, rather than the times of year when higher or lower TGM concentrations are expected (as shown in Figure 2). Due to the presence of many overlapping points, Figure 5b shows the same back trajectory starting points, but illustrated as point density (see Section 3.5 for further details on point density mapping). Figure 5b indicates that most 72-hour back trajectory starting points were located in northern Alberta, northern Saskatchewan, northern British Columbia, and southern Northwest Territories where the highest point density is mapped in red. Only a limited number of back trajectories started outside western Canada, where the lowest point density is mapped in blue in Figure 5b; grey regions in Figure 5b indicate where no back trajectory starting points exist within a 700 km radius. Subsequently, a boundary condition to this back trajectory analysis was set to exclude regions with less than 1% of the total point density (*i.e.*, fewer than 170 back trajectory starting points within a 700 km radius neighborhood), represented as the dashed lines in Figure 5a–c. Regions outside this dashed line were deemed to have an insufficient number of back trajectory starting points to draw any useful conclusions. Finally, Figure 5c shows the point density of back trajectory starting points weighted with de-trended/seasonally adjusted TGM concentrations and normalized by the point density shown in Figure 5b, resulting in a normalized de-trended/seasonally adjusted TGM concentration map (see Section 3.5). Red and blue areas in Figure 5c correspond to areas leading to the highest and lowest normalized de-trended/seasonally adjusted TGM concentration observations, respectively. The grey areas in Figure 5c are outside the boundary limits of the analysis due to an insufficient number of back trajectory starting points.

Figure 5c shows that the lowest normalized de-trended/seasonally adjusted TGM concentration observations primarily arrive via arctic regions of northern Canada. This region has minimal land disturbance or industry, so lower TGM concentration observations from air passing through this region are not surprising. Note that this back trajectory analysis cannot clearly distinguish between lower TGM concentrations originating from arctic regions as a result of arctic atmospheric mercury depletion events (AMDEs) [1,3,5,27–31] as opposed to as a result of generally cleaner air (*i.e.*, air with lower TGM concentrations). However, given that during the study period, arctic air parcels are often associated with unseasonably lower TGM concentrations, this possibility is explored further in Section 2.4 below.

Higher normalized de-trended/seasonally adjusted TGM concentration observations can be traced back to regions further southeast and west from the monitoring station, suggesting the role of transport from more populated areas of North America and Asian influenced trans-Pacific transport. That said, high TGM concentration may be introduced into an air parcel anywhere along a back trajectory, and the intention of this analysis is not to identify sources of atmospheric mercury, but rather to highlight the importance of long-range transport when dealing with atmospheric mercury as a global pollutant. Indeed, it is unlikely there are any significant sources of atmospheric mercury in the areas shaded red in Figure 5c, but only that the air has passed through those regions on its way to the monitoring station where higher atmospheric mercury concentrations were observed. Note that Figure 5c uses de-trended/seasonally adjusted TGM concentrations; thus, red or blue regions in this map indicate the starting points for the unseasonably high or low TGM concentrations for any given time of year and cannot be related to the seasonal trend in TGM concentrations illustrated in Figure 2. In other words, Figure 5c does not indicate whether back trajectories in the spring or fall tend to originate from regions shaded red or blue, respectively. The intermediate regions in Figure 5c between extremes do not show any consistently high or low values as exhibited in the red or blue regions, but rather a relatively similar number of high and low values (as can be seen in Figure 5a). The normalized de-trended/seasonally adjusted TGM concentrations approach 0 $ng \cdot m^{-3}$ in these regions. For example, normalized de-trended/seasonally adjusted TGM concentrations are approximately 0 $ng \cdot m^{-3} \cdot km^{-2}$ in the region near the sampling station because the high and low de-trended/seasonally adjusted TGM concentrations associated with these starting points cancel each other in the weighted point density calculations.

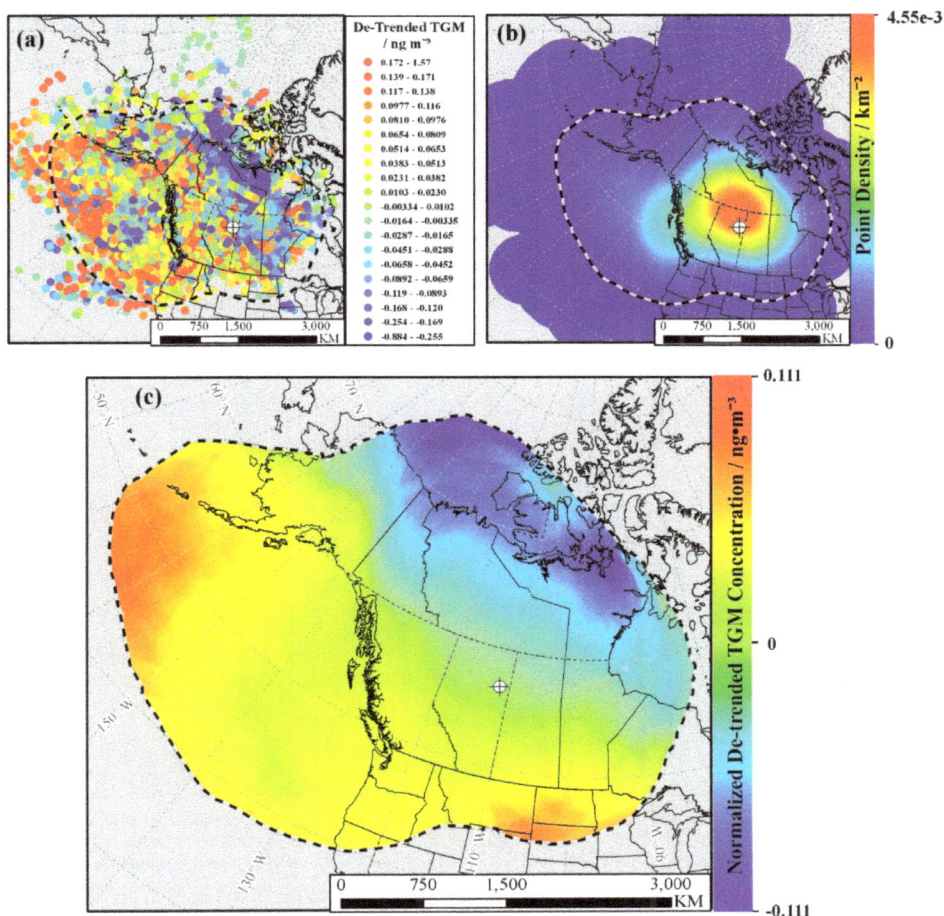

Figure 5. HYSPLIT 72-hour back trajectory analysis. Panel (**a**) shows all back trajectory starting points colour-coded to the corresponding de-trended/seasonally adjusted TGM concentrations; panel (**b**) shows the point density of back trajectory starting points; panel (**c**) shows the normalized de-trended/seasonally adjusted TGM concentrations as described in the text. Dashed lines indicate the boundary limits of the analysis as described in the text; grey areas contained insufficient or no back trajectory starting points. Fort McMurray is indicated in northern Alberta with the cross-hair symbol. Data observed with forest fire smoke were excluded from this analysis.

2.4. Mercury Chemistry Implications

In this dataset, low concentrations of TGM are often concurrently observed with low concentrations of ozone and higher RH, as indicated with the PCA results

13

above. Furthermore, unseasonably low TGM concentrations are often traced back to arctic regions as illustrated in Figure 5c. These associations suggest the possible role of atmospheric mercury chemistry, particularly oxidation of GEM to form GOM [3,5,27,32]. GOM can then undergo deposition faster than GEM, thereby reducing the ambient TGM concentration in an area. Previous studies have provided detailed discussions on this type of mercury chemistry, most notably in terms of arctic AMDEs during spring-time polar sunrises [1,3,5,27–31]. AMDEs are associated with reduced GEM concentrations, reduced ozone concentrations, and increasing solar radiation. The PCA results above, however, show a positive correlation between TGM concentrations and solar radiation (*i.e.*, reduced TGM concentrations with decreasing solar radiation), which suggests this reduction in TGM concentration is not likely a locally occurring process from oxidation by ozone. Since, in this dataset, lower TGM concentrations are associated with lower ozone and lower levels of solar radiation, the link between reduced TGM concentrations and reduced ozone concentrations is more likely an association with cleaner air passing over the monitoring station rather than a result of local GEM oxidation.

To further support this argument, Figure 6 shows the diel and monthly fractional distributions of observations of "high", "average", and "low" TGM concentrations, with divisions at the 67th- and 33rd-percentiles of de-trended/seasonally adjusted TGM concentrations (described above in Section 2.2 and illustrated in Figure 1) excluding data affected by forest fire smoke. Figure 6a shows the "low" TGM concentration observations typically occurred in the early morning or overnight while "high" TGM concentration observations typically occurred in the mid-day. The frequency of "average" TGM concentration observations is consistent throughout the day as indicated in Figure 6a. This is not to say that "average" TGM concentration observations do not exhibit a diel trend, but rather that "average" TGM concentrations can be observed with the same frequency at any time of day. In contrast, Figure 6b shows that the "high", "average", and "low" TGM concentration subsets do not have any clear seasonal trend. Figure 6b emphasizes that "low" TGM concentration observations are distributed more or less equally over the course of the year, suggesting "low" TGM concentrations originating from arctic regions are more likely due to generally cleaner air than to AMDEs. Had the "low" TGM concentration observations been a result of local/regional AMDEs, one would expect to see the highest frequency of "low" TGM concentration observations during daylight hours. Likewise, had the "low" TGM concentrations been a result of arctic air transported into the region after arctic AMDEs, one would expect to see the highest frequency of "low" TGM concentration observations during the spring. Note that Figure 3 also indicates that lower TGM concentration observations are not particularly common in the spring.

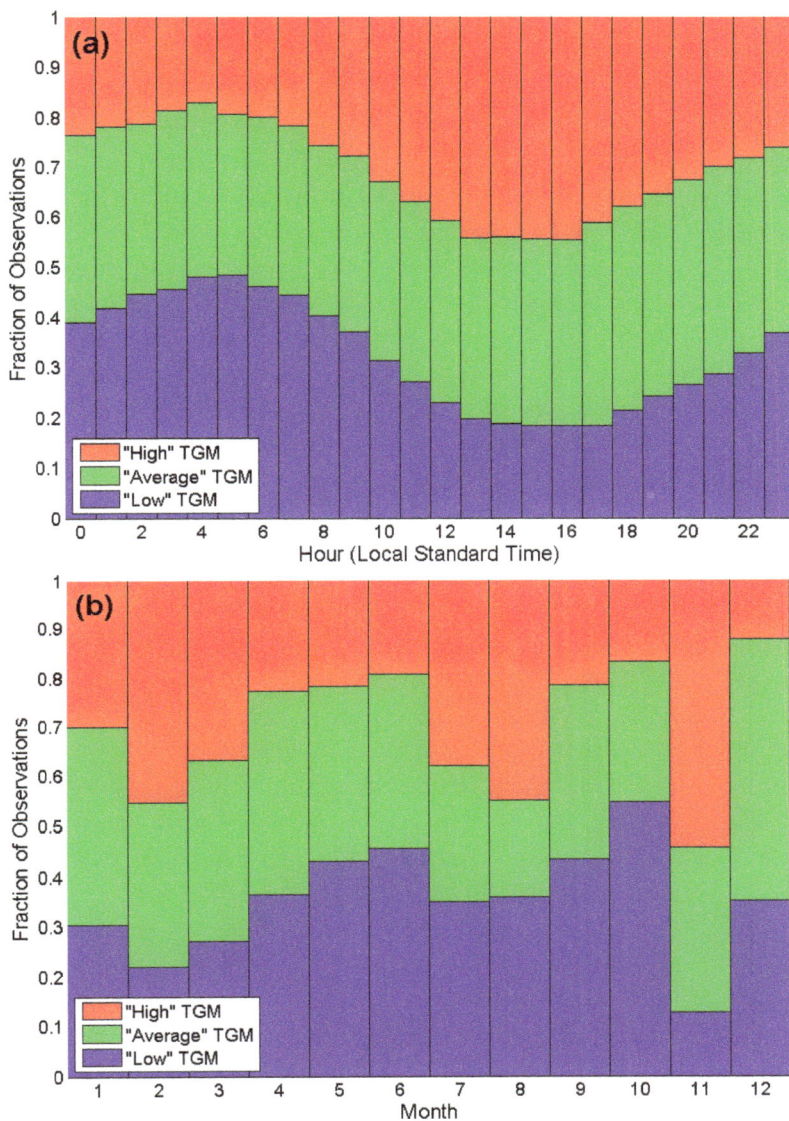

Figure 6. (a) Diel and (b) monthly fractional distribution of "high", "average", and "low" de-trended/seasonally adjusted TGM concentrations as defined in the text. These plots exclude data affected by forest fire smoke.

Figure 6a indicates again that, in general, local daytime oxidation by ozone is an unviable reaction mechanism for TGM depletion observed at Patricia McInnes. The diel trend in Figure 6a also supports conclusions that TGM is driven by diel trends affecting underlying surface-air flux and long-range transport of global atmospheric mercury. In other words, a diel trend is superimposed on the TGM concentration time series that undergoes lower frequency oscillations as air with higher or lower TGM concentrations move over the monitoring station. The back trajectory analysis above also supports this conclusion, in that the lowest TGM values came with air that has passed through arctic regions of northern Canada, and are thus presumably cleaner than air originating from more populated regions. GEM oxidation chemistry may still be occurring locally, but it is difficult to speculate to what extent it occurs, if at all, without speciated mercury measurements providing GOM concentration data. Also, this dataset cannot provide any insight to the role of particulate bound mercury in the area. Mercury speciation measurements will form a future component of the Joint Canada-Alberta Implementation Plan for Oil Sands Monitoring, and that future work may provide further information to more definitively understand any local mercury chemistry taking place in the region with associated effects on local mercury deposition.

3. Experimental Section

3.1. Site Description

Fort McMurray—Patricia McInnes station (56°45′7′′N, 111°28′34′′W; elevation 362 m; see Figure 7) is an ambient air quality monitoring station operated by WBEA and located in a neighbourhood in the northwest quadrant of Fort McMurray, Alberta (2011 population: 61,374). As a community-based station, as opposed to an air permitting approval-based station, Patricia McInnes station is surrounded by housing to the immediate north, south, and east (325°–0°–245°), and a recreational park and woodland to the immediate west (245°–325°). Fort McMurray is located in the Regional Municipality of Wood Buffalo on the Athabasca River approximately 30 km south of the oil sands developments and approximately 380 km north-northeast of Edmonton, Alberta. The area is within the Boreal Plains Ecozone of Canada in which balsam fir/poplar, white/black spruce, white birch, and trembling aspen are common [33]. Under the Köppen climate classification system, Fort McMurray is categorized as humid continental climate bordering on subarctic with large seasonal temperature differences ranging from the mean winter minimum of $-20.7\,°C$ to the mean summer maximum of $22.5\,°C$ [34].

3.2. Methods

Ambient total gaseous mercury (TGM) concentrations are measured as part of Environment Canada's contribution to the Canada-Alberta Joint Oil Sands Monitoring Plan with the assistance of WBEA. Environment Canada operates Tekran Model 2537 Mercury Vapor Analyzers in continuous 5-minute intervals following the

Canadian Atmospheric Mercury Measurement Network (CAMNet) TGM guidelines for instrument configuration and maintenance [7,35]. Measurements have been ongoing at WBEA's Patricia McInnes station since October 2010 and continue today with technical support from the WBEA staff. This study considers data from 21 October 2010 through 31 May 2013, inclusively. The quality controlled TGM data is publicly available on the Canada-Alberta Joint Oil Sands Environmental Monitoring Information Portal [36].

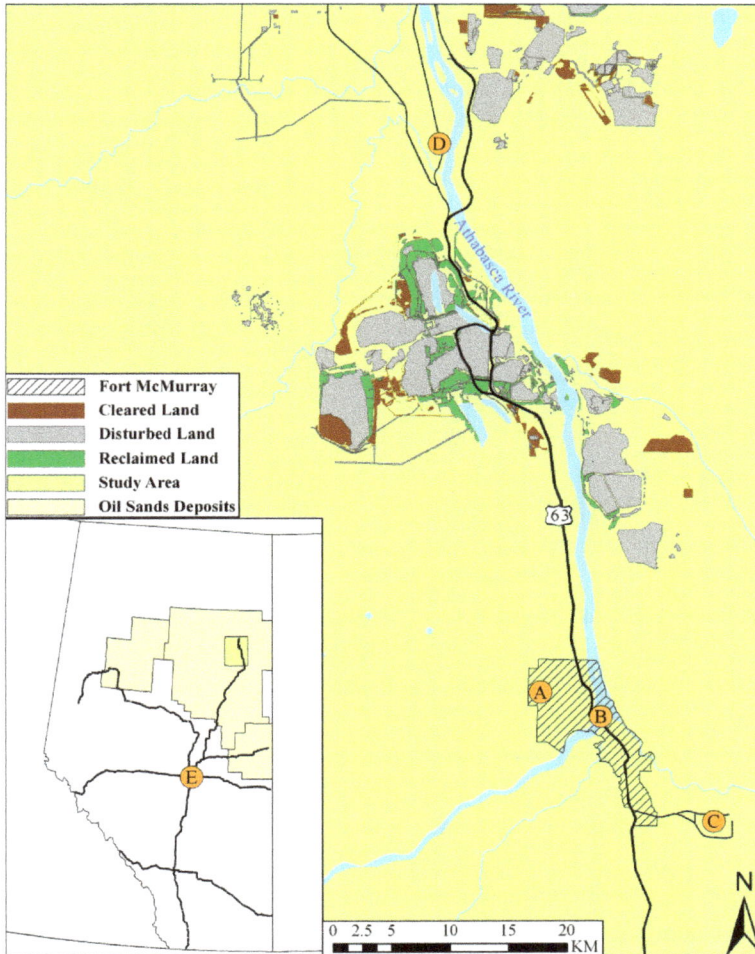

Figure 7. Map of the area showing Patricia McInnes Station (A), Athabasca Valley Station (B), Fort McMurray Airport (C), and Fort McKay (D). Inset map of Alberta shows study area in relationship to Edmonton (E). Fort McMurray and oil sands deposits and production areas are shaded according to the legend [37].

Table 6 lists TGM and other measured parameters used in this study, with station, operator, and source details. Air quality parameters include nitrous oxide (NO), nitrogen dioxide (NO_2), oxides of nitrogen (NO_X), ozone (O_3), fine particulate matter ($PM_{2.5}$), sulphur dioxide (SO_2), total hydrocarbon (THC), total reduced sulphur (TRS), ammonia (NH_3), and carbon monoxide (CO). Meteorological parameters include temperature (T), relative humidity (RH), wind speed and direction, solar radiation, and snow depth. Most of these variables were chosen due to the fact they were measured at Patricia McInnes station. Solar radiation, snow depth, and RH measured at nearby stations were included due to associations with TGM concentration noted in previous work [19]; CO was included as a forest fire tracer as noted below. The data source of each parameter, as listed in Table 6, provides corresponding details on instrumentation; all public data has been quality controlled by the respective operator and provided as hourly averages from continuous instrumentation. Wind speed and direction are hourly vector averages. Measurement probes and sampling inlets are approximately 2–3 m above ground at each station.

Elevated concentrations of $PM_{2.5}$ and CO were used as tracers to determine the presence of forest fire smoke in the vicinity of the monitoring station, corroborated with archived forecasts of ground-level $PM_{2.5}$ concentrations due to forest fire smoke [38].

3.3. Data Quality Assurance/Quality Control

Raw TGM data is ingested into the Research Data Management and Quality Control System (RDMQ) developed by Environment Canada [8,42] where every 5-minute measurement is manually scrutinized and flagged. As supplementation to RDMQ, raw and quality controlled air quality and meteorological data from WBEA's Patricia McInnes air monitoring station provided various parameters with which to compare. These parameters, alongside instrument diagnostics, are compared to the flagged values in RDMQ to determine if the measurement is valid. After acceptance of a valid data point (measured over 5 min), averages are calculated for hours where three or more data points exist. If less than three data points exist in a given hour, the hour is marked as missing. A large gap in the dataset exists from 21 May 2011 to 20 October 2011. Measurements were made during this period; however, significant forest fires north of Fort McMurray and associated heavy smoke caused particles to bypass plugged particulate filters and contaminate the instrument's gold traps. This caused lower measurements to be recorded during the 5-minute collection phase in that not all of the mercury in the ambient air was able to be pre-concentrated on the contaminated gold traps. The issue was discovered during the QC process after an increase in TGM concentrations was observed upon replacement of the gold traps in October 2011. As a result, data was invalidated back to the date in May where flow issues were first observed, and it is believed smoke first affected the gold traps.

Invalidated data were not included in the analysis, statistical tests, or interpretation of results in this study.

Table 6. Summary of data sources.

Parameter	Station	Latitude	Longitude	Elevation	Operator	Source
TGM					Environment Canada	JOSM[a]
NO NO$_2$ NO$_X$ O$_3$ PM$_{2.5}$ SO$_2$ THC TRS Temperature Wind Speed Wind Direction	Fort McMurray—Patricia McInnes	56.75°N	111.48°W	362 m	Wood Buffalo Environmental Association	CASA[b]
NH$_3$						WBEA[c]
CO	Fort McMurray—Athabasca Valley	56.73°N	111.39°W	260m	Wood Buffalo Environmental Association	CASA[b]
Solar Radiation	Fort McKay	57.19°N	111.64°W	269	Wood Buffalo Environmental Association	CASA[b]
RH	Fort McMurray Airport	56.65°N	111.22°W	369m	Environment Canada	EC[d]
Snow Depth						

Notes: [a] Joint Oil Sands Monitoring Data Portal website [36]. [b] Clean Air Strategic Alliance website [39]. [c] Wood Buffalo Environmental Associations website [40].[d] Environment Canada National Climate Data and Information Archive [41].

3.4. Statistical Analyses

Principal component analyses, descriptive statistics, trend analysis, t-tests, and de-trending/seasonal adjustments were carried out with Matlab software. Several other parameters measured at Patricia McInnes and neighbouring monitoring stations (as summarized in Table 6) were used for PCA. Components identified with PCA were varimax rotated to identify a handful of the most important factors that explain the most variability in the dataset. For a detailed description of PCA, see Hair *et al.* [22]. By comparing how TGM concentrations load on each factor for various data subsets, one can gain an insight into which factors have the most influence over TGM concentration.

3.5. HYSPLIT Back Trajectory Analysis

The online version of the Hybrid Single Particle Lagrangian Integrated Trajectory (HYSPLIT) model [24,25] was applied using default meteorological settings, including archived meteorological data from the National Centers for

Environmental Prediction's Global Data Assimilation System using model vertical velocity for vertical motion. Archived meteorological data was provided on a 1° grid with 23 vertical levels along the pressure gradient. A separate 72-hour back trajectory was obtained for each hourly observation of de-trended/seasonally adjusted TGM concentration measured at Patricia McInnes station, excluding data associated with forest fire smoke for a total of 17,020 unique trajectories. The objective behind running back trajectories in this study was to better understand the source region(s) of air parcels impinging upon the monitor, not specifically to track regional air flows. Since the TGM monitor samples below the mixing height nearly all the time, it is important that the end point of the back trajectories also be below the mixing height. For this reason, back trajectories arrived at the monitoring station 50 m above ground level, to be generally less than the minimum mixing height at any given time of day or year [43], but also high enough to avoid model surface artifacts. The 72-hour back trajectories were chosen to provide a balance between sufficient spatial distribution of back trajectory starting points while minimizing errors associated with longer back trajectories.

Three figures are presented to show the results of the HYSPLIT back trajectory analysis. Figure 5a presents each individual back trajectory's starting point and is colour coded based on the associated de-trended/seasonally adjusted TGM concentrations. Note that the points in Figure 5a were layered by observation—as opposed to layered by de-trended/seasonally adjusted TGM concentrations—to provide a sense of randomization to prevent visually biasing either high or low de-trended/seasonally adjusted TGM concentrations in this panel. Figure 5b uses the point density tool in ArcMap 10 and is defined by a 10 km × 10 km grid focused over western Canada. The point density tool determines the density of data points within a user defined distance from the center of a given grid cell—a neighbourhood. For Figure 5b, the point density tool is applied to the number of back trajectories within a 700 km radius neighbourhood from the center of each grid cell. The trajectory point density map may be conceptualized as a population density map where the number of back trajectory starting points is substituted for population. Figure 5b does not take into account the TGM concentrations associated with each back trajectory. A 700 km radius neighbourhood was used to provide a balance between smoothing the map of fine details that can overemphasize individual starting points without losing larger-scale features. For further details on point density mapping, see [44,45].

Figure 5c uses the combination of Figures 5a and 5b to create a de-trended/ seasonally adjusted TGM concentration weighted point density map, normalized by point density. In Figure 5c, each back trajectory starting point was weighted with its corresponding de-trended/seasonally adjusted TGM concentration. The de-trended/seasonally adjusted TGM concentration weighted points were then normalized by the number of trajectories in the grid cell, as determined by Figure 5b.

20

Normalizing by the number of trajectories in the grid cell avoids biasing the map in regions where there are fewer back trajectory starting points. For each $10\,\text{km} \times 10\,\text{km}$ grid cell, this process is mathematically defined as the weighted point density (in units of $\text{ng}\cdot\text{m}^{-3}\cdot\text{km}^{-2}$) divided by the point density (in units of km^{-2}), to result in the normalized TGM concentrations (in units of $\text{ng}\cdot\text{m}^{-3}$) as presented in Figure 5c.

4. Conclusions

Ambient concentrations of total gaseous mercury were measured at WBEA's Fort McMurray—Patricia McInnes air quality monitoring station from 20 October 2010 through 31 May 2013. The results show that TGM concentrations are comparable to that measured at other stations in the province of Alberta. TGM concentration undergoes seasonal and diel trends, with high values in the spring and midday, and low values in the fall and early morning. Higher TGM concentrations were generally a result of forest fire smoke transported to the sampling area and other long-range transport via the southeast and west, whereas lower TGM concentrations were generally a result of cleaner air transported into the region via the arctic. There was no correlation between concentrations of TGM and other anthropogenic pollutants (*i.e.*, NO_X and SO_2), despite the fact that the station is within an urban area and near the industrialized center of the Canadian oil sands region. Although reduced TGM concentrations were associated with reduced ozone concentrations, the data do not suggest that mercury oxidation chemistry via ozone plays a major role in the area. This work provides a critical component for future work to better understand mercury deposition and cycling within the Canadian oil sands region.

Acknowledgments: The authors thank WBEA staff responsible for local site maintenance, operations, and logistics. This study forms a part of the Joint Canada-Alberta Implementation Plan for Oil Sands Monitoring.

Conflicts of Interest: The authors declare no conflict of interest.

References

1. AMAP/UNEP. *Technical Background Report for the Global Mercury Assessment 2013*; Arctic Monitoring and Assessment Programme: Oslo, Norway, 2013.
2. Lin, C.J.; Pehkonen, S.O. The chemistry of atmospheric mercury: A review. *Atmos. Environ.* **1999**, *33*, 2067–2079.
3. Schroeder, W.H.; Munthe, J. Atmospheric mercury—An overview. *Atmos. Environ.* **1998**, *32*, 809–822.
4. Percy, K.E.; Hansen, M.C.; Dann, T. Air Quality in the Athabasca Oil Sands Region 2011. In *Alberta Oil Sands: Energy, Industry and the Environment*, 1st ed.; Percy, K.E., Ed.; Elsevier: Oxford, UK, 2012; pp. 47–91.

5. Lindberg, S.; Bullock, R.; Ebinghaus, R.; Engstrom, D.; Feng, X.; Fitzgerald, W.; Pirrone, N.; Prestbo, E.; Seigneur, C. A synthesis of progress and uncertainties in attributing the sources of mercury in deposition. *Ambio* **2007**, *36*, 19–32.

6. Driscoll, C.T.; Mason, R.P.; Chan, H.M.; Jacob, D.J.; Pirrone, N. Mercury as a global pollutant: Sources, pathways, and effects. *Environ. Sci. Tech.* **2013**, *47*, 4967–4983.

7. Mazur, M.; Mintz, R.; Lapalme, M.; Wiens, B. Ambient air total gaseous mercury concentrations in the vicinity of coal-fired power plants in Alberta, Canada. *Sci. Total Environ.* **2009**, *408*, 373–381.

8. Temme, C.; Blanchard, P.; Steffen, A.; Banic, C.; Beauchamp, S.; Poissant, L.; Tordon, R.; Wiens, B. Trend, seasonal and multivariate analysis study of total gaseous mercury data from the Canadian atmospheric mercury measurement network (CAMNet). *Atmos. Environ.* **2007**, *41*, 5423–5441.

9. Selin, N.E.; Jacob, D.J.; Park, R.J.; Yantosca, R.M.; Strode, S.; Jaeglé, L.; Jaffe, D. Chemical cycling and deposition of atmospheric mercury: Global constraints from observations. *J. Geophys. Res.* **2007**, *112*, D02308.

10. National Pollutant Release Inventory. Available online: http://www.ec.gc.ca/npri (accessed on 24 July 2013).

11. Percy, K.E. *Alberta Oil Sands: Energy, Industry and the Environment*, 1st ed.; Percy, K.E., Ed.; Elsevier: Oxford, UK, 2012; pp. 1–496.

12. Timoney, K.P.; Lee, P. Does the Alberta Tar Sands Industry Pollute? The Scientific Evidence. *Open Conserv. Biol. J.* **2009**, *3*, 65–81.

13. Kelly, E.N.; Schindler, D.W.; Hodson, P.V.; Short, J.W.; Radmanovich, R.; Nielsen, C.C. Oil sands development contributes elements toxic at low concentrations to the Athabasca River and its tributaries. *Proc. Natl. Acad. Sci. USA* **2010**, *107*, 16178–16183.

14. Wiklund, J.A.; Hall, R.I.; Wolfe, B.B.; Edwards, T.W.; Farwell, A.J.; Dixon, D.G. Has Alberta oil sands development increased far-field delivery of airborne contaminants to the Peace-Athabasca Delta? *Sci. Total Environ.* **2012**, *433*, 379–382.

15. *Joint Canada-Alberta Implementation Plan for Oil Sands Monitoring*; Government of Alberta Government of Canada: Edmonton, AB, Canada, 2012. Available online: http://ec.gc.ca/scitech/D0AF1423--351C-4CBC-A990--4ADA543E7181/COM1519_ Final%20OS%20Plan_02.pdf (accessed on 14 June 2013).

16. Kellerhals, M.; Beauchamp, S.; Belzer, W.; Blanchard, P.; Froude, F.; Harvey, B.; McDonald, K.; Pilote, M.; Poissant, L.; Puckett, K.; *et al.* Temporal and spatial variability of total gaseous mercury in Canada: Results from the Canadian Atmospheric Mercury Measurement Network (CAMNet). *Atmos. Environ.* **2003**, *37*, 1003–1011.

17. Blanchard, P.; Froude, F.A.; Martin, J.B.; Dryfhout-Clark, H.; Woods, J.T. Four years of continuous total gaseous mercury (TGM) measurements at sites in Ontario, Canada. *Atmos. Environ.* **2002**, *36*, 3735–3743.

18. Poissant, L.; Pilote, M.; Beauvais, C.; Constant, P.; Zhang, H. A year of continuous measurements of three atmospheric mercury species (GEM, RGM and Hg) in southern Québec, Canada. *Atmos. Environ.* **2005**, *39*, 1275–1287.

19.   Eckley, C.S.; Parsons, M.T.; Mintz, R.; Lapalme, M.; Mazur, M.; Tordon, R.; Elleman, R.; Graydon, J.A.; Blanchard, P.; St Louis, V. Impact of closing Canada's largest point-source of mercury emissions on local atmospheric mercury concentrations. *Environ. Sci. Tech.* **2013**, *47*, 10339–10348.

20.   Mathworks. Seasonal Adjustment. In *Econometrics Toolbox User Guide*, R2013a ed.; Mathworks: Natick, MA, USA, 2013; pp. 2–51.

21.   Findley, D.F.; Monsell, B.C.; Bell, W.R.; Otto, M.C.; Chen, B.C. New Capabilities and Methods of the X-12-ARIMA Seasonal Adjustment Program. *J. Bus. Econ. Stat.* **1998**, *16*, 127–152.

22.   Hair, J.F.; Anderson, R.E.; Tatham, R.L.; Black, W.C. *Multivariate Data Analysis*, 5th ed.; Prentice Hall: Upper Saddle River, NJ, USA, 1998.

23.   Andreae, M.O.; Merlet, P. Emission of trace gases and aerosols from biomass burning. *Glob. Biogeochem. Cy.* **2001**, *15*, 955–966.

24.   Draxler, R.R.; Rolph, G.D. *HYSPLIT (Hybrid Single-Particle Lagrandian Integrated Trajectory) Model Access via NOAA ARL READY*; NOAA Air Resources Laboratory: College Park, MD, USA, 2013. Available online: http://www.arl.noaa.gov/HYSPLIT.php (accessed on 24 July 2013).

25.   Rolph, G.D. *Real-time Environmental Applications and Display sYstem (READY) Website*; NOAA Air Resources Laboratory: College Park, MD, USA. Available online: http://www.ready.noaa.gov (accessed on 24 July 2013).

26.   Fleming, Z.L.; Monks, P.S.; Manning, A.J. Review: Untangling the influence of air-mass history in interpreting observed atmospheric composition. *Atmos. Res.* **2012**, *104–105*, 1–39.

27.   Skov, H.; Christensen, J.H.; Goodsite, M.E.; Heidam, N.Z.; Jensen, B.; Wåhlin, P.; Geernaert, G. Fate of elemental mercury in the arctic during atmospheric mercury depletion episodes and the load of atmospheric mercury to the arctic. *Environ. Sci. Tech.* **2004**, *38*, 2373–2382.

28.   Steffen, A.; Douglas, T.; Amyot, M.; Ariya, P.; Aspmo, K.; Berg, T.; Bottenheim, J.; Brooks, S.; Cobbett, F.; Dastoor, A.; *et al.* A synthesis of atmospheric mercury depletion event chemistry in the atmosphere and snow. *Atmos. Chem. Phys.* **2008**, *8*, 1445–1482.

29.   Lu, J.Y.; Schroeder, W.H.; Barrie, L.A.; Steffen, A.; Welch, H.E.; Martin, K.; Lockhart, L.; Hunt, R.V.; Boila, G.; Richter, A. Magnification of atmospheric mercury deposition to polar regions in springtime: The link to tropospheric ozone depletion chemistry. *Geophys. Res. Lett.* **2001**, *28*, 3219–3222.

30.   Lu, J.Y.; Schroeder, W.H. Annual time-series of total filterable atmospheric mercury concentrations in the Arctic. *Tellus* **2004**, *56*, 213–222.

31.   Sprovieri, F.; Pirrone, N.; Landis, M.; Stevens, R. Atmospheric mercury behavior at different altitudes at Ny Alesund during Spring 2003. *Atmos. Environ.* **2005**, *39*, 7646–7656.

32.   Ebinghaus, R.; Kock, H.H.; Temme, C.; Einax, J.W.; Löwe, A.G.; Richter, A.; Burrows, J.P.; Schroeder, W.H. Antarctic Springtime Depletion of Atmospheric Mercury. *Environ. Sci. Tech.* **2002**, *36*, 1238–1244.

33. Parks Canada. *National Parks System Plan*, 3rd ed. 2009. Available online: http://www.pc.gc.ca/docs/v-g/nation/sec2/~/media/docs/v-g/pn-np/SysPlan_e.ashx (accessed on 11 June 2013).

34. Canadian Climate Normals. Available online: http://climate.weather.gc.ca/climate_normals/index_e.html (accessed on 13 August 2013).

35. Steffen, A.S.; Schroeder, B. *Standard Operating Procedures for Total Gaseous Mercury Measurements—Canadian Atmospheric mercury Measurement Network (CAMNet)*; Environment Canada: Toronto, ON, Canada, 1999.

36. Canada-Alberta Oil Sands Environmental Monitoring Information Portal. Available online: http://www.jointoilsandsmonitoring.ca (accessed on 8 August 2013).

37. Alberta Environment and Sustainable Resource Development. *Oil Sands Landcover Status 2011*; AESRD: Edmonton, AB, Canada, 2013.

38. Bluesky Forecast Archive. Available online: http://www.env.gov.bc.ca/epd/bcairquality/bluesky/data (accessed on 16 August 2013).

39. Clean Air Strategic Alliance Data Warehouse. Available online: http://www.casadata.org (accessed on 8 August 2013).

40. Wood Buffalo Environmental Association. Available online: http://www.wbea.org (accessed on 8 August 2013).

41. National Climate Data and Information Archive. Available online: http://climate.weather.gc.ca (accessed on 8 August 2013).

42. Steffen, A.; Scherz, T.; Olson, M.; Gay, D.; Blanchard, P. A comparison of data quality control protocols for atmospheric mercury speciation measurements. *J. Environ. Monit.* **2012**, *14*, 752–765.

43. Davies, M.J.E. Air Quality Modeling in the Athabasca Oil Sands Region. In *Alberta Oil Sands: Energy, Industry and the Environment*, 1st ed.; Percy, K.E., Ed.; Elsevier: Oxford, UK, 2012; pp. 267–309.

44. How Point Density Works. Available online: http://help.arcgis.com/en/arcgisdesktop/10.0/help/index.html#/How_Point_Density_works/009z00000013000000/ (accessed on 11 November 2013).

45. Silverman, B.W. *Density Estimation for Statistics and Data Analysis*; Chapman and Hall: London, UK, 1986.

Seasonal and Diurnal Variations of Total Gaseous Mercury in Urban Houston, TX, USA

Xin Lan, Robert Talbot, Patrick Laine, Barry Lefer, James Flynn and Azucena Torres

Abstract: Total gaseous mercury (THg) observations in urban Houston, over the period from August 2011 to October 2012, were analyzed for their seasonal and diurnal characteristics. Our continuous measurements found that the median level of THg was 172 parts per quadrillion by volume (ppqv), consistent with the current global background level. The seasonal variation showed that the highest median THg mixing ratios occurred in summer and the lowest ones in winter. This seasonal pattern was closely related to the frequency of THg episodes, energy production/consumption and precipitation in the area. The diurnal variations of THg exhibited a pattern where THg accumulated overnight and reached its maximum level right before sunrise, followed by a rapid decrease after sunrise. This pattern was clearly influenced by planetary boundary layer (PBL) height and horizontal winds, including the complex sea breeze system in the Houston area. A predominant feature of THg in the Houston area was the frequent occurrence of large THg spikes. Highly concentrated pollution plumes revealed that mixing ratios of THg were related to not only the combustion tracers CO, CO_2, and NO, but also CH_4 which is presumably released from oil and natural gas operations, landfills and waste treatment. Many THg episodes occurred simultaneously with peaks in CO, CO_2, CH_4, NO_x, and/or SO_2, suggesting possible contributions from similar sources with multi-source types. Our measurements revealed that the mixing ratios and variability of THg were primarily controlled by nearby mercury sources.

Reprinted from *Atmosphere*. Cite as: Lan, X.; Talbot, R.; Laine, P.; Lefer, B.; Flynn, J.; Torres, A. Seasonal and Diurnal Variations of Total Gaseous Mercury in Urban Houston, TX, USAr. *Atmosphere* **2014**, *5*, 399–419.

1. Introduction

Mercury is a toxic environmental pollutant [1]. It is mobilized from deep reservoirs in the Earth to the atmosphere, where it then deposits to terrestrial systems and water bodies. Mercury in the water can be transformed into methylmercury, a much more toxic form that accumulates in fish and shellfish [2]. Humans are exposed to mercury poisoning mainly by consuming contaminated seafood. In these processes, the atmosphere serves as a major pathway for mercury transport from sources to receptors. Thus, it is important to quantify and characterized mercury in the atmosphere. In the atmosphere, mercury exists in three chemical forms: gaseous elemental mercury (GEM = Hg°), gaseous oxidized mercury (GOM), and particulate bound mercury (PBM). Gaseous elemental mercury is the most abundant chemical from, which accounts for about 95% of total atmospheric mercury [3–5].

Atmospheric mercury is emitted from both natural and anthropogenic sources. Natural sources include volcanoes and geothermal areas, mercury enriched soils, wild fires, and the ocean [6–12]. Major anthropogenic sources include coal combustion, industrial and commercial boilers, electric arc furnaces, cement production, and waste treatment facilities [13]. The National Emission Inventory (NEI) from the U.S. Environmental Protection Agency (EPA) reports that 59 tons out of 61 tons of total mercury emissions in the U.S. are from stationary sources. Combustion of fossil fuels is considered as the major anthropogenic source of atmospheric mercury; coal-fired utility boilers alone account for 49% of anthropogenic emissions [13]. The sources of atmospheric mercury are just beginning to be characterized and quantified; large uncertainties remain in various mercury sources, such as on-road vehicles, oil refineries, and other industrial facilities [14–17].

Ambient mercury levels have been assessed through careful measurements. A 10-year (1995–2005) measurement at 11 sites from the Canadian Atmospheric Mercury Measurement Network (CAMNet) reported that the averaged THg (THg = GEM + GOM) concentration was 1.58 ng·m^{-3} (177 ppqv) [18]. Long-term measurements have also been conducted at two European background sites [19]. The 6-year (1998–2004) mean of THg concentration at these coastal sites were of 1.72 ng·m^{-3} (193 ppqv) and 1.66 ng·m^{-3} (186 ppqv) at Mace Head, Ireland and Zingst, Germany, respectively. In a recent review, Sprovieri *et al.* [20] concluded that the current background concentration of atmospheric mercury was 1.5–1.7 ng·m^{-3} (168–190 ppqv) in the Northern Hemisphere.

The temporal and spatial variations of atmospheric mercury are of critical importance as it can help to understand the physical transformations of mercury species or chemical sources and sinks. A study of CAMNet data showed a common diurnal pattern for most rural sites with minimum concentrations right before sunrise and maximum concentrations around solar noon, which was attributed to nighttime depletion in the lowermost atmosphere [21]. A recent report on the U.S. Atmospheric Mercury Network (AMNet) found similar pattern [22]. GEM dissolution into dew at night and re-volatilization in the early morning were considered to be dominant factors controlling this diurnal pattern, which was similar to the results for New Hampshire reported by Mao and Talbot [23,24].

In urban areas, however, the mixing ratios of atmospheric mercury, and its seasonal and diurnal variation patterns are very different from rural sites, due to complex anthropogenic emissions, topography and meteorology. The averaged GEM concentration in urban Detroit was reported to be 2.5 ng·m^{-3} (280 ppqv) for 2004 [25]. The seasonal variation pattern showed the highest seasonal GEM concentration in summer and the lowest concentration in winter, which was different from most rural site measurements [21,22]. A study in urban Birmingham, Alabama, reported averaged GEM concentration of 2.12 ng·m^{-3} (237 ppqv) for the 2005–2008 period [26]. Measurements in Salt Lake City showed that the median GEM concentration was 226 ppqv for the 2009–2010 period [22], which was also higher than many rural site measurements.

In general, four different diurnal patterns are frequently reported in the literature, concerning the relative importance of surface emission rate *versus* the deposition rate: (1) Mercury steadily decreases at night and reaches its minimum level just before sunrise, then gradually increases to reach a maximum level at noon or in the early afternoon. The decrease at night may be related to dry deposition or uptake by wet surfaces. This pattern was observed frequently at rural and remote areas, such as some of the CAMNET and AMNet sites. (2) Mercury accumulates overnight under the influence of a low nocturnal boundary layer, and reaches its maximum before sunrise, followed by a rapid decrease after sunrise and a daily minimum in the afternoon. This pattern most commonly occurs in urban sites where strong local and regional emissions are predominant [27–29]. (3) Mercury rapidly increases right after sunrise, followed by a gradual decrease in the later hours. The rapid increase after sunrise was due to the erosion of the residual layer with elevated mercury concentrations brought down to the surface. This pattern likely occurs in rural or suburban areas, and some urban areas with special topographies [26,30–32]. (4) Mercury exhibits very small diurnal variation at elevated sites above the nocturnal boundary layer [33], and in the marine environment [23,34].

This study characterized the seasonal and diurnal variations of total gaseous mercury (THg = GEM + GOM) in the Houston area, and investigated the factors governing those changes. Understanding mercury pollution in the Houston area is complex and unique because of its distinct industrial emissions and changing meteorological conditions. Approximately 400 refineries surround the Galveston Bay in Houston, and a multitude of other industrial facilities are distributed across the region. Numerous mercury sources were reported in the 2008 EPA NEI [35] in this area (Figure 1). The Formosa Plastics Company, located to the southwest of Houston, was reported to have 1.068 metric tons of mercury emissions per year. The W.A. Parish plant, one of the largest coal-fired electrical generating plants in the U.S., is located southwest of Houston and emits 265 kg/yr. Several mercury sources, such as oil refineries and waste treatment facilities, are congregated in the Houston Ship Channel area, the vicinity to the northeast and east of our monitoring site. To the southeast of Houston, Texas City also has industrial facilities that are reported as mercury sources. The Houston-The Woodlands-Sugar Land is the fifth-largest metropolitan area in the U.S. with over 6 million people [36]. There is a dense highway/roadway network with heavy traffic in the metropolitan area. In addition to these large number of potential mercury sources, the meteorological conditions, especially the bay and sea breezes complicate the regional and local transport of mercury.

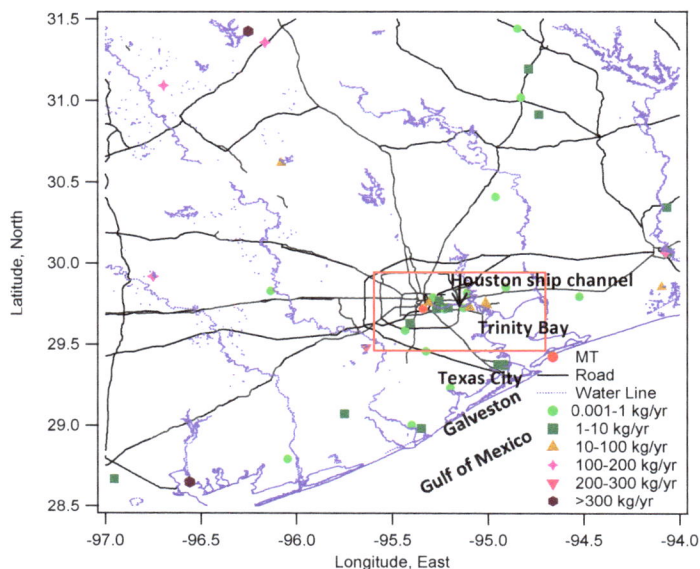

Figure 1. Facility emission sources around MT site from 2008 NEI facility data. A satellite image of the red box area is provided in the supporting material (Figure S1) for details of Houston Ship Channel area.

Atmospheric mercury was observed in the Houston area for the first time during the Texas Air Quality Study II (TexAQS II, 2005–2006). Measurements on the Moody Tower (MT) observing site on the University of Houston campus captured extremely high concentrations of PBM (79 pg·m^{-3}) downwind of the Galveston Bay refinery complex [37]. During TexAQS II concentrated plumes of GEM (up to 28,000 ppqv or 250 ng·m^{-3}) were observed repeatedly in the Houston Ship Channel area and once in the Beaumont-Port Arthur area [38]. However, no continuous measurements were reported from this metropolitan area to quantify the average ambient levels and temporal variations. In fact, few long-term ambient mercury measurements are available for urban environments. Our group has been conducting measurements of atmospheric mercury in the Houston area since August 2011. This study is our first report based on 14 months of continuous measurements. Our aim is to provide important information on mercury pollution in a heavily polluted urban area to help evaluate the regional mercury budget, and further facilitate regional modeling and policy-making processes.

2. Results and Discussion

2.1. Seasonal and Diurnal Variations

The complete time series of THg is presented in Figure 2. The median level of THg in Houston was 172 ppqv (181 ± 63 ppqv for mean ± S.D.) during our observation period (see Figure S2 and Table S1 for more statistical details), which was in agreement with the current background level in the Northern Hemisphere [20], but slightly lower than other urban sites with mercury levels higher than 2 ng·m^{-3} (224 ppqv) [22,25,28,39,40]. The majority of THg observations fell within the range of 148 ppqv (10th percentile) to 215 ppqv (90th percentile). The maximum THg mixing ratio, however, was as high as 4876 ppqv, exceeding 25 times the current global background level. The minimum level was 80 ppqv, which occurred with southerly wind that brought cleaner air from the Gulf of Mexico into urban Houston. A prominent feature of THg in the Houston area was the frequent occurrence of large THg spikes (Figure 2b). From 14-month measurements, we documented 81 well-developed spikes with THg levels higher than 300 ppqv, 34 of which were higher than 500 ppqv. Extremely large peaks with THg levels higher than 1000 ppqv were observed 12 times, and six of them were higher than 3000 ppqv. The time scale of elevated mercury in these spikes ranged from 30 minutes to a few hours. As a consequence of the THg spikes, the standard deviations of our data reached 63 ppqv, which was comparable with other urban sites measurements, such as Salt Lake City (95 ppqv [22]) and Reno (45–90 ppqv [39]).

Seasonal median levels of THg were 178 ppqv (179 ± 54 ppqv for mean ± S.D.), 161 ppqv (161 ± 80 ppqv), 172 ppqv (172 ± 26 ppqv), and 185 ppqv (186 ± 32 ppqv) for fall 2011, winter 2011, spring 2012 and summer 2012, respectively. The monthly median THg values are displayed in Figure 3. Unlike many other ambient mercury measurements, which show higher mercury in winter time [21,22], high THg in Houston area occurred in the warm seasons (June to October). It is obvious that the frequent occurrences of large THg spikes during the warm seasons, especially in August, September and October, contributed to the elevated THg levels. The great enrichments in episodic THg spikes suggested that the large pollutant plumes originated from the nearby industrial/urban emission sources. Mercury emissions from anthropogenic sources are closely linked to energy production, especially from coal-fired power plants, the largest anthropogenic mercury source in U.S. [13]. Enhanced energy production is expected in the Houston area in summer and fall. The high ambient air temperature requires energy to operate air-conditioning units. As a consequence, the ambient THg levels may increase. To illustrate this point, the state-level energy data achieved from Energy Information Administration (EIA) is presented in Figure 3. The monthly total energy produced in warm months was 50%–100% higher than in the cold months (The correlation coefficient between monthly median THg and monthly median energy production is 0.735, which is statistically significant ($p \leq 0.005$)). Besides the fluctuations in mercury sources, we noticed some changes in potential sinks. Mercury deposition in rain water has been commonly reported [41–43], indicating rainfall as an important removal mechanism for atmospheric mercury. The precipitation in Houston was observed to be consistently higher in the period from December 2011 to March 2012 (Figure 3), which coincided with low wintertime THg.

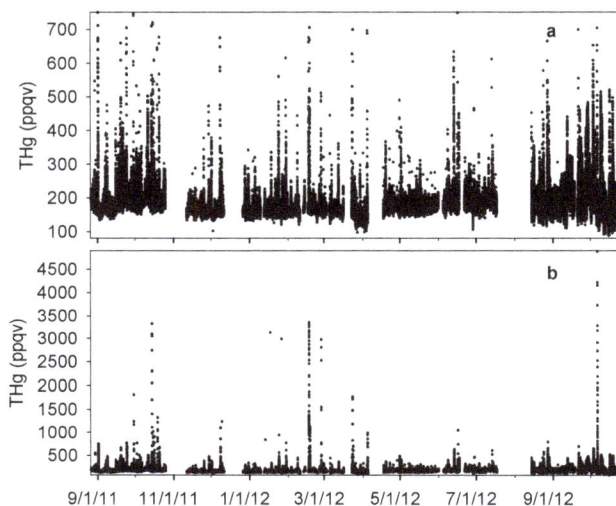

Figure 2. Complete time series of THg from MT measurements. (**a**) and (**b**) show the same data with different ranges in y axis.

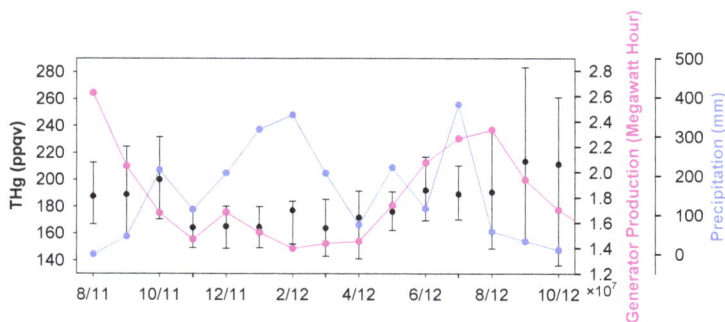

Figure 3. Monthly medians of THg, energy production and precipitation.

30

Precipitous day-to-day peaks and valleys variations were also observed as another outstanding feature of THg signal. Hourly median THg mixing ratios are displayed in Figure 4a. Despite the differences in variation amplitudes, diurnal variation patterns in fall, winter and spring were generally similar. Their diurnal patterns show a period with increasing THg levels starting around 16:00–18:00 local standard time (LST) and lasting for 2–4 hours, followed by a period with relatively constant THg levels. A sharp enhancement of THg levels started at around 04:00 LST, and then the THg reached a daily maximum at 07:00–09:00 LST. The summer diurnal pattern, however, appeared to be very different from other seasons. The THg level in summer gradually increased throughout the whole night and reached its maximum at about 07:00 LST, indicating an accumulation process of THg. Summertime had the highest THg mixing ratios during a day, and the largest diurnal variation amplitude (about 30 ppqv) among the four seasons. To conclude, the THg levels were higher at night than during the daytime. The maximum levels appeared right before sunrise, followed by rapid decreases after sunrise, and daily minimum shortly after noon. Similar diurnal patterns were observed in other urban cities, such as Guiyang, China [27], Detroit, U.S. [28], and Toronto, Canada [29].

It is interesting to note that the THg mixing ratio reached its diurnal maximum at about 07:00 LST in fall, spring and summer, while it was two hours later in winter. This phenomenon was related to the diurnal development of the PBL (Figure 4b). The PBL heights started increasing at about 06:00–07:00 LST for spring and summer and two hours later for winter. The rapid enhancements of PBL height in early morning can facilitate air mixing in three-dimensions and dilute the ambient THg within the boundary layer, and thus caused striking decreases of THg at the same time. The rate of decrease in summer was the largest (7.0 ppqv·h^{-1}), compared to those in winter (1.9 ppqv·h^{-1}), spring (2.6 ppqv·h^{-1}), and fall (4.6 ppqv·h^{-1}). The high PBL height in daytime then contributed to the low THg mixing ratios, especially in the afternoon. This phenomenon suggested that the residual layer was not a significant source for THg in Houston; instead, the industrial/urban emissions from the surface were more prominent.

However, some features of the diurnal THg variations cannot only be explained by the PBL height propagations, for example, the summertime THg mixing ratios were the highest during all times of the day even though the daytime PBL heights were the highest. It was observed that the diurnal variations of horizontal wind speeds were anti-correlated with THg levels most of the time (Figure 4c, median wind speeds were calculated using scalar values). High horizontal wind speeds can enhance horizontal mixing of polluted air with cleaner ambient air, and effectively advect the pollutants away from the urban area. The summertime wind speed was the lowest of all seasons, and the especially low wind speeds at night supported the accumulation of THg and caused high summertime THg mixing ratios. In spring,

31

low wind speeds were observed at 04:00–07:00 LST, which probably contributed to the high THg mixing ratios in this period.

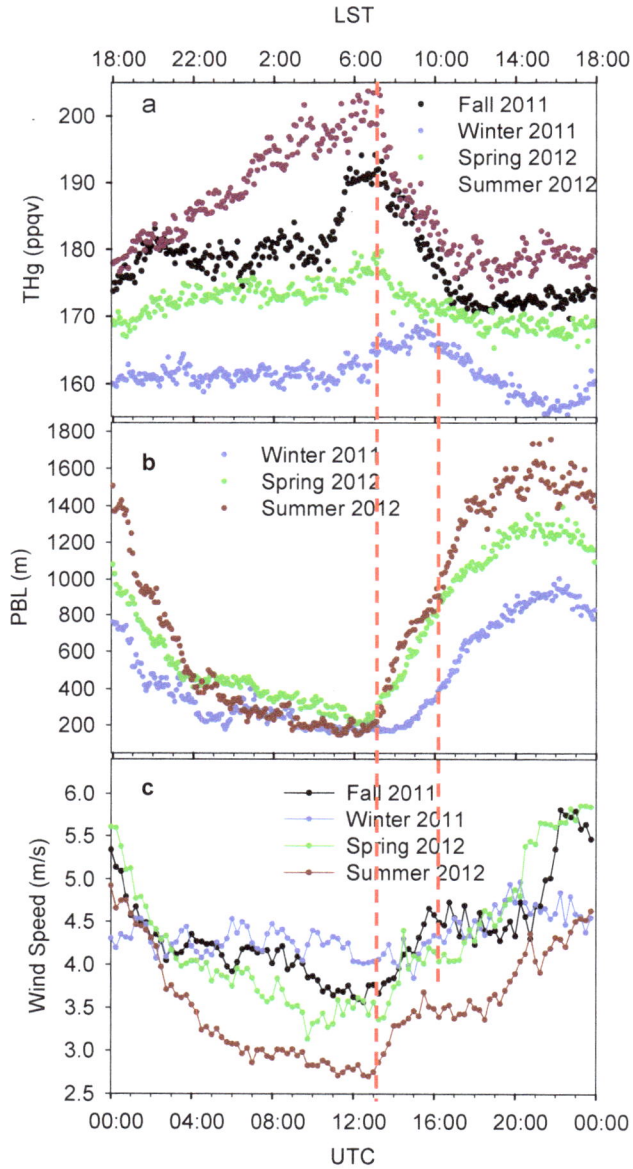

Figure 4. Seasonally diurnal variations of THg (**a**), PBL height (**b**) and horizontal wind speed (**c**). Data shows the median values with 5 min. time interval. The top x axis shows time in LST and the bottom x axis shows time in UTC. LST = UTC − 6:00.

The energy production/consumption, precipitation, PBL height, and horizontal wind speed played important roles in the seasonal and diurnal variations of THg; however, we cannot quantify the contribution of each factor from our observations. Regional modeling with a good emission inventory and dynamic processing is necessary for further investigation.

It is unclear whether photochemical reactions play an important role in determining THg mixing ratios in the Houston area. Previous research suggests that the main sink of GEM in the atmosphere is oxidation to GOM [44]. The GOM then can attach to particles and be transformed to PBM. Both GOM and PBM are easily removed from the air via wet and dry deposition [21,45], which will eventually reduce the mixing ratio of THg. From our observations, the diurnal THg mixing ratios remained constant from 12:00 LST to 16:00 LST in fall, spring and summer. During these times, photochemical reactions should be actively changing due to the fluctuations of solar radiation and variations in the abundance of oxidants. The effect of photochemical reactions may be obscured due to complex local emissions and meteorological conditions.

Natural sources can also influence the seasonal and diurnal variations of ambient THg levels; however, we are unable to quantify the contributions of natural emissions in metropolitan Houston due to a lack of direct measurements. However, a large portion of vegetation in the Houston area is evergreen; with no snow in winter, we expect small contributions from natural emissions to the THg seasonal variations.

2.2. Wind Induced Influences

In addition to the wind speed, the wind direction also exerts considerable impacts on THg variations, suggesting the importance of local/regional transport. In Figure 5, the percentage values on the R axis shows the frequency (%) of THg coming from a certain range (22.5°) of wind directions. The predominant wind directions in our observed period were south and southeast directions (120°–190°) (Figure 5a), which were from the Gulf of Mexico. The overall frequency from these directions accounts for about 40%, after summing up the percentage values from 120°–190° directions in Figure 5a. Cleaner air masses with especially low THg levels (<150 ppqv) were observed from these directions. However, the refinery facilities emissions in Texas City may also advected to Houston area as we obtained about 6% of air masses with THg levels higher than 200 ppqv from the same directions.

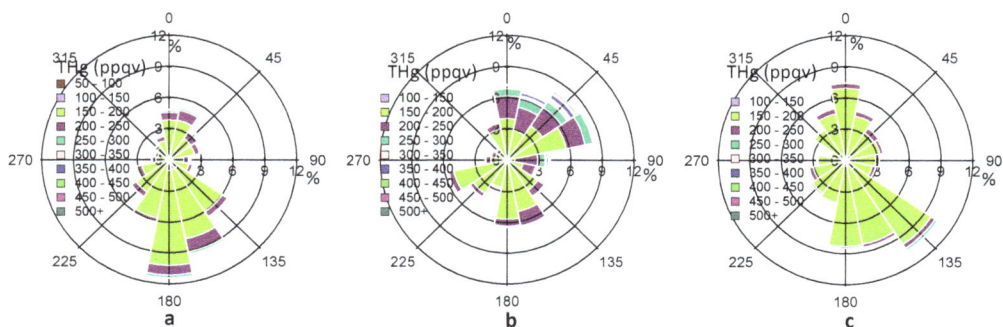

Figure 5. Complete THg *versus* wind direction (**a**), THg *versus* wind direction in HMP (**b**), THg *versus* wind direction in LMP (**c**). The color scale shows the ranges of THg mixing ratios. The percentage values on the R axis shows the frequency of THg coming from a certain range (22.5°) of wind directions.

Wind direction in Houston varied diurnally. From Figure 4a, we noticed elevated THg mixing ratios at about 05:00–11:00 LST, and lower mercury mixing ratios at about 12:00–17:00 LST. To evaluate the impacts of shifting wind direction, we defined a daily high mercury period (HMP) of 05:00–11:00 LST and a daily low mercury period (LMP) of 12:00–17:00 LST. Figure 5b,c depict the wind rose for HMP and LMP in 2011 fall, respectively. In the HMP, air masses with high THg levels (200 ppqv) mainly came from north, northeast, and east sectors, where urban and industrial influences were remarkable, especially in the Houston Ship Channel area. In the LMP, air masses mainly came from the south and southeast sectors, where cleaner marine air seemed to dominate. The median THg level in HMP was statistically higher than that in LMP (14 ppqv higher, p ≤ 0.001). Analysis of other seasons (not shown) also suggested similar influences from shifting source regions. It was also observed that consistent northerly winds with high wind speeds (>4 m/s, occurred in December) yielded THg levels ≥ 150 ppqv, and CO levels ≥ 125 ppbv. In comparison, consistent southerly winds with high wind speeds (occurred in April) produced THg levels as low as 120 ppqv and CO levels as low as 85 ppbv. The large discrepancies between southerly air and northerly air highlight the significance of local urban/industrial influences on elevated THg mixing ratios in the Houston area.

The sea breeze exerted significant and complex impacts on THg levels in the Houston area. The sea breeze is driven by diurnally uneven heating in coastal areas, which produces warmer temperatures over land than over water during the day, and cooler land temperatures at night [46]. As a result, a sea breeze is produced when the air flows from the sea to the land at low altitude (<500 m). The low level wind vector then rotates through a clockwise (Northern Hemisphere) cycle under the influence of the Coriolis force [46,47]. As the sea breeze passes through the shoreline, it can bring cleaner marine air to inland areas that can dilute urban/industrial emissions. The sea breeze can also bring back ashore the aged polluted air that was once transported offshore by other wind systems. A detailed analysis on the wind system in Houston area reported that the afternoon sea breeze was responsible for large O_3 enrichment events, by bringing back aged, polluted air masses to the urban area [46].

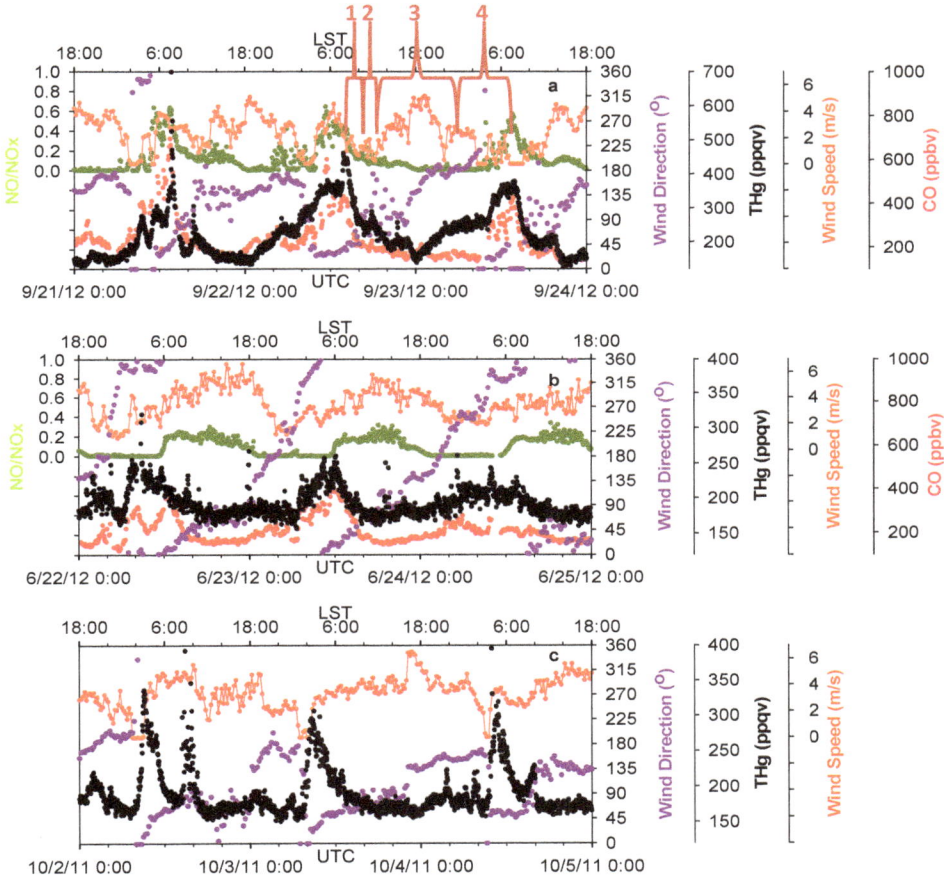

Figure 6. Time series of THg, CO, NO/NO$_x$, wind speed and wind direction. (**a,b,c**) show the influences of three common wind patterns on THg levels.

Our observations found similar effects from the bay breezes (from Trinity Bay and Galveston Bay, southeast of MT) and the sea breeze (from the Gulf of Mexico, south of MT). Figure 6a presents an example of sea breeze impacts on THg diurnal variations. Four stages of THg variations with corresponding wind shifts are marked for illustration. In stage one, THg mixing ratios decreased significantly, partially because of the increase in PBL height, and also the northerly winds that were

35

controlling this area and pushing the urban pollutants offshore. In the second stage, the bay breeze and sea breeze built up and intruded inland as southerly winds. The bay breeze and sea breeze then moved into urban Houston, where they met the northerly flow and formed a sea breeze front. With the influence of a sea breeze front, the horizontal wind speeds decreased to less than 2 m/s, causing a stagnant period, which would not favor vigorous air diffusion or transport. In addition, the polluted flow transported offshore then returned back to Houston with the southerly winds. We observed slightly elevated THg and CO in this period. It is known that NO/NO_x can serve as an indicator for the relative age of air masses, because of the short life time of NO. The NO/NO_x values in this stage were less than 0.3, indicating the influences of aged air masses in which NO was oxidized. In this stage, a convergent zone formed where sea breeze front was concentrated. As a result, the updraft can bring polluted air to higher altitudes and enable long-distance transport of THg. In the third stage, the sea breeze moved farther inland and cleaner marine air behind the sea breeze front controlled air quality in the Houston area. The THg levels were observed to be the lowest of the day. In the final stage, the sea breeze faded out at night and northerly wind again controlled this area. Significant enhancement of THg occurred within the nocturnal boundary layer, along with high CO and NO/NO_x (~0.6), indicating the influences of fresh urban/industrial emissions. These four stages occurred frequently day-to-day during the warm season.

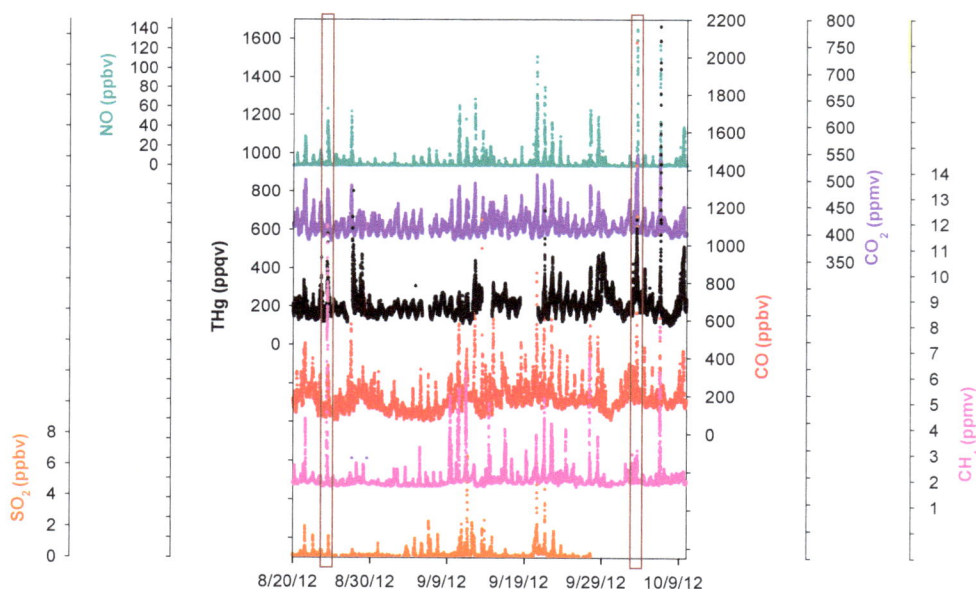

Figure 7. Time series of THg, CO, CO_2, NO, SO_2, and CH_4. The two boxes are comparisons of two episodes with different CH_4 mixing ratios.

Another diurnal pattern of THg that also involved the sea breeze is depicted in Figure 6b. Clockwise rotation of wind directions diurnally were accompanied with obvious THg diurnal changes in which daily low THg mixing ratios occurred when winds were southeast and southerly flow dominated. The diurnally 360° changes in wind direction occurred in March through October, but not in November through February. This pattern was documented on about 30 days in our 14-month data span. Another interesting feature we observed in THg diurnal patterns was that THg levels tended to peak right after abrupt changes in wind direction (45 degree change, normally from south winds to north winds), with few exceptions. This condition occurred on 90 days over 14 months, while the wind shifts occurred mostly around 02:00–04:00 LST (Figure 6c). Please note that while sea breeze constantly corresponds with relatively low levels of THg in the afternoon and early evening in Figure 6b,c, the aged plumes were not always observed as in Figure 6a, which may due to the differences in the strength (relative to northerly) and the daily propagation of the sea breeze.

2.3. Relationship of THg with Key Trace Gases

An advantage of this study was that multiple trace gas species were sampled at the same location (MT) along with THg with high time resolutions. Investigating the relationships between THg and other key trace gases can provide important information concerning mercury sources. Figure 7 displaced the time series of SO_2, NO, CO, CO_2, CH_4, and THg. A significant feature we noticed was the common co-occurrence of THg peaks with SO_2, NO, CO, CO_2, CH_4 peaks in pollution plumes. A zoom-in picture depicting the maximum THg spike observed during this study is presented in Figure 8. This episode corresponded with air masses coming from the northeast of our sampling site, which is the direction of Houston Ship Channel. This highly concentrated pollution plume with THg mixing ratios as large as 4876 ppqv was related to combustion tracers, such as CO, CO_2, and NO. In addition, this THg peak occurred coincidentally with a peak in CH_4, which is presumably released from oil and natural gas operations, landfills, and waste treatment facilities. In this episode significant enhancements of NO, THg, CO, CO_2, and CH_4 started at 04:30 LST and peaked at 06:45–07:45 LST. The maximum levels of each species in this plume were 4876 ppqv for THg, 1053 ppbv for CO, 45 ppbv for NO, 549 ppmv for CO_2 and 8.0 ppmv for CH_4. The close correspondence between THg and other species in the THg episodes is clearly evident, and thus it is possible that THg came from similar sources as some of the other trace gases. However, the peaks signatures varied greatly in different pollution plumes. Extreme THg peaks sometimes were in conjunction with exceptional large peaks of CH_4 (e.g., Figure 7 left box, 11.8 ppmv), but sometimes with slightly elevated CH_4 (e.g., Figure 7 right box, 3.1 ppmv). Correlations of THg with other trace gases were diffuse overall;

correlation coefficients (R values) between THg and those species were low (R < 0.43, see Table S2) albeit their correlations were statistically significant (95%), suggesting the possible contributions from diverse source types in the Houston area.

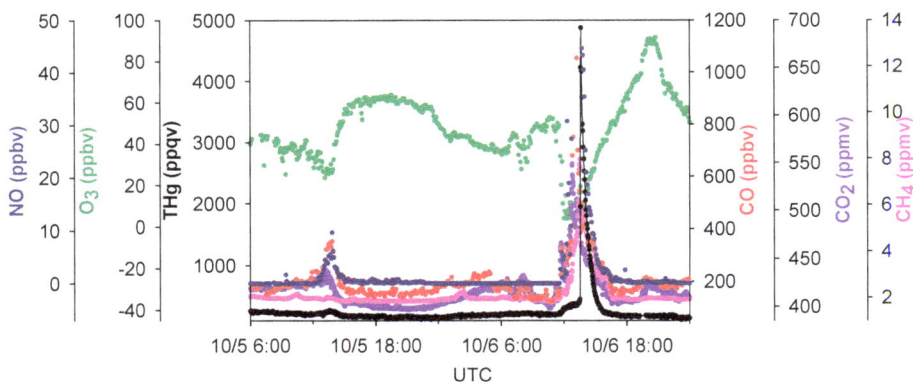

Figure 8. Time series of THg, CO, CO_2, NO, O_3 and CH_4 in the largest THg plume.

Detailed analyses of diurnal variations of THg and other species can also provide valuable information on mercury sources and sinks. Figure 9 exhibits the summertime diurnal variations of THg, NO, NO_2, CO, O_3, CO_2, and CH_4. Interestingly, THg, CO_2 and CH_4 had similar diurnal patterns; however, it is still difficult to determine whether these patterns were resulting from meteorological forcing, such as PBL height and winds, and/or their similar emission sources. The diurnal patterns of NO, NO_2, and CO, were ubiquitously distinguished from those of THg, CO_2, and CH_4. Vehicle emissions are believed to be the dominant sources of NO, NO_2, and CO in urban areas [48,49]. We noticed that the mixing ratios of CO and NO_2 started increasing simultaneously at 03:30 LST (04:30 Local Daylight Saving Time) and reached their maximums at 06:00–07:00 LST. The mixing ratios of NO started increasing after sunrise, about two hours later compared to NO_2 and CO. Ozone titration was also observed at the same time when elevated CO and NO_2 occurred (note that O_3 photochemistry in this site is limited by VOCs, instead of NO_x [50]). We believe these were traffic signals; NO was emitted by vehicles subsequently titrating O_3 and being converted to NO_2, but without sunlight NO_2 builds up. The diurnal pattern of THg tracked those of CO_2 and CH_4, instead of CO and NO_2, suggested that vehicle emissions may not exert outstanding impacts on THg levels in summer. The importance of vehicle emissions as a THg source is yet to be further investigated.

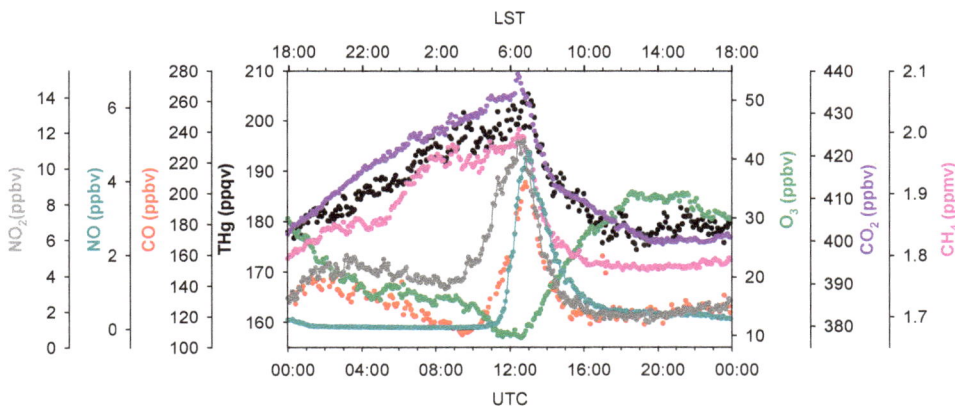

Figure 9. Diurnal variations of THg, CO, CO_2, NO, NO_2, O_3 and CH_4 in summer 2012. Data shows the median values within a 5 min. time interval.

2.4. Mercury Episodes

As we reported previously, 81 spikes were documented with THg mixing ratios greater than 300 ppqv. We calculated the enhancement ratios (ER) in these plumes to retrieve information about point sources. Enhancement ratios are obtained by dividing the excess species (THg in this study) concentrations measured in a pollution plume by the excess concentration of a reference gas, for example, CO_2, CO and CH_4 in this study. Enhancement ratios are commonly expressed in molar ratios, and the ambient background levels of each gas must be subtracted to get the "excess" values. For example, the ER for THg relative to CO is:

$$ER_{THg/CO} = \frac{\Delta THg}{\Delta CO} = \frac{THg_{plume} - THg_{background}}{CO_{plume} - CO_{background}} \tag{1}$$

Enhancement ratio is a different parameter compared to emission ratio because the measurement is conducted from a downwind location instead of at the source. Since THg, CO, CO_2 and CH_4 are not highly reactive species, it is possible that ERs remain close to emission ratios of a specific source in short-distant transport. Further we may assume that pollution plumes coming from the same direction were from the same emission source if they have similar ERs. For ER calculation, we characterized the air masses with the lower 25th percentiles of CO mixing ratios as background air [23], and the corresponding medians values for this part of data were used as the background levels so as to minimize the possible influence of local sinks. Those values were 172 ppqv for THg, 112.6 ppbv for CO, 1.79 ppmv for CH_4, and 403.3 ppmv for CO_2. Figure 10 presents the ER values for $\Delta THg/\Delta CO$, $\Delta THg/\Delta CO_2$

39

and $\Delta THg/\Delta CH_4$, *versus* wind directions. The CO_2 measurements are only available from March 2012 and CH_4 measurements from June 2012. Thus, fewer episodes were available for $\Delta THg/\Delta CO_2$ and $\Delta THg/\Delta CH_4$ compared to $\Delta THg/\Delta CO$.

The patterns of $\Delta THg/\Delta CO$ and $\Delta THg/\Delta CO_2$ *versus* wind direction shared some similarities (Figure 10). Pollution plumes with THg mixing ratios higher than 500 ppqv tended to produce higher ERs, compared to those with THg mixing ratios between 300 ppqv and 500 ppqv. A large percentage of THg episodes with high CO or CO_2 mixing ratios originated from the 30°–120° direction, the Houston Ship Channel area. Note that, from Figure 5a, air masses from this direction only accounted for about 17% of the total air mass measured at MT. This means that most THg episodes originated from directions that were the least favored wind directions.

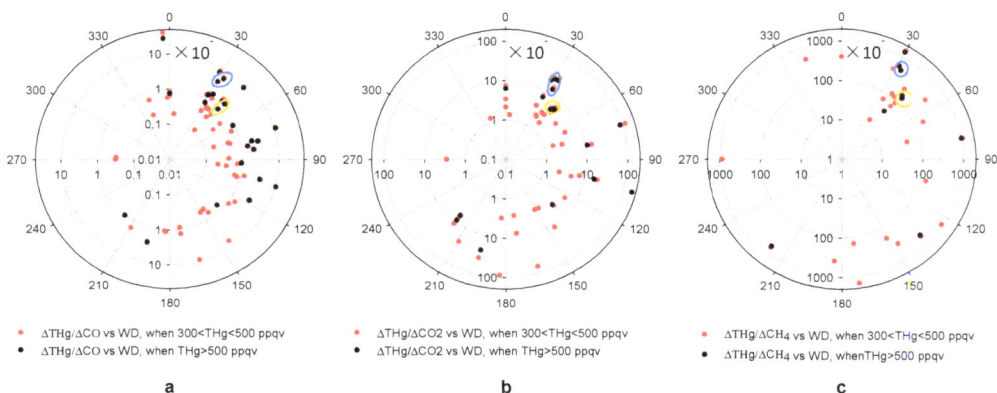

Figure 10. Enhancement ratios *versus* wind direction in high THg episodes. (**a,b,c**) show $\Delta THg/\Delta CO$, $\Delta THg/\Delta CO_2$ and $\Delta THg/\Delta CH_4$ *versus* wind direction, respectively. The yellow circles point out two ERs potentially from the same source at ~40° direction and the blue circles represent another two episodes from a source at ~30° direction.

For now, we are unable to correlate these pollution plumes with specific sources due to the absence of detailed and accurate source signatures. The EPA NEI [13] provides annual total emissions for THg and CO, and the Greenhouse Gases (GHGs) Emissions Data [51] also provides annual total emissions of CO_2 and CH_4 for significant emission sources. We attempted to compare our ER values with emission ratios calculated from the EPA inventory. In the northeast direction from MT, the direction with a large number of THg episodes, we found only 2 point sources within a short distance (<80 km) from MT that were documented in the 2008 EPA NEI as mercury sources. From the ER *versus* wind direction plot, we noticed 2 episodes with similar $\Delta THg/\Delta CO$, $\Delta THg/\Delta CO_2$ and $\Delta THg/\Delta CH_4$ values from the 40° direction (Figure 10, yellow circles). Considering the large elevations of THg, CO, CO_2 and CH_4

in these episodes, we suspect that they could be attributed to local sources. A waste treatment facility located about 5 km away from MT in that direction was reported by the 2008 NEI and 2010 GHGs Emissions Data as a source for THg, CO, CO_2 and CH_4. The other point source that was likely associated with another 2 episodes with similar $\Delta THg/\Delta CO$, $\Delta THg/\Delta CO_2$ and $\Delta THg/\Delta CH_4$ values were from the 30° direction (Figure 10, blue circles). We found an iron and steel casting facility from EPA NEI 9 km away from MT. Table 1 shows the comparisons between ER values calculated from observations and emission ratios calculated from the EPA inventory (please note that this iron and steel casting facility is not listed in the 2010 GHGs Emissions Data, which may be due to the fact that the 2010 GHGs Emissions Data does not include sources that have annual emissions of less than 25,000 metric tons of CO_2e. For the emission ratios calculations, we assume the CO_2 and CH_4 emissions from this facility were 25,000 metric tons of CO_2e). The differences between these two ratios were significant, up to a few orders of magnitude. It is highly possible that the EPA emission inventory may have inaccurate estimations of the total emissions. It is also possible that some emission sources may not be documented in the EPA inventory.

Table 1. Comparisons between ER values derived from MT observations and emission factors calculated from EPA Inventory.

Facility	$\Delta THg/\Delta CO$		$\Delta THg/\Delta CO_2$		$\Delta THg/\Delta CH_4$	
	EPA	Observation	EPA	Observation	EPA	Observation
30°, Iron and steel casting (blue circles)	1.77×10^{-5}	8.9×10^{-7}~1.49×10^{-6}	5.04×10^{-7}	5.3×10^{-9}	9.78×10^{-3}	1.40×10^{-7}
40°, Waste treatment (yellow circles)	5.70×10^{-8}	3.7×10^{-6}~5.4×10^{-6}	1.77×10^{-10}	1.3×10^{-8}~2.3×10^{-8}	1.35×10^{-9}	5.0×10^{-7}~5.8×10^{-7}

3. Experimental Section

Continuous THg measurements were conducted atop Moody Tower (29.71°N, 95.34°W), an elevated site (75 m above ground level) located on the main campus of the University of Houston (Figure 1). This urban site is located about 2–4 km away from major freeways and industrial facilities, and 4–5 km away from downtown Houston. The campus is surrounded by vegetation covered land surfaces. It is 35 km west of the Galveston Bay and 70 km northwest of Galveston, TX and the Gulf of Mexico. The data reported in this study was acquired from 26 August 2011 to 25 October 2012.

Total gaseous mercury was measured with a Tekran 2537A automated instrument. The Tekran instrument uses cold vapor atomic florescence (CVAF) detection in sequential dual channels with 2.5 min. time interval resolution, yielding a detection limit of 5–10 ppqv. A Teflon filter (1 μm pore size) was used to remove aerosols at the entrance of the inlet, and was changed weekly. The sample line was

41

6 m long and composed of Silco-Steel Sulfinert coated (Restek) stainless steel. We conducted extensive testing of the passing efficiency of GOM through Silco Steel Sulfinert tubing using permeation sources of $HgBr_2$ and $HgCl_2$. This tubing passed these two mercury species with 100% efficiency. Heating of the tubing was not required to attain this high passing efficiency. We also tested teflon filters and found 100% passing efficiency at the flow rates used for sampling. Instrument calibration was checked weekly using the internal mercury permeation source. The internal standard was verified every six months using a Tekran 2505 unit. Judging from the inter-comparison with 2 other co-located Tekran instruments, the overall precision of this instrument was $\pm$ 10% [34]. The accuracy of the mercury measurement was estimated to be $\pm$ 5%. In this work, mercury is reported using mixing ratios in ppqv so that it is directly comparable with other trace gases. Under Standard Temperature and Pressure conditions, 1 ng$\cdot$m^{-3} equals to 112 ppqv. Except for the time series analysis, the THg data was averaged to 5 min. for various comparisons.

Carbon monoxide was measured using a Thermal Environmental (TE) CO Analyzer (48C-TLE). Ozone was measured with a TE O_3 U.V. photometric analyzer (49C). Sulfur dioxide was measured using a TE 43C instrument. The CO, O_3, and SO_2 analyzers shared the same inlet, whose sample lines were around 10–12 m long. Nitric oxide, nitrogen dioxide and NO_x were measured using the TE NO-NO_2-NO_x chemiluminescence analyzer (42C) with a blue light converter to photolyze NO_2 to NO. The inlet of NO-NO2-NOx analyzer was 5 m away from the inlet of inlets of CO, O_3, and SO_2 analyzers, and the sample line was 7.5 m long. These inlets used PFA Teflon tubing, and both of them were about 6 m away from the THg inlet. To maintain accurate measurements, these four instruments were adjusted to zero and the span checked on a daily basis. Multi-point calibration was conducted every two weeks. The original time resolution of CO, O_3, SO_2, NO, NO_2, and NO_x was 10 s, and these data were average to 5 min. in this study. Carbon dioxide was measured using a LI-COR LI-7000 analyzer and CH_4 was measured with a PICARRO G2132-i instrument. Both instruments calibration was checked monthly using standards obtained from Scott-Marrin with $\pm$1% NIST traceable quality. The sample lines of LI-COR and PICARRO instruments were about 6 m long, and their inlets were co-located with THg inlet. The original time resolutions of CO_2 and CH_4 were 1 min. and 1.5 s, respectively. Except for the time series analysis, both dataset were average to 5 min. for comparison to other species.

Basic meteorological parameters, including ambient temperature, wind speed, wind direction, precipitation, relative humidity, and pressure were monitored with a collection of Campbell Scientific sensors with 10 s resolution and averaged to 5 min. for this study.

Planetary boundary layer (PBL) height was measured at another site (29.72°N, 95.34°W), also located on the main campus of University of Houston. This was a ground level site, about 700 m away from Moody Tower site. A Vaisala CL31 Ceilometer was used to measure PBL height with 5 m height resolution, and PBL height was reported every 5 minute. The PBL height measurements from this instrument agreed very well with our radiosonde and ozonesonde measurements. More details concerning the performance and algorithm settings of this instrument are discussed in Haman *et al.* [52].

4. Conclusions

This study provides the first assessment of mercury pollution in the highly populated and industrial city of Houston, Texas. Our continuous measurements show that the median THg level was consistent with the current global background level; however, a predominant feature of THg in the Houston area was the frequent occurrence of large peaks ($\geq$300 ppqv). The primary focus of this study has been placed on characterizing the seasonal and diurnal variation patterns of THg in Houston. We found that the highest seasonal median THg mixing ratios occurred in summer and the lowest ones occurred in winter. The seasonal patterns of THg were closely linked to the frequency of THg episodes, and also the energy production/consumption and precipitation (rainfall) in the area. Diurnally, the THg levels were higher at night than at daytime. The maximum THg levels appeared right before sunrise, followed by rapid decreases after sunrise. The diurnal variations were clearly under the influence of PBL height and horizontal winds, including the complex sea breeze system in the Houston area. Our results combined with those of Banta *et al.* [46] suggest a very active influence of the sea breeze in the Houston area, which can either increase or decrease the urban THg levels over a day depending on the position of the sea breeze front.

Special emphasis has been placed on the THg episodes. Highly concentrated pollution plumes originated from urban Houston agglomeration revealed that THg levels were related to the combustion tracers such as CO, CO_2, and NO. The occurrences of THg peaks also coincided with peaks in CH_4, which was presumably released from oil and natural gas operations, landfill, and waste treatments. The frequent co-occurrences of THg episodes with peaks in CO, CO_2, CH_4, NO_x, and/or SO_2 suggested possible contributions from similar sources. We noticed that the majority of the THg episodes were from the Houston Ship Channel Area. When our observed ERs values were compared with the emission ratios calculated from the annual total emissions documented by EPA inventory, we found that the EPA 2008 NEI and 2010 GHGs Emissions Data may not reflect the actual emissions from industrial facilities.

Future work is warranted to quantify the contributions of each factor on THg variations. This study established a strong connection between PBL height and THg mixing ratios, which suggested the possible influence of vertical mixing on

THg variations. The influence of local/regional transport was also convoluted by the sea breeze system. We suggest using atmospheric mercury modeling to better understand the complexity and quantify the relative impacts of these factors. The modeling should benefit from good simulations of PBL development and sea breeze circulation, as well as a good estimation of mercury emissions. Consider the current uncertainty in EPA emission inventory, we also recommend further investigations on urban/industrial emissions using our mobile laboratory.

Acknowledgments: We would like to thank Huiting Mao for helpful review of the manuscript. We also thank Zuoyuan Sun for his help in the QA/QC process of our CO, NO_x, and O_3 data. Financial support for this work was provided by the National Science Foundation under grant # ATM1141713.

Author Contributions: Xin Lan (data analysis, manuscript writing), Robert Talbot (measurements, manuscript writing), Patrick Laine (data collection), Barry Lefer (data collection), James Flynn (data collection, data quality assurance and quality control) and Azucena Torres (data collection).

Conflicts of Interest: The authors declare no conflict of interest.

References and Notes

1.   United States Environmental Pretention Agency. Mercury Study Report to Congress 1997. Available online: www.epa.gov/mercury/report.html (accessed on 07 March 2014).

2.   United States Environmental Pretention Agency. Health Effect of Mercury. Available online: http://www.epa.gov/hg/effects.htm (accessed on 07 March 2014).

3.   Lindberg, S.E.; Stratton, W.J. Atmospheric mercury speciation: Concentrations and behavior of reactive gaseous mercury in ambient air. *Environ. Sci. Technol.* **1998**, *32*, 49–57.

4.   Valente, R.J.; Shea, C.; Humes, K.L.; Tanner, R.L. Atmospheric mercury in the great smoky mountains compared to regional and global levels. *Atmos. Environ.* **2007**, *41*, 1861–1873.

5.   Choi, H.D.; Holsen, T.M.; Hopke, P.K. Atmospheric mercury (Hg) in the adirondacks: Concentrations and sources. *Environ. Sci. Technol.* **2008**, *42*, 5644–5653.

6.   Ferrara, R.; Mazzolai, B.; Lanzillotta, E.; Nucaro, E.; Pirrone, N. Volcanoes as emission sources of atmospheric mercury in the mediterranean basin. *Sci. Total Environ.* **2000**, *259*, 115–121.

7.   Sigler, J.M.; Lee, X. Recent trends in anthropogenic mercury emission in the northeast united states. *J. Geophys. Res.: Atmos.* **2006**.

8.   Xin, M.; Gustin, M.S. Gaseous elemental mercury exchange with low mercury containing soils: Investigation of controlling factors. *Appl. Geochem.* **2007**, *22*, 1451–1466.

9.   Friedli, H.R.; Radke, L.F.; Lu, J.Y.; Banic, C.M.; Leaitch, W.R.; MacPherson, J.I. Mercury emissions from burning of biomass from temperate north american forests: Laboratory and airborne measurements. *Atmos. Environ.* **2003**, *37*, 253–267.

10. Friedli, H.R.; Radke, L.F.; Prescott, R.; Hobbs, P.V.; Sinha, P. Mercury emissions from the august 2001 wildfires in washington state and an agricultural waste fire in oregon and atmospheric mercury budget estimates. *Glob. Biogeochem. Cy.* **2003**.

11. Mason, R.P.; Sheu, G.R. Role of the ocean in the global mercury cycle. *Glob. Biogeochem.Cy.* **2002**.

12. Pirrone, N.; Hedgecock, I.M.; Cinnirella, S.; Sprovieri, F. Overview of major processes and mechanisms affecting the mercury cycle on different spatial and temporal scales. In Proceedings of European Research Course on Atmospheres (ERCA 9)—From the Global Mercury Cycle to the Discoveries of Kuiper Belt Objects, Grenoble, France, 11 January–12 February 2010; EDP Sciences: Grenoble, France, 2010; pp. 3–33.

13. United States Environmental Pretention Agency. 2008 National Emission Inventory Technical Support Document. Available online: http://www.epa.gov/ttnchie1/net/2008inventory.html (accessed on 07 March 2014).

14. Conaway, C.H.; Mason, R.P.; Steding, D.J.; Flegal, A.R. Estimate of mercury emission from gasoline and diesel fuel consumption, san francisco bay area, california. *Atmos. Environ.* **2005**, *39*, 101–105.

15. Won, J.H.; Park, J.Y.; Lee, T.G. Mercury emissions from automobiles using gasoline, diesel, and lpg. *Atmos. Environ.* **2007**, *41*, 7547–7552.

16. Landis, M.S.; Lewis, C.W.; Stevens, R.K.; Keeler, G.J.; Dvonch, J.T.; Tremblay, R.T. Ft. Mchenry tunnel study: Source profiles and mercury emissions from diesel and gasoline powered vehicles. *Atmos. Environ.* **2007**, *41*, 8711–8724.

17. Wilhelm, S.M. Estimate of mercury emissions to the atmosphere from petroleum. *Environ. Sci. Technol.* **2001**, *35*, 4704–4710.

18. Temme, C.; Blanchard, P.; Steffen, A.; Banic, C.; Beauchamp, S.; Poissant, L.; Tordon, R.; Wiens, B. Trend, seasonal and multivariate analysis study of total gaseous mercury data from the canadian atmospheric mercury measurement network (CAMNet). *Atmos. Environ.* **2007**, *41*, 5423–5441.

19. Kock, H.H.; Bieber, E.; Ebinghaus, R.; Spain, T.G.; Thees, B. Comparison of long-term trends and seasonal variations of atmospheric mercury concentrations at the two european coastal monitoring stations mace head, ireland, and zingst, germay. *Atmos. Environ.* **2005**, *39*, 7549–7556.

20. Sprovieri, F.; Pirrone, N.; Ebinghaus, R.; Kock, H.; Dommergue, A. A review of worldwide atmospheric mercury measurements. *Atmos. Chem. Phys.* **2010**, *10*, 8245–8265.

21. Kellerhals, M.; Beauchamp, S.; Belzer, W.; Blanchard, P.; Froude, F.; Harvey, B.; McDonald, K.; Pilote, M.; Poissant, L.; Puckett, K.; *et al.* Temporal and spatial variability of total gaseous mercury in canada: Results from the canadian atmospheric mercury measurement network (CAMNet). *Atmos. Environ.* **2003**, *37*, 1003–1011.

22. Lan, X.; Talbot, R.; Castro, M.; Perry, K.; Luke, W. Seasonal and diurnal variations of atmospheric mercury across the us determined from amnet monitoring data. *Atmos. Chem. Phys.* **2012**, *12*, 10569–10582.

23. Mao, H.; Talbot, R. Speciated mercury at marine, coastal, and inland sites in new england—Part 1: Temporal variability. *Atmos. Chem. Phys.* **2012**, *12*, 5099–5112.

24. Mao, H.; Talbot, R.; Hegarty, J.; Koermer, J. Speciated mercury at marine, coastal, and inland sites in new england—Part 2: Relationships with atmospheric physical parameters. *Atmos. Chem. Phys.* **2012**, *12*, 4181–4206.

25. Liu, B.; Keeler, G.J.; Dvonch, J.T.; Barres, J.A.; Lynam, M.M.; Marsik, F.J.; Morgan, J.T. Urban-rural differences in atmospheric mercury speciation. *Atmos. Environ.* **2010**, *44*, 2013–2023.

26. Nair, U.S.; Wu, Y.; Walters, J.; Jansen, J.; Edgerton, E.S. Diurnal and seasonal variation of mercury species at coastal-suburban, urban, and rural sites in the southeastern united states. *Atmos. Environ.* **2012**, *47*, 499–508.

27. Feng, X.B.; Shang, L.H.; Wang, S.F.; Tang, S.L.; Zheng, W. Temporal variation of total gaseous mercury in the air of guiyang, china. *J. Geophys. Res.: Atmos.* **2004**.

28. Liu, B.; Keeler, G.J.; Dvonch, J.T.; Barres, J.A.; Lynam, M.M.; Marsik, F.J.; Morgan, J.T. Temporal variability of mercury speciation in urban air. *Atmos. Environ.* **2007**, *41*, 1911–1923.

29. Song, X.; Cheng, I.; Lu, J. Annual atmospheric mercury species in downtown toronto, Canada. *J. Environ. Monit.* **2009**, *11*, 660–669.

30. Snyder, D.C.; Dallmann, T.R.; Schauer, J.J.; Holloway, T.; Kleeman, M.J.; Geller, M.D.; Sioutas, C. Direct observation of the break-up of a nocturnal inversion layer using elemental mercury as a tracer. *Geophys. Res. Lett.* **2008**.

31. Weiss-Penzias, P.; Gustin, M.S.; Lyman, S.N. Observations of speciated atmospheric mercury at three sites in nevada: Evidence for a free tropospheric source of reactive gaseous mercury. *J. Geophys. Res.: Atmos.* **2009**.

32. Peterson, C.; Gustin, M.; Lyman, S. Atmospheric mercury concentrations and speciation measured from 2004 to 2007 in Reno, Nevada, USA. *Atmos. Environ.* **2009**, *43*, 4646–4654.

33. Sigler, J.M.; Mao, H.; Sive, B.C.; Talbot, R. Oceanic influence on atmospheric mercury at coastal and inland sites: A springtime noreaster in new england. *Atmos. Chem. Phys.* **2009**, *9*, 4023–4030.

34. Mao, H.; Talbot, R.W.; Sigler, J.M.; Sive, B.C.; Hegarty, J.D. Seasonal and diurnal variations of hg degrees over new england. *Atmos. Chem. Phys.* **2008**, *8*, 1403–1421.

35. United States Environmental Pretention Agency. 2008 National Emission Inventory. Available online: http://www.epa.gov/ttnchie1/net/2008inventory.html (accessed on 7 March 2014).

36. United States Census. Annual Estimates of the Population. Available online: http://www.census.gov/popest/data/metro/totals/2012/ (accessed on 7 March 2014).

37. Brooks, S.; Luke, W.; Cohen, M.; Kelly, P.; Lefer, B.; Rappenglueck, B. Mercury species measured atop the moody tower tramp site, houston, texas. *Atmos. Environ.* **2010**, *44*, 4045–4055.

38. Final Rapid Science Synthesis Report: Findings from the Second Texas Air Quality Study (TexAQS II); A Report to the Texas Commission on Environmental Quality. Available online: http://aqrp.ceer.utexas.edu/docs/RSSTFinalReportAug31.pdf (accessed on 7 March 2014).

39. Stamenkovic, J.; Lyman, S.; Gustin, M.S. Seasonal and diel variation of atmospheric mercury concentrations in the reno (Nevada, USA) airshed. *Atmos. Environ.* **2007**, *41*, 6662–6672.

40. Engle, M.A.; Tate, M.T.; Krabbenhoft, D.P.; Schauer, J.J.; Kolker, A.; Shanley, J.B.; Bothner, M.H. Comparison of atmospheric mercury speciation and deposition at nine sites across central and eastern north america. *J. Geophys. Res.: Atmos.* **2010**.

41. Pehkonen, S.O.; Lin, C.J. Aqueous photochemistry of mercury with organic acids. *J. Air Waste Manage.* **1998**, *48*, 144–150.

42. Keeler, G.J.; Gratz, L.E.; Al-Wali, K. Long-term atmospheric mercury wet deposition at underhill, Vermont. *Ecotoxicology* **2005**, *14*, 71–83.

43. Lombard, M.A.S.; Bryce, J.G.; Mao, H.; Talbot, R. Mercury deposition in southern new hampshire, 2006–2009. *Atmos. Chem. Phys.* **2011**, *11*, 7657–7668.

44. Selin, N.E. Global biogeochemical cycling of mercury: A review. *Annu. Rev. Environ. Resour.* **2009**, *34*, 43–63.

45. Poissant, L.; Pilote, M.; Xu, X.H.; Zhang, H.; Beauvais, C. Atmospheric mercury speciation and deposition in the bay St. Francois wetlands. *J. Geophys. Res.: Atmos.* **2004**.

46. Banta, R.M.; Seniff, C.J.; Nielsen-Gammon, J.; Darby, L.S.; Ryerson, T.B.; Alvarez, R.J.; Sandberg, S.R.; Williams, E.J.; Trainer, M. A bad air day in houston. *Bull. Amer. Meteor. Soc.* **2005**, *86*, 657–669.

47. Darby, L.S. Cluster analysis of surface winds in houston, texas, and the impact of wind patterns on ozone. *J. Appl. Meteorol.* **2005**, *44*, 1788–1806.

48. Hao, J.; Wu, Y.; Fu, L.; He, K.; He, D. Motor vehicle source contributions to air pollutants in beijing. *Environ.Sci.* **2001**, *22*, 1–6. (in Chinese).

49. Parrish, D.D. Critical evaluation of us on-road vehicle emission inventories. *Atmos. Environ.* **2006**, *40*, 2288–2300.

50. Luke, W.T.; Kelley, P.; Lefer, B.L.; Flynn, J.; Rappengluck, B.; Leuchner, M.; Dibb, J.E.; Ziemba, L.D.; Anderson, C.H.; Buhr, M. Measurements of primary trace gases and noy composition in houston, texas. *Atmos. Environ.* **2010**, *44*, 4068–4080.

51. United States Environmental Pretention Agency. Greenhouse Gas Reporting Program: 2010 Data Sets. Available online: http://www.epa.gov/climate/ghgreporting/ghgdata/2010data.html (accessed on 07 March 2014).

52. Haman, C.L.; Lefer, B.; Morris, G.A. Seasonal variability in the diurnal evolution of the boundary layer in a near-coastal urban environment. *J. Atmos. Ocean. Technol.* **2012**, *29*, 697–710.

Temporal Variability of Atmospheric Total Gaseous Mercury in Windsor, ON, Canada

Xiaohong Xu, Umme Akhtar, Kyle Clark and Xiaobin Wang

Abstract: Atmospheric Total Gaseous Mercury (TGM) concentrations were monitored in Windsor, Ontario, Canada, during 2007 to 2011, to investigate the temporal variability of TGM. Over five years, the average concentration was $2.0 \pm 1.3 \, \text{ng/m}^3$. A gradual decrease in annual TGM concentrations from $2.0 \, \text{ng/m}^3$ in year 2007 to $1.7 \, \text{ng/m}^3$ in 2009 was observed. The seasonal means show the highest TGM concentrations during the summer months ($2.4 \pm 2.0 \, \text{ng/m}^3$), followed by winter ($1.9 \pm 1.4 \, \text{ng/m}^3$), fall ($1.8 \pm 0.81 \, \text{ng/m}^3$), and spring ($1.7 \pm 0.73 \, \text{ng/m}^3$). Diurnal patterns in summer, fall, and winter were similar. A different diurnal pattern was observed in spring with an early depletion in the morning. The TGM concentrations were lower on weekends ($1.8 \pm 0.77 \, \text{ng/m}^3$) than on weekdays ($2.0 \pm 1.5 \, \text{ng/m}^3$), suggesting 10% of TGM in Windsor was attributable to emissions from industrial sectors in the region. Directional TGM concentrations also indicated southwesterly air masses were TGM enriched due to emissions from coal-fired power plants and industrial facilities. Correlation and principal component analysis identified that combustion of fossil fuel, ambient temperature, wind speed, synoptic systems, and O_3 concentrations influenced TGM concentrations significantly. Overall, inter-annual, seasonal, day-of-week, and diurnal variability was observed in Windsor. The temporal patterns were affected by anthropogenic and surface emissions, as well as atmospheric mixing and chemistry.

Reprinted from *Atmosphere*. Cite as: Xu, X.; Akhtar, U.; Clark, K.; Wang, X. Temporal Variability of Atmospheric Total Gaseous Mercury in Windsor, ON, Canada. *Atmosphere* **2014**, *5*, 536–556.

1. Introduction

Mercury (Hg) has been extensively used since ancient times in medical, agricultural, industrial, and scientific purposes because of its unique chemical and physical properties. Major anthropogenic sources of mercury emissions include coal-fired power plants, manufacturing of metal and chemical products, and waste incineration. Total atmospheric mercury consists of 97%–99% elemental mercury ($Hg°$), with the remaining 1%–3% encompassing particulate mercury and reactive gaseous mercury (RGM) [1,2]. Gaseous elemental mercury is slow-reacting, highly volatile, and sparsely soluble in water, leading to long atmospheric residence times, approximately 0.5–2 years [2]. Extensive use of mercury over the last several centuries, its long residence time—thus, global distribution by large-scale atmospheric circulations, and reemission of historical depositions resulted in an increased concentration of atmospheric mercury and consequent depositions by a factor of three to five compared with preindustrial periods [3]. In the northern hemisphere, current background atmospheric mercury concentrations range between 1.5 and 1.7 ng/m^3 [2]. Atmospheric mercury concentrations at these levels are not likely to affect human health. However, deposition of atmospheric mercury to aquatic surfaces and consequent bioaccumulation of mercury compounds in aquatic food webs at high concentration are highly toxic. The rate of mercury accumulation in an aquatic system is believed to be proportional to atmospheric mercury concentration [4].

Study of atmospheric mercury leads to better understanding of the impact of emission, chemistry, and deposition on mercury cycles among air, water, and land. A number of studies have been performed on atmospheric mercury concentrations in different rural [5–13] and urban [5,11,13–20] sites. Results of these studies indicated elevated mercury concentration and deposition in urban areas compared to rural locations. Different temporal patterns, *i.e.*, seasonal and diurnal variability, were also observed at urban sites. Temporal variability in urban settings was site specific, whereas most of the rural areas had a general pattern of a spring high and a fall low. The difference in concentration and variability between urban and rural sites observed could be due to differences in local sources, surface characteristics, meteorological conditions and presence of other pollutants in urban and rural sites [16]. To advance our understanding in regards to mercury emission, transport, transformation, and deposition processes, further studies are required in urban areas to facilitate source identification through measurements of atmospheric mercury along with a variety of air pollutants.

Windsor (Ontario, Canada) is an industrial city along the Canada-USA border (Figure 1). It is located downwind of several industrial states, including Michigan, Ohio, and Indiana, thus, experiencing trans-boundary air pollution. The combined effects from local anthropogenic sources and trans-boundary pollution have resulted in occasional poor air quality in Windsor [21]. Windsor is also at the heart of the Great Lakes region. Therefore, ambient mercury levels have significant implications on fish contamination. The objective of this study is to investigating the effects of emission, transport, chemistry, and deposition processes on temporal variability of Total Gaseous Mercury (TGM) concentrations in Windsor.

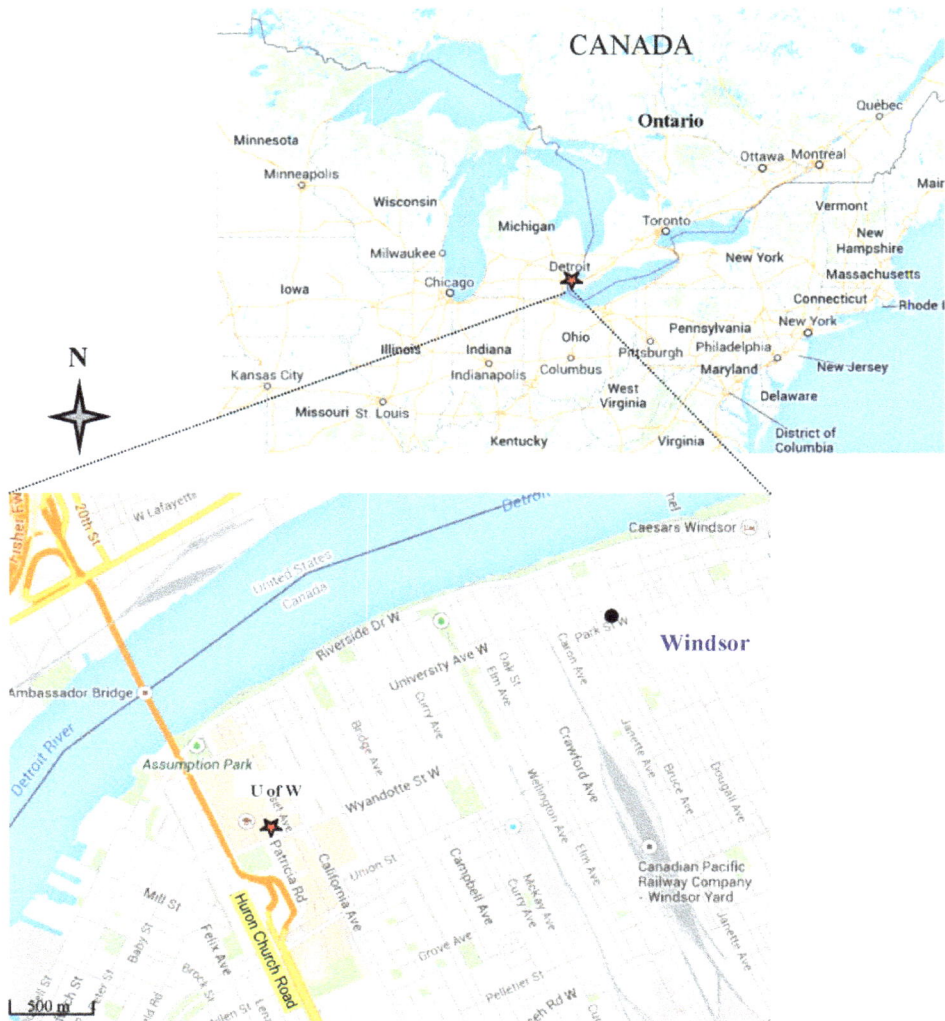

Figure 1. Map of sampling location (✹) at the University of Windsor, Ontario, Canada. (● represents Windsor Downtown Air Quality Station. Base maps adapted from Google Maps).

2. Methodology

2.1. Sampling Site and Instrumentation

The sampling site was located on the University of Windsor campus (42°18.27'N, 83°3.98'W) (Figure 1). The site is on the north side of Wyandotte St. West (27 m) and opposite to the entrance roadway to the Ambassador Bridge—connecting Windsor and Detroit, Michigan in the US. The site is also close to Huron Church Road (approximately 200 m west), which is the major roadway for entering/exiting the Ambassador Bridge. Heavy local and border-crossing traffic in the nearby area of the sampling site is due to the Ambassador Bridge and University of Windsor traffic. A monitoring campaign of 13 sites in 2006 revealed a rather homogeneous spatial distribution of TGM except for high concentrations near the Ambassador Bridge [22]. Hence, there are no significant local or urban sources of pollution that may bias TGM measurements at this site.

A Tekran® 2537A mercury vapor analyzer (Tekran Inc., Toronto, ON, Canada) was used to measure TGM concentration in the ambient air at 5 min intervals. Ambient air was collected at a height of 5 m above ground level. Hourly averaged TGM concentrations were calculated for the study period of 2007–2011, because the meteorological parameters and other air pollutants were available at hourly intervals. Quality Assurance and Quality Control procedures can be found in [22].

2.2. Meteorological Data and Other Air Pollutants

Hourly meteorological parameters for the years 2007–2011 were collected from the Environment Canada website [23], including surface air temperature, relative humidity (RH), wind speed, wind direction, and atmospheric pressure. These parameters were measured at Windsor International Airport, located approximately 10 km northwest of the sampling site.

Hourly ambient concentrations of carbon monoxide (CO), sulfur dioxide (SO_2), nitrogen oxide (NO), nitrogen dioxide (NO_2), nitrogen oxides (NO_x), ozone (O_3), and fine particulate matter ($PM_{2.5}$) were downloaded from the Ministry of Environment Ontario website [24]. These data were collected at Windsor Downtown Air Quality Station, which is 2 km east of the sampling site (Figure 1).

2.3. Data Analysis

For seasonal analysis, the four seasons are winter—December, January, and February, spring—March, April, and May, summer—June, July, and August, and fall—September, October, and November. Day-of-week variability was analyzed for weekdays, Saturdays and Sundays. Diurnal variability was determined for the study period and by season. The analysis of variance (ANOVA) was used to check the statistical difference in mean concentrations between hour-of-day, day-of-week, seasons, and years. Tukey's test was used for comparisons of multiple means for various temporal scales.

To study the inter-relationships, Pearson correlation coefficients were calculated among twelve parameters. The parameters are hourly TGM concentrations, meteorological conditions (temperature, relative humidity, wind speed, and atmospheric pressure) and other air pollutants (CO, SO_2, NO, NO_2, NO_x, O_3, and $PM_{2.5}$). Furthermore, Spearman rank correlation coefficients and Kendall rank correlation coefficients were calculated because those two methods are less sensitive to outliers. Principal component analysis (PCA) with varimax rotation was conducted to identify the factors influencing TGM concentrations using the same twelve parameters as in the correlation analysis.

Wind rose was generated with WRPLOT View (Lakes Environmental, Waterloo, ON, Canada), using hourly wind speed and direction data. Hourly TGM and wind directions were used to produce pollutant rose at 10 degree intervals and percentile values of 5, 25, 50, 75, and 95th for concentrations in each of the 36 directions, using Grapher (R7, Golden Software, CO, USA).

All statistical analysis were performed at a confidence interval of 95% ($\alpha = 0.05$), using MINITAB (R14, Minitab Inc., State College, Pennsylvania, PA, USA). The exception is PCA for which MATLAB (R2013a, MathWorks, Natick, Massachusetts, USA) was used.

3. Results and Discussion

3.1. Temporal Variability

3.1.1. Inter-Annual Variability

During 2007–2011, the TGM concentrations were in the range of 0.3–57 ng/m^3; the average concentration was 2.0 ng/m^3 with a standard deviation of 1.3 ng/m^3. This concentration was higher than the reported background value of 1.5–1.7 ng/m^3 for Hg$^\circ$ in the Northern Hemisphere, which constitutes 97%–99% of TGM [2]. The observed TGM concentration was also higher than the average value of 1.58 ng/m^3 observed at all CAMNet rural sites during 1995–2005 [8] and at Egbert, rural Ontario during 2000–2009 [25], and 1.45 ng/m^3 observed at Fort McMurray, Alberta, Canada during 2010–2013 [20]. However, it was close to 2.2–2.5 ng/m^3 observed in other urban sites, e.g., Toronto, Detroit, Connecticut, Alabama, and Nova Scotia [5,11,13,15,16].

A gradual decrease in annual TGM concentrations from years 2007 to 2009 was observed (2007, 2.0 ± 1.8 ng/m^3; 2008, 1.9 ± 0.99 ng/m^3; 2009, 1.7 ± 1.15 ng/m^3). TGM measurements in 2010 and 2011 were excluded from annual mean calculations due to uneven seasonal coverage, *i.e.*, data were missing for substantial parts of certain seasons. ANOVA and Tukey's tests indicate that the annual concentrations in 2007, 2008, and 2009 were statistically different.

The gradual decrease in TGM concentrations during 2007–2009 is consistent with decreasing trends observed at Egbert (a rural Ontario site, 400 km northeast of Windsor) and other mid-latitude or sub-arctic Canadian sites [25], and two background sites in New England [10] during the same time period. It is also consistent with the decreasing trends derived with land and cruise observations in both the Northern Hemisphere and Southern Hemisphere during 1996–2009 [26]. A 2007–2009 time series of TGM or elemental mercury in localities near Windsor and with industrial impact are not available in the literature yet. Nonetheless, the observed declines of mercury concentrations in precipitation at Mercury Deposition Network sites in the Northeast ($-4.1 \pm 0.49\%$/year) and Midwest ($-2.7 \pm 0.68\%$/year) US sites during 2004–2010 [27] are supportive of our finding. The gradual decrease in TGM concentrations is likely because of anthropogenic mercury emission reductions in the US [27] and Canada [28].

3.1.2. Seasonal Variability

As shown in Figure 2, monthly TGM concentrations were high in the coldest (January) and warmest (May–August) months, and low in the transition months. A low p-value (<0.05) from ANOVA suggests statistically significant differences in the monthly means.

Seasonal patterns of TGM, SO_2, NO_x, O_3, wind speed, and ambient temperature are presented in Figure 3. The highest seasonal TGM concentration was observed in summer (2.4 ± 2.0 ng/m^3), followed by winter (1.9 ± 1.4 ng/m^3), while low in fall (1.8 ± 0.81 ng/m^3) and spring (1.7 ± 0.73 ng/m^3). Results of ANOVA and Tukey's test indicate the differences in all four means were statistically significant. A lower variation of TGM concentration was found during the spring and fall seasons, while higher variation was found in summer and winter.

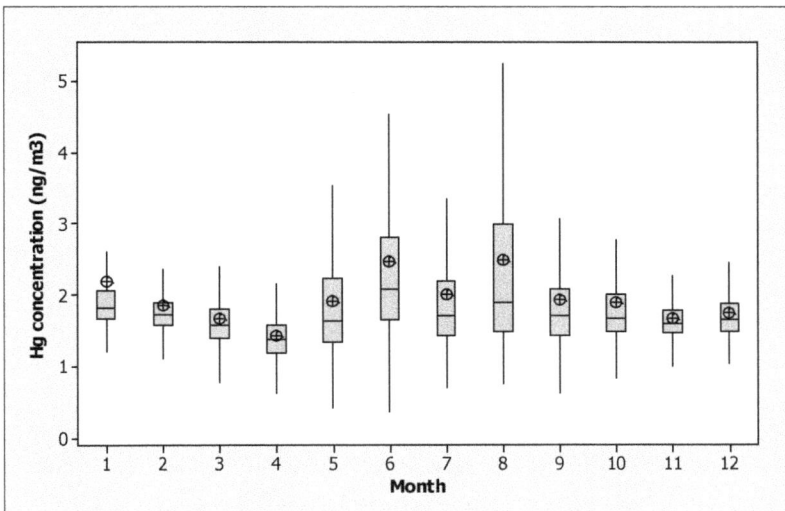

Figure 2. Monthly TGM concentrations during 2007–2011 in Windsor.

As expected, ambient temperatures were highest in summer (22.6 ± 4.5 °C) and lowest in winter (−2.9 ± 5.4 °C). Wind speed was high in winter (17.6 ± 9.6 km/h) and spring (16.9 ± 9.6 km/h), and low in summer (12.0 ± 7.6 km/h). O_3 concentrations were high in spring (32.6 ± 14 ppb), peaked in summer (35.6 ± 18 ppb), but low in fall (20.7 ± 14 ppb) and winter (17.9 ± 10 ppb). An opposite seasonal trend was observed for NO_x (Figure 3f) and CO (not shown). SO_2 concentrations were right skewed. While the medians were similar throughout the four seasons, the mean value was higher in winter (4.9 ± 7.2 ppb), indicating strong influences of air masses with enriched SO_2, especially during the cold season.

Overall, summer in Windsor is characterized by high temperature leading to high Hg reemission, more coal consumption, and lower wind speed thus less dispersion. On the other hand, higher concentrations of photo-chemical oxidants, such as O_3, would result in transformation of Hg° and sequential deposition. The elevated TGM concentrations observed were the net effect of all above-mentioned processes.

The seasonal patterns of higher summer concentrations, in comparison with other seasons observed in this study, are quite similar to most of the studies conducted at urban sites in North America [5,15,16]. Winter TGM concentration in Windsor was relatively high—similar as observed in Toronto, Canada [15]. However, the pattern in Windsor was different from the spring high and fall low seasonal pattern observed in North American rural sites [5,8,11,29].

Difference in meteorological conditions, as well as atmospheric chemistry between urban and rural sites, could lead to elevated TGM concentrations in summer [16]. A study conducted in several urban sites in New York [14] suggested that the emissions from urban surfaces could elevate urban mercury concentration. For example, higher mercury fluxes during summer from soil, grass, and pavement were observed at an urban site [30]. Another possible reason for high concentrations in summer could be the increased electricity demand resulting in more coal combustion. During 2007–2011, the highest quarterly US national coal consumption in the electric power sector was recorded in July–September, ranging 26%–28% (average 27%) of the annual values [31]. Though there is no coal fired-power plant in Windsor, US states to the south and west of Windsor, e.g., Indiana, Illinois, Ohio, and Michigan, are the largest mercury point sources in North America [32]. Five coal-fired power plants located in Ontario, Canada, also emit one third of the total mercury emissions in Ontario [33]. Thus, transportation of airborne mercury from regional sources could affect TGM concentrations. In addition to higher emissions, lower wind speeds during the summer cause less dilution of atmospheric mercury, resulting in a build-up of mercury concentrations.

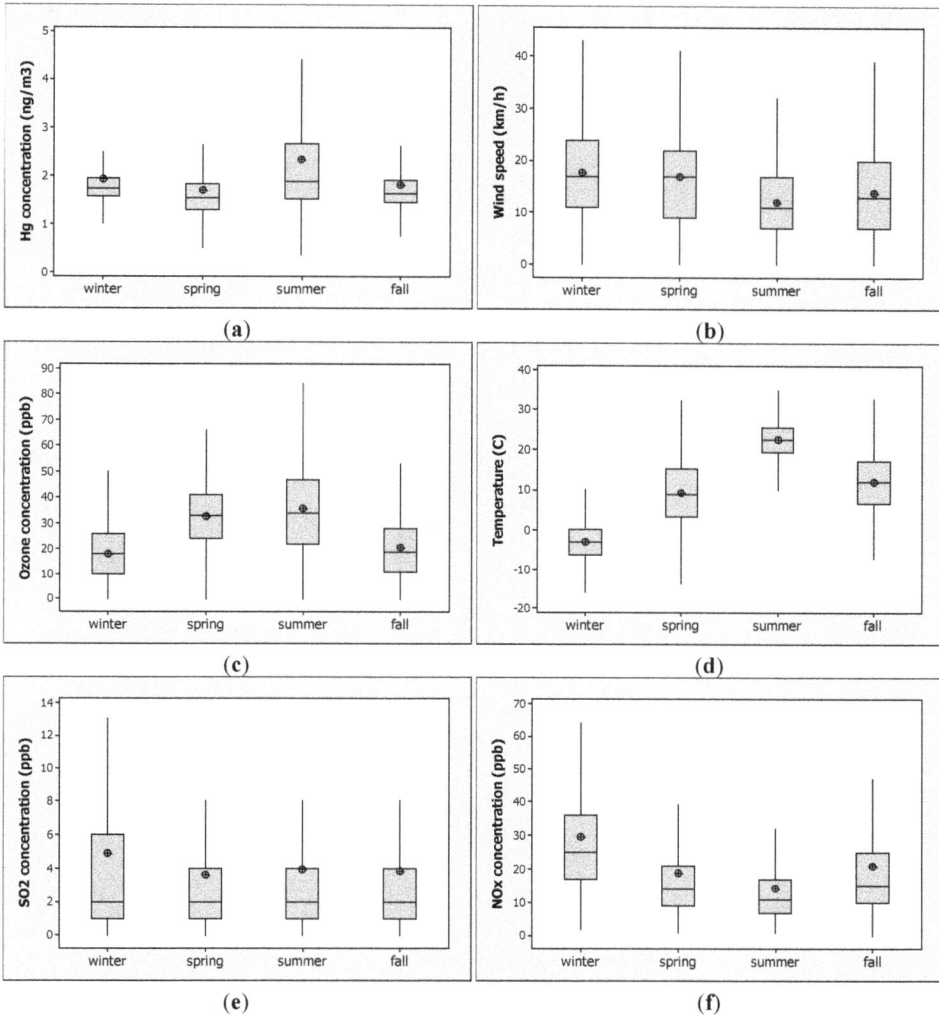

Figure 3. Box plot of seasonal (**a**) TGM, (**b**) wind speed, (**c**) O_3, (**d**) temperature, (**e**) SO_2, (**f**) NO_x, during 2007–2011 in Windsor.

A possible reason for higher concentrations in winter could be the increase in heating demand, resulting in more coal combustion [15]. Low atmospheric oxidant, e.g., O_3 concentrations, and low removal rate of atmospheric mercury in colder seasons, could also result in high concentrations [34]. For fall and spring, lower ambient temperatures compared to the summer, and less coal consumption than in winter/summer may lead to lower concentrations.

3.1.3. Day-of-Week Variability

Weekdays had higher levels of TGM and higher variability (2.0 ± 1.5 ng/m^3) when compared to Saturdays (1.8 ± 0.71 ng/m^3) and Sundays (1.8 ± 0.77 ng/m^3), shown in Figure 4, because of more emissions from anthropogenic sources, as pointed out by Brooks *et al.* [35]. The differences between Saturdays and Weekdays, as well as between Sundays and Weekdays, were statistically significant.

The difference between weekday and weekend means were 0.2 ng/m^3, implying at least 10% of observational TGM in Windsor was attributable to mercury emissions by the industrial sector in this study region. This estimate is consistent with simulation results of a global model [36]. They found North American emissions accounted for approximately 0.3–0.4 ng/m^3 of the total mercury observed at the non-Arctic CAMNet sites.

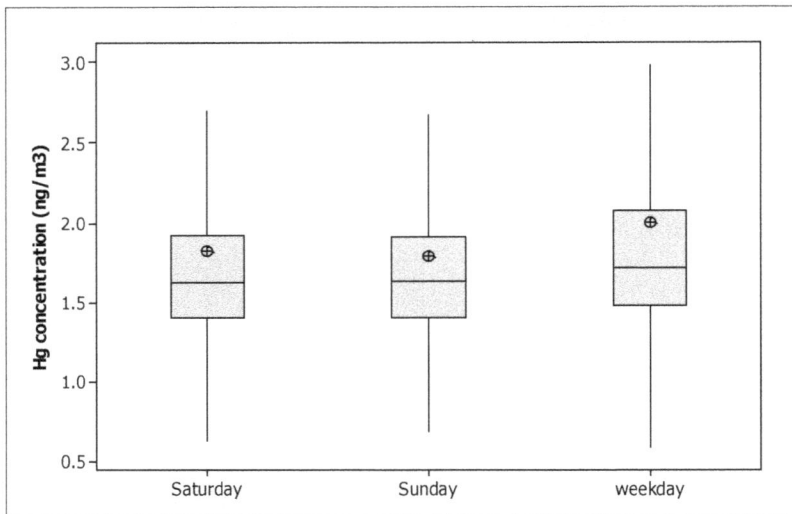

Figure 4. Box plot of day-of-week TGM during 2007–2011 in Windsor.

3.1.4. Diurnal Variability

Figure 5a depicts the diurnal distribution of hourly TGM concentration during the study period. From hours 00:00 to 14:00, there was a gradual increase in TGM concentrations, followed by a comparatively steep decrease when ambient temperature and O$_3$ concentrations peaked, *i.e.*, at 14:00. A small spike in concentration occurred at hour 19:00, which was followed by another slow and gradual concentration increase for the remainder of the day. Results of ANOVA indicate there were significant differences in means among the hours. However, all

Tukey groups had overlapped due to large within-hour variances as opposed to between-hour variances.

After reviewing a number of studies reporting urban TGM diurnal cycles, it was found that each study had reported a site-specific diurnal trend [11,13,15,16,19,20,22,34]. In our study, decreasing concentration in the afternoon, followed by an increasing concentration at night was observed. This diurnal cycle is similar to that reported in Detroit [16]. The similarity in concentration trends is likely due to similar emissions and meteorological characteristics in these two neighboring cities.

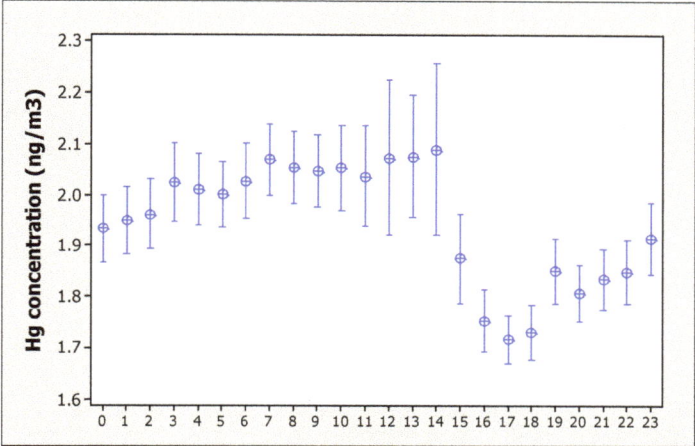

(a)

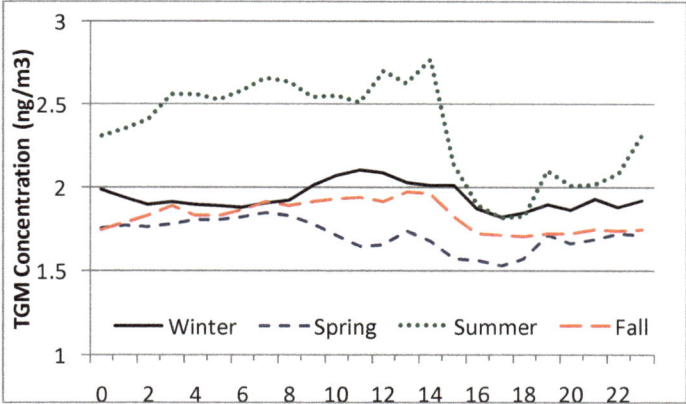

(b)

Figure 5. Hour-of-day TGM (a) during 2007–2011 (circles are means and bars represent 95% confidence intervals of means), and (b) by season.

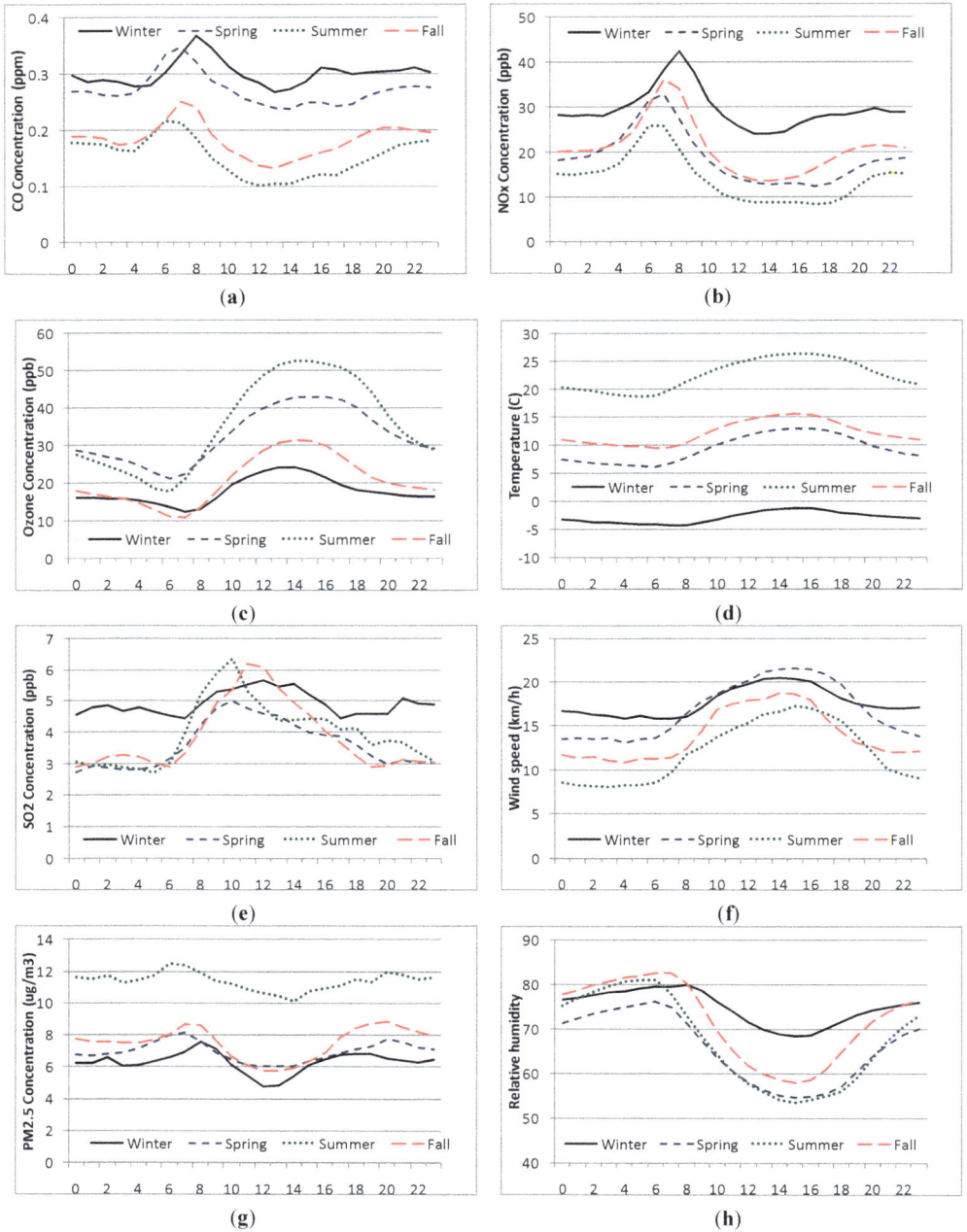

Figure 6. Seasonally averaged hour-of-day trends of meteorological conditions and air quality. (**a**) CO; (**b**) NO$_x$; (**c**) O$_3$; (**d**) Temperature; (**e**) SO$_2$; (**f**) Wind speed; (**g**) PM$_{2.5}$; (**h**) Relative humidity.

As shown in Figure 5b, in summer, winter, and fall the TGM diurnal patterns were similar to the entire study period (Figure 5a). The increases before sunrise track that of CO and NO_x, reflecting nocturnal inversions. After sunrise, TGM followed the diurnal temperature cycle (Figure 6). There was a depletion in the afternoon that continued until early evening. The diurnal pattern was significantly different in spring compared to other seasons; TGM depletion started in the morning instead of the afternoon.

In winter, summer, and fall, the TGM concentration followed the diurnal temperature trend from morning until noon. In this time period, the increase in TGM could be governed by solar radiation/temperature driven surface emissions and enhanced biological activities, *i.e.*, foliar emission especially in summer and fall [15,37]. Regardless of season, a decrease in afternoon concentration was observed when the O_3 concentrations were high (Figure 6c). This supports the mechanism of TGM depletion by oxidation. An increased level of photo-chemically originated oxidants (e.g., O_3) in the afternoon could lead to oxidation of $Hg°$ to RGM which is then rapidly deposited, resulting in a decrease in overall TGM concentrations [34]. Several studies in urban sites [7,38] have reported high RGM concentrations between solar noon and late evening hours, causing an overall decrease in TGM concentrations similar to this study. In the afternoon there were also increased surface reemissions. However, the strong presence of increased vertical mixing and oxidation resulted in an overall downward trend. We hypothesize that the early depletion in spring morning is due to an increase in the atmospheric oxidation rate, following O_3 concentrations, and lower rates of reemissions, following ambient temperature and owing to reduced vegetative covers.

In summer, the rate of TGM increase from evening to morning and the rate of decrease in afternoon were higher compared to other seasons (Figure 5b). Formation of photo-chemical oxidants such as O_3 in the presence of high solar radiation and high temperature is greatest in the summer, which could lead to a steeper TGM decline in the afternoon during this season. Large variability in the summer could be due to greater diurnal variations in temperature and mixing height, a higher rate of uptake/emission by vegetation, increased surface emissions, and enhanced oxidation in warmer seasons [8,13,37,39]. In spite of afternoon concentration depletion, nighttime TGM levels were elevated when compared with other seasons. Higher anthropogenic emissions due to cooling demands and significant surface emissions may lead to the overall higher TGM concentration in summer, as observed in seasonal variability.

Overall, TGM temporal variability analysis indicates influence by environmental conditions, such as meteorological parameters, presence of oxidants, atmospheric chemistry, and local/regional sources. To understand the impact of these processes on temporal variability of TGM, relationships between TGM concentrations and meteorological parameters as well as other air pollutants are further investigated in the following two sections.

3.2. Environmental Factors Influencing TGM

To further investigate the relationships on an hourly basis, Pearson correlation coefficients for hourly TGM and other air contaminant measurements, and meteorological parameters are presented in Table 1. Cross-correlation matrixes of all 12 valuables can be found in Figure S1 (Supplemental Materials).

Table 1. Pearson correlation coefficients of hourly Total Gaseous Mercury (TGM) concentration with other parameters (all significant at $p < 0.05$ except shaded cells).

Parameter	2007–2011 (N = 23,467)	Winter (N = 5303)	Spring (N = 6041)	Summer (N = 5512)	Fall (N = 6612)
SO_2	−0.008	0.046	0.031	−0.045	−0.039
NO	0.083	0.084	0.188	0.112	0.162
NO_2	0.106	0.117	0.288	0.130	0.146
NO_x	0.105	0.109	0.269	0.142	0.177
CO	0.100	0.105	0.318	0.178	0.190
O_3	−0.068	−0.090	−0.227	−0.106	−0.092
$PM_{2.5}$	0.074	0.067	0.248	−0.021	0.053
Temperature	0.090	−0.084	0.144	−0.042	0.112
Relative humidity	0.078	−0.007	0.251	0.108	0.006
Wind speed	−0.083	0.025	−0.216	−0.051	−0.068
Pressure	−0.073	−0.022	−0.115	−0.076	−0.063

When the five-year dataset was considered, hourly TGM had weak but significant positive correlations with NO, NO_2, NO_x, CO, $PM_{2.5}$, temperature, and relative humidity; weak negative correlations with O_3, wind speed, and pressure. A New England study also found positive correlations with NO_y and CO in winter [40]. The average SO_2 concentration over the study period was 4.1 ppb, with higher seasonal levels in winter (4.9 ppb) and summer (4.0 ppb). The much anticipated positive correlation with SO_2 due to coal combustion [41,42] was absent. In other words, mercury emissions from coal-fired power plants may not significantly increase TGM in regions a few hours downwind [39] due to the depositional removal of more soluble and reactive oxidized Hg from the plume as it is advected to the Windsor measurements site. Monitoring of speciated mercury is needed to further investigate the Hg-sulfur relation. Another possible reason is the TGM-SO_2 association was damped by the long-term record. Our result is similar to that of a short-term 2007 study in Windsor [22], and to the findings in New England [40] and in Seoul, Korea [19]. Among the four seasons, the relationships were similar but strongest in spring, while winter TGM was only correlated with one (temperature) out of four weather conditions.

Focusing on coefficients >0.2 in spring, the direction (*i.e.*, positive or negative) of the correlation is indicative. Positive correlations, although weak, of TGM with NO, NO_2, NO_x, CO, and $PM_{2.5}$, suggest fossil fuel combustion as the common source. Negative correlation with wind speed indicates a decrease in concentration at high wind speeds because of dilution of air pollutants, including mercury. Anti-correlation with O_3 points toward oxidation of $Hg°$. TGM was positively correlated with RH. However, the correlation was negative when RH $\geq$ 99%, although not significant ($r = -0.267$, $p > 0.05$). High ambient RH leads to condensation of water vapor. Removal of gaseous mercury from the atmosphere is enhanced due to rapid oxidation of elemental mercury in the aqueous phase [1]. Thus, high water contents could lead to enhanced oxidation, resulting in deposition of atmospheric mercury.

Similar results were observed with the Spearman rank correlations (Table 2) and Kendall rank correlations (Table S1). In comparison with the Pearson correlations (Table 1), the differences among the four seasons were less pronounced when rank correlations were employed, especially between TGM and other pollutants. The results indicate that the relationships in spring were less affected by extreme concentration values. Nonetheless, the directions of the correlations were largely unaltered by rank correlations. Therefore, the conclusions regarding emission sources and weather conditions influenced temporal variability of TGM significantly remain the same.

Although most correlations listed in Tables 1 and 2 are statistically significant, they are of little practical significance because coefficient values were low. The contrast between moderate to strong correlations on the seasonally averaged diurnal trends and the weak correlations on the basis of hourly data is striking. Figure 6 shows strong correlations among seasonally averaged hour-of-day temperature, RH, wind speed, O_3, $PM_{2.5}$, and SO_2 concentrations. However, those associations were all lost on an hourly basis, except for the anti-correlation between temperature and RH in summer (Figure S1). On the other hand, the lack of strong association between hourly TGM and other parameters, particularly in winter [17], is not unexpected. One possible reason is the large variability of each parameter, which is governed by different atmospheric physical and chemical processes of different time and special scales. Therefore, the apparent associations in seasonally averaged trends were diminished by substantial scattering in the hourly data. Another explanation is the existence of a rather stable and homogeneous distribution of TGM over North America, especially in the winter [29]. Consequently, factors modulating TGM levels above and beyond the diurnal cycles are difficult to identify. In-depth analysis of episodic events as in [29] may provide insightful relationships, which could be easily marginalized in long-term time series analysis.

Table 2. Spearman rank correlation coefficients of hourly TGM concentration with other parameters (all significant at $p < 0.05$ except shaded cells).

Parameter	2007–2011 (N = 23,467)	Winter (N = 5303)	Spring (N = 6041)	Summer (N = 5512)	Fall (N = 6612)
SO_2	0.018	0.110	0.089	−0.082	−0.015
NO	0.205	0.382	0.165	0.228	0.245
NO_2	0.305	0.349	0.398	0.336	0.271
NOx	0.311	0.399	0.381	0.356	0.309
CO	0.249	0.378	0.378	0.439	0.341
O_3	−0.263	−0.294	−0.384	−0.273	−0.224
$PM_{2.5}$	0.180	0.252	0.247	0.029	0.091
Temperature	0.065	−0.135	0.044	−0.115	0.071
Relative humidity	0.212	0.017	0.348	0.257	0.120
Wind speed	−0.186	−0.076	−0.265	−0.183	−0.122
Pressure	−0.142	−0.050	−0.145	−0.172	−0.120

The results of PCA over the five-year study period and by season are shown in Table 3. When analyzed by year, similar results were obtained as with the full dataset, as shown in Table S2. For comparison, the four or five factors with largest eigenvalues (≥ 0.93) were retained. These factors explained 71%–80% of the total variances. The interpretation of each factor was based on the loadings and outcomes of other PCA studies of atmospheric mercury [16,17,19,20,22,39,42]. For example, in Table 3a, Factor 1 had strong positive loadings for NO, NO_x, CO, and a weak positive loading for TGM. All these compounds have a common source, the combustion of fossil fuel. The impact of each factor on TGM was evaluated by TGM loading in that factor.

For the study period of 2007–2011, the dominant factor is Fossil Fuel Combustion, followed by Diurnal Trend (large loading of temperature), Photochemistry (large loadings of O_3 and RH with opposite signs) and $PM_{2.5}$, Synoptic Systems (large loading of pressure), and Industrial Sulfur. The low loadings of TGM (<0.15) in the Photochemistry factor indicate its relatively small impact on TGM. The concurrent strong loadings of NO, NO_2, NO_x in the same factor (Table 3) are because of strong correlations among the three compounds (Figure S1).

The seasonal variation of PCA results is interesting. The two common factors influencing TGM variability are Fossil Fuel Combustion and Diurnal Trend. The impact of combustion on TGM has also been identified in most PAC studies in urban settings [16,17,19,22,39,42], with the exception of Fort McMurray, Alberta [20]. In winter (Table 3b), the Transport factor (large loading of wind speed) seems to be very influential to TGM (loading = 0.65), likely due to higher wind speeds (Figure 3). Photochemistry was only significant in spring (Table 3c), which is in agreement with a comparatively stronger correlation between TGM and O_3 than in other seasons (Table 1). In summer (Table 3d), Synoptic Systems (loading of pressure = 0.70) was a lead factor for TGM (loading = −0.49). The opposite signs suggest stagnant air during high-pressure systems not leading to TGM build-up, while high winds during low-pressure systems causing transport of TGM. In addition, summer is when TGM was only associated with two factors (TGM loading > 0.15), verses three or four in other seasons, making it more challenging to pin-point the major mechanisms driving TGM variability in the warm months. The opposite loadings of SO_2 and TGM in spring, fall and winter under Industrial Sulfur contradict a 2007 short-term PCA study in Windsor, where a factor of Coal-fired Power Generation with positive loadings of SO_2 and TGM was identified [22]. A lack of strong association between TGM and Industrial Sulfur was also reported in Toronto, Ontario [17], Rochester, New York [39], and Fort McMurray, Alberta [20].

Overall, our correlation and PCA analyses suggest that the temporal variability of TGM was affected by man-made and surface emissions, atmospheric mixing and reactions, and aqueous chemistry. Future studies should conduct onsite measurement of mercury species, solar radiation, mercury fluxes, and potential oxidant/reductant concentrations to further examine the impact of environmental factors.

Table 3. Principal component analysis (PCA) factor loadings (bold numbers indicate loadings > 0.4). (**a**) 2007–2011; (**b**) Winter; (**c**) Spring; (**d**) Summer; (**e**) Fall.

(a)

Parameter	Fossil Fuel Combustion	Diurnal Trend&PM$_{2.5}$	Photo-Chemistry	Synoptic Systems	Industrial Sulfur
TGM	0.26	0.20	0.11	0.21	**0.67**
SO$_2$	0.17	0.21	0.11	0.18	**−0.63**
NO	**0.48**	−0.06	0.11	−0.05	0.12
NO$_2$	0.39	0.05	−0.16	−0.08	−0.13
NO$_x$	**0.50**	−0.01	−0.01	−0.07	0.01
CO	**0.43**	−0.01	−0.02	0.17	−0.02
O$_3$	−0.15	0.24	**0.48**	0.06	0.00
PM$_{2.5}$	0.14	**0.62**	−0.05	0.00	−0.24
Temperature	−0.17	**0.56**	0.14	0.00	0.17
Relative humidity	−0.09	0.08	**−0.75**	0.11	0.00
Wind speed	0.01	−0.35	0.28	**0.59**	−0.16
Pressure	0.07	−0.15	0.22	**−0.72**	−0.09
Variance (%)	34.8	14.1	11.2	10.5	7.8
Eigen value	4.18	1.70	1.35	1.26	0.93

Table 3. *Cont.*

(b)

Parameter	Fossil Fuel Combustion	Diurnal Trend	Transport	Industrial Sulfur
TGM	0.28	−0.10	0.65	−0.27
SO$_2$	0.06	−0.05	0.06	**0.87**
NO	0.40	−0.04	0.06	−0.07
NO$_2$	0.40	−0.06	−0.09	0.09
NO$_x$	**0.45**	−0.06	0.00	−0.01
CO	0.39	0.04	0.07	−0.03
O$_3$	−0.31	−0.22	0.15	−0.14
PM$_{2.5}$	0.30	0.12	−0.09	0.19
Temperature	−0.12	**0.56**	−0.02	0.08
Relative humidity	0.10	**0.60**	−0.10	−0.13
Wind speed	−0.18	−0.06	**0.57**	0.27
Pressure	0.05	**−0.49**	**−0.44**	−0.03
Variance (%)	38.9	15.9	9.9	8.0
Eigen value	4.67	1.90	1.19	0.96

(c)

Parameter	Fossil Fuel Combustion	Diurnal Trend&PM$_{2.5}$	Synoptic Systems &Photo-Chemistry	Transport &Industrial Sulfur
TGM	0.02	0.33	−0.30	−0.37
SO2	0.31	0.18	0.20	**0.49**
NO	**0.41**	−0.08	0.04	0.06
NO2	**0.43**	0.02	0.00	−0.07
NOx	**0.47**	−0.03	0.02	−0.01
CO	0.38	0.14	−0.14	0.04
O3	−0.27	0.29	0.28	0.16
PM$_{2.5}$	0.24	**0.51**	0.01	−0.01
Temperature	−0.16	**0.65**	0.04	−0.04
Relative humidity	0.05	−0.14	**−0.60**	−0.10
Wind speed	−0.10	−0.10	−0.15	**0.68**
Pressure	0.11	−0.20	**0.62**	−0.34
Variance (%)	38.1	14.7	12.5	8.8
Eigen value	4.58	1.76	1.50	1.05

Table 3. *Cont.*

(d)

Parameter	Fossil Fuel Combustion	PM$_{2.5}$ &Industrial Sulfur	Synoptic Systems	Photo-Chemistry &Diurnal Trend
TGM	0.19	−0.12	**−0.49**	−0.05
SO2	0.08	**0.45**	0.13	−0.09
NO	**0.59**	−0.15	0.02	−0.15
NO2	0.36	0.21	−0.06	0.18
NOx	**0.54**	0.05	−0.03	0.04
CO	0.24	0.25	−0.36	0.07
O3	−0.19	0.27	0.02	−0.39
PM$_{2.5}$	−0.04	**0.67**	−0.04	0.06
Temperature	−0.05	0.27	−0.14	**−0.48**
Relative humidity	−0.20	0.10	−0.25	**0.59**
Wind speed	−0.03	−0.19	−0.18	**−0.45**
Pressure	0.20	0.06	**0.70**	0.05
Variance (%)	34.4	17.4	10.9	9.0
Eigen value	4.12	2.09	1.31	1.08

(e)

Parameter	Fossil Fuel Combustion	Diurnal Trend&PM$_{2.5}$	Synoptic Systems	Relative Humidity	Industrial Sulfur
TGM	0.24	0.15	−0.16	−0.16	**0.70**
SO$_2$	0.20	0.14	−0.20	−0.25	**−0.61**
NO	**0.46**	−0.08	0.06	−0.14	0.12
NO$_2$	**0.42**	0.02	0.06	0.10	−0.17
NO$_x$	**0.51**	−0.05	0.07	−0.06	0.01
CO	0.40	0.06	−0.17	0.12	0.05
O$_3$	−0.21	**0.41**	0.04	−0.31	0.01
PM2.5	0.18	**0.58**	0.02	0.22	−0.23
Temperature	−0.09	**0.63**	0.00	−0.06	0.14
Relative humidity	−0.05	−0.01	−0.03	**0.75**	0.01
Wind speed	−0.08	−0.19	**−0.60**	−0.28	−0.09
Pressure	0.02	−0.10	**0.73**	−0.26	−0.07
Variance (%)	33.3	15.6	11.1	9.8	8.9
Eigen value	4.00	1.87	1.33	1.17	1.07

3.3. Directional TGM Concentrations

During this study, the prevailing winds were from the south to west (180–270°) section (Figure 7a), consistent with the long-term weather record in Windsor. Annual TGM roses in 2010 and 2011 were not generated due to uneven seasonal coverage in those two years. The TGM roses over 2007–2009 are shown in Figure 7b; a longer bar indicates a higher inter-percentile concentration. Higher TGM concentrations were associated with winds from the southwest (220–260°), while air masses from the east to south (90–200°) contained lower mercury concentrations. Similar patterns were observed on an annual basis in 2007 and 2008 (Figure 8). In 2009, the distribution of directional TGM concentrations was rather uniform, with the exception of extreme values, *i.e.*, 98th percentiles, by westerly and northerly winds (250° clockwise to 20°). Higher TGM associated with southwesterly air flow is due to coal-fired power generation and other industrial facilities located in the southwest region of Windsor. These sources have been identified through a back-trajectory and potential source contribution function investigation using the annual and seasonal TGM in 2007 [18].

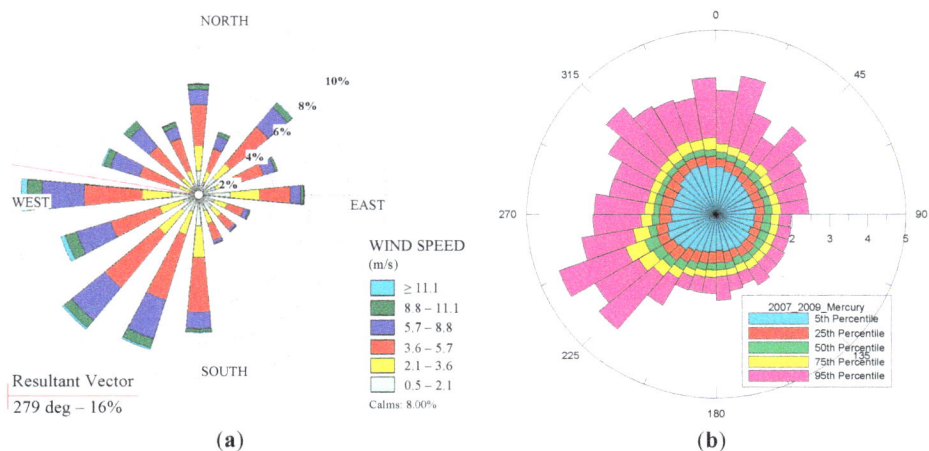

Figure 7. (a) Wind rose and (b) TGM concentration rose (ng/m^3) during 2007–2009 in Windsor.

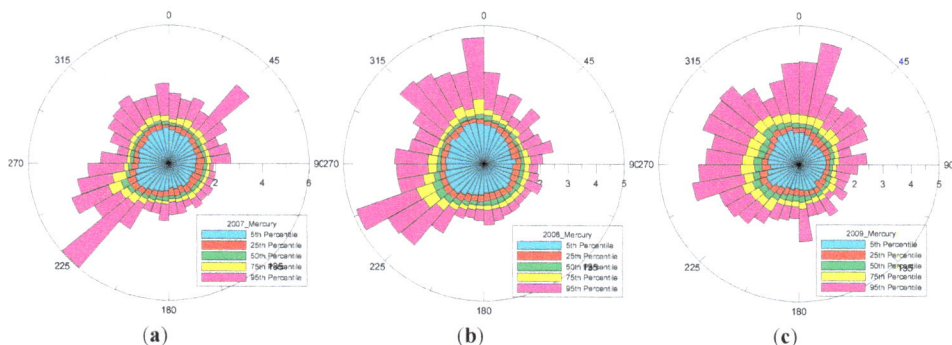

Figure 8. TGM concentration rose in (a) 2007, (b) 2008, and (c) 2009.

The directional TGM distributions were similar in winter and spring, when air mass from the southwest brought in higher concentrations. In summer and fall, regions north of Windsor also contributed to elevated TGM levels (Figure 9). Further analysis is recommended to identify air mass conditions associated with those episodes.

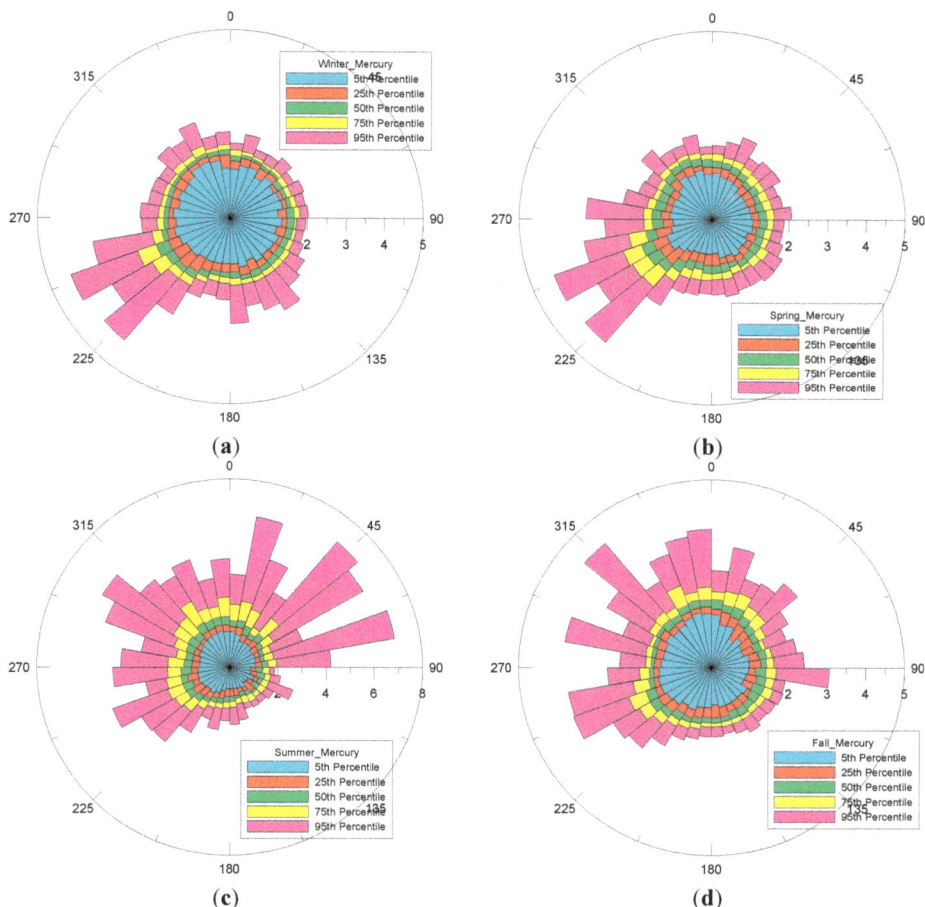

Figure 9. TGM concentration rose in (**a**) winter, (**b**) spring, (**c**) summer and (**d**) fall.

4. Conclusions

This study monitored TGM concentrations from 2007 to 2011 in Windsor, which is an industrialized border city affected by local emissions and regional transport of air pollutants. The average concentration was 2.0 ± 1.3 ng/m^3, which is higher than the concentration observed in all CAMNet sites in Canada (1995–2005), but close to the concentrations observed in other urban sites in North America. Different temporal aspects of TGM concentration were investigated. A gradual decrease in annual TGM concentrations from year 2007–2009 was observed (2.0–1.7 ng/m^3). TGM exhibited seasonality with the highest concentrations in summer, relatively high in winter, and low in spring and fall. High surface emissions in summer and an elevated anthropogenic mercury release from regional sources in summer and winter due to increased power consumption, most likely resulted in the elevated TGM concentrations in these two seasons. Weekday levels were higher than weekend levels by 10%, attributable to industrial activities in the region. On an annual basis, a distinctive diurnal pattern was observed. TGM concentration was relatively high overnight, followed by a continuous increase in the morning, a decline in the afternoon, minima in the early evening, and finally rising again. Diurnal patterns in winter, summer, and fall were similar to the annual pattern. However, spring diurnal trends were characterized by an early depletion due to the strong effect of oxidation loss, in conjunction with very little gain from vegetative reemission.

Environmental conditions, including temperature, relative humidity, wind speed, and O$_3$ concentrations influenced temporal variability of TGM significantly. The associations between TGM and the study parameters were moderate to strong on seasonally averaged hour-of-day trends. The aforementioned relationships effectively explained the TGM diurnal cycles due to anthropogenic emissions, solar radiation/temperature driven surface emissions and mixing during the daytime, photochemical reactions in the afternoon, and atmospheric inversions at night. However, the analysis of hourly TGM and meteorological parameters, as well as SO$_2$, NO, NO$_2$, NO$_x$, CO, O$_3$, and PM$_{2.5}$ concentrations, did not yield any strong correlations. Our findings draw attention to the limitations imposed on the investigative power of correlation analysis using long-term hourly measurements.

The PCA results indicate combustion of fossil fuel as the major source of TGM. Directional analysis shows the southwesterly air masses were associated with higher TGM levels due to emissions from coal-fired power plants and industrial facilities. Diurnal cycles driven primarily by solar radiation (thus, temperature) and regional transport of air containments also significantly contributed to temporal variability of TGM. The impact of photochemistry, *i.e.*, reduction of ambient Hg° by photochemical oxidation to RGM, was more clear to attribute in the spring because there are less confounding factors, e.g., reemission of Hg°.

Acknowledgments: We are grateful to Tekran Inc. (Toronto, ON, Canada) for its kind contribution toward the purchasing of the Mercury Analyzer and for technical support. The authors would like to thank Bill Middleton and Tianzhu Zhang at University of Windsor for technical assistance. The funding of this research work was provided by the Natural Sciences and Engineering Research Council of Canada (NSERC) and the University of Windsor.

Author Contributions: Xiaohong Xu, project PI, did literature review of some papers, analysis of variance (ANOVA), and wrote the manuscript. Umme Akhtar conducted data collection in 2007 and a few months in 2008, did literature review of some papers, and drafted some passages. Kyle Clark conducted compilation and screening of all data collected in 2007–2011, did general statistics, plotted most charts, and performed English correction of the manuscript. Xiaobin Wang conducted Principal Component Analysis (PCA), and did literature review of some papers.

Conflicts of Interest: The authors declare no conflict of interest.

References

1. Poissant, L. Field observations of total gaseous mercury behaviour: Interactions with ozone concentration and water vapour mixing ratio in air at a rural site. *Water Air Soil Pollut.* **1997**, *97*, 341–353.

2. Lindberg, S.E.; Bullock, R.; Ebinghaus, R.; Engstrom, D.; Feng, X.; Fitzgerald, W.; Pirrone, N.; Prestbo, E.; Seigneur, C.A. Synthesis of progress and uncertainties in attributing the sources of mercury in deposition. *Ambio* **2007**, *36*, 19–32.

3. Krabbenhoft, D.P.; Branfireun, B.A.; Heyes, A. Biogeochemical cycles affecting the speciation, fate and transport of mercury in the environment. In *Mercury: Sources, Measurements, Cycles, and Effects*; Parsons, M.B., Percival, J.B., Eds.; Short Course Series; Mineralogical Association of Canada: Ottawa, ON, Canada, 2005; Volume 34, pp. 139–156.

4. Swain, E.B.; Jakus, P.M.; Rice, G.; Lupi, F.; Maxon, P.A.; Pacyna, J.M.; Penn, A.; Speigel, S.J.; Veiga, M.M. Socioeconomic consequences of mercury use and pollution. *Ambio* **2007**, *36*, 45–61.

5. Nadim, F.; Perkins, C.; Liu, S.; Carley, R.J.; Hoag, J.E. Long-term investigation of atmospheric mercury contamination in Connecticut. *Chemosphere* **2001**, *45*, 1033–1043.

6. Han, Y.J.; Holsen, T.M.; Lai, S.O.; Hopke, P.K.; Yi, S.M.; Liu, W.; Pagano, J.; Falanga, L.; Milligan, M.; Andolina, C. Atmospheric gaseous mercury concentrations in New York State: Relationships with meteorological data and other pollutants. *Atmos. Environ.* **2004**, *38*, 6431–6446.

7. Lynam, M.M.; Keeler, G.J. Automated speciated mercury measurements in Michigan. *Environ. Sci. Technol.* **2005**, *39*, 9253–9262.

8. Temme, C.; Blanchard, P.; Steffen, A.; Banic, C.; Beauchamp, S.; Poissant, L.; Tordon, R.; Wiens, B. Trend, seasonal and multivariate analysis study of total gaseous mercury data from the Canadian Atmospheric Mercury Measurement Network (CAMNet). *Atmos. Environ.* **2007**, *41*, 5423–5441.

9. Ci, Z.J.; Zhang, X.S.; Wang, Z.W.; Niu, Z.C. Atmospheric gaseous elemental mercury (GEM) over a coastal/rural site. *Atmos. Environ.* **2011**, *45*, 2480–2487.

10. Mao, H.; Talbot, R. Long-term variation in speciated mercury at marine, coastal and inland sites in New England: Part I Temporal Variability. *Atmos. Chem. Phys.* **2012**, *12*, 5099–5112.

11. Nair, U.S.; Wu, Y.L.; Walters, J.; Jansen, J.; Edgerton, E.S. Diurnal and seasonal variation of mercury species at coastal-suburban, urban, and rural sites in the southeastern United States. *Atmos. Environ.* **2012**, *47*, 499–508.

12. Kolker, A.; Engle, M.A.; Peucker-Ehrenbrink, B.; Geboy, N.J.; Krabbenhoft, D.P.; Bothner, M.H.; Tate, M.T. Atmospheric mercury and fine particulate matter in coastal New England: Implications for mercury and trace element sources in the northeastern United States. *Atmos. Environ.* **2013**, *79*, 760–768.

13. Cheng, I.; Zhang, L.M.; Mao, H.T.; Blanchard, P.; Tordon, R.; Dalziel, J. Seasonal and diurnal patterns of speciated atmospheric mercury at a coastal-rural and a coastal-urban site. *Atmos. Environ.* **2014**, *82*, 193–205.

14. Capri, A.; Chen, Y.F. Gaseous elemental mercury fluxes in New York City. *Water Air Soil Pollut.* **2002**, *140*, 371–379.

15. Denis, M.S.; Song, X.; Lu, J.Y.; Feng, X. Atmospheric gaseous elemental mercury in downtown Toronto. *Atmos. Environ.* **2006**, *40*, 4016–4024.

16. Liu, B.; Keeler, G.J.; Dvonch, J.T.; Barres, J.A.; Lynam, M.M.; Marsik, F.J.; Morgan, J.T. Temporal variability of mercury speciation in urban air. *Atmos. Environ.* **2007**, *41*, 1911–1923.

17. Cheng, I.; Lu, J.; Song, X.J. Studies of potential sources that contributed to atmospheric mercury in Toronto, Canada. *Atmos. Environ.* **2009**, *43*, 6145–6158.

18. Xu, X.; Akhtar, U.S. Identification of potential regional sources of atmospheric total gaseous mercury in Windsor, Ontario, Canada using hybrid receptor modeling. *Atmos. Chem. Phys.* **2010**, *10*, 7073–7083.

19. Kim, K.H.; Shon, Z.H.; Nguyen, H.T.; Jung, K.; Park, C.G.; Bae, G.N. The effect of man made source processes on the behavior of total gaseous mercury in air: A comparison between four urban monitoring sites in Seoul Korea. *Sci. Total Environ.* **2011**, *409*, 3801–3811.

20. Parsons, M.T.; McLennan, D.; Lapalme, M.; Mooney, C.; Watt, C.; Mintz, R. Total gaseous mercury concentration measurements at Fort McMurray, Alberta, Canada. *Atmosphere* **2013**, *4*, 472–493.

21. Ministry of the Environment Ontario (MOE). Air Quality in Ontario 2011 Report. Available online: https://dr6j45jk9xcmk.cloudfront.net/documents/1118/70-air-quality-in-ontario-2011-report-en.pdf (accessed on 25 June 2014).

22. Akhtar, U.S. Atmospheric Total Gaseous Mercury Concentration Measurement in Windsor: A Study of Variability and Potential Sources. Master's Thesis, University of Windsor, Windsor, ON, Canada, 2008.

23. Environment Canada. Climate Data. Available online: http://www.climate.weatheroffice.ec.gc.ca/climateData/hourlydata_e.html (accessed on 1 August 2013).

24. Ministry of Environment Ontario (MOE). Historical Air Pollutant Data. Available online: http://www.airqualityontario.com/history/index.php. (accessed on 1 August 2013).

25. Cole, A.S.; Steffen, A.; Pfaffhuber, K.A.; Berg, T.; Pilote, M.; Poissant, L.; Tordon, R.; Hung, H. Ten-year trends of atmospheric mercury in the high Arctic compared to Canadian sub-Arctic and mid-latitude sites. *Atmos. Chem. Phys.* **2013**, *13*, 1535–1545.

26. Slemr, F.; Brunke, E.G.; Ebinghaus, R.; Kuss, J. Worldwide trend of atmospheric mercury since 1995. *Atmos. Chem. Phys.* **2011**, *11*, 4779–4787.

27. Zhang, Y.; Jaeglé, L. Decreases in mercury wet deposition over the United States during 2004–2010: Roles of domestic and global background emission reductions. *Atmosphere* **2013**, *4*, 113–131.

28. Environment Canada. NPRI. Heavy Metals and Persistent Organic Pollutants (National: 1990–2012). Available online: http://www.ec.gc.ca/inrp-npri/default.asp?lang=en\&n=0EC58C98 (accessed on 1 December 2013).

29. Mao, H.; Talbot, R.; Hegarty, J.; Koermer, J. Long-term variation in speciated mercury at marine, coastal and inland sites in New England: Part II Relationships with atmospheric physical parameters. *Atmos. Chem. Phys.* **2012**, *12*, 4181–4206.

30. Gabriel, M.C.; Williamson, D.G.; Zhang, H.; Brooks, S.; Lindberg, S. Diurnal and seasonal trends in total gaseous mercury flux from three urban ground surfaces. *Atmos. Environ.* **2006**, *40*, 4269–4284.

31. United States Department of Energy (USDOE). Table 32. USA Coal Consumption by End-Use Sector, 2007–2013. Available online: http://www.eia.gov/coal/production/quarterly/pdf/t32p01p1.pdf (accessed on 25 June 2014).

32. Keating, M. *Mercury and Midwest Power Plants*; Clean Air Task Force: Boston, MA, USA, 2003.

33. Ontario Clean Air Alliance. An OCAA Air quality Report, OPC: Ontario's Pollution Giant. Available online: http://www.cleanairalliance.org/resource/opgiant.pdf (accessed on 25 June 2014).

34. Stamenkovic, J.; Lyman, S.; Gustin, M.S. Seasonal and diel variation of atmospheric mercury concentrations in the Reno (NV, USA) Airshed. *Atmos. Environ.* **2007**, *41*, 6662–6672.

35. Brooks, S.; Luke, W.; Cohen, M.; Kelly, P.; Lefer, B.; Rappenglück, B. Mercury species measured atop the Moody Tower TRAMP site, Houston, Texas. *Atmos. Environ.* **2010**, *44*, 4045–4055.

36. Durnford, D.; Dastoor, A.; Figueras-Nieto, D.; Ryjkov, A. Long range transport of mercury to the Arctic and across Canada. *Atmos. Chem. Phys.* **2010**, *10*, 6063–6086.

37. Kellerhals, M.; Beauchamp, S.; Belzer, W.; Blanchard, P.; Froude, F.; Harvey, B.; McDonald, K.; Pilote, M.; Poissant, L.; Puckett, K.; *et al.* Temporal and spatial variability of total gaseous mercury in Canada: Results from the Canadian Atmospheric Mercury Measurement Network (CAMNet). *Atmos. Environ.* **2003**, *37*, 1003–1011.

38. Lindberg, S.E.; Stratton, W.J. Atmospheric mercury speciation: Concentrations and behaviour of reactive gaseous mercury in ambient air. *Environ. Sci. Technol.* **1998**, *32*, 49–57.

39. Huang, J.Y.; Choi, H.D.; Hopke, P.K.; Holsen, T.M. Ambient mercury sources in Rochester, NY: Results from Principle Components Analysis (PCA) of mercury monitoring network data. *Environ. Sci. Technol.* **2010**, *44*, 8441–8445.

40. Mao, H.; Talbot, R.; Sigler, J.M.; Sive, B.C.; Hegarty, J.D. Seasonal and diurnal variation in Hg° over New England. *Atmos. Chem. Phys.* **2008**, *8*, 1403–1421.

41. Kim, K.H.; Kim, M.Y. Some insights into short-term variability of total gaseous mercury in urban air. *Atmos. Environ.* **2001**, *35*, 49–59.
42. Lynam, M.M.; Keeler, G.J. Source-receptor relationships for atmospheric mercury in urban Detroit, Michigan. *Atmos. Environ.* **2006**, *40*, 3144–3155.

A Survey of Mercury in Air and Precipitation across Canada: Patterns and Trends

Amanda S. Cole, Alexandra Steffen, Chris S. Eckley, Julie Narayan,
Martin Pilote, Rob Tordon, Jennifer A. Graydon, Vincent L. St. Louis,
Xiaohong Xu and Brian A. Branfireun

Abstract: Atmospheric mercury (Hg) measurements from across Canada were compiled and analysed as part of a national Hg science assessment. Here we update long-term trends of Hg in air and precipitation, and present more extensive measurements on patterns and trends in speciated Hg species (gaseous elemental mercury—GEM, reactive gaseous mercury—RGM, and total particulate mercury on particles <2.5 μm—TPM$_{2.5}$) at several sites. A spatial analysis across Canada revealed higher air concentrations and wet deposition of Hg in the vicinity of local and regional emission sources, and lower air concentrations of Hg at mid-latitude maritime sites compared to continental sites. Diel and seasonal patterns in atmospheric GEM, RGM and TPM$_{2.5}$ concentrations reflected differences in patterns of anthropogenic emissions, photo-induced surface emissions, chemistry, deposition and mixing. Concentrations of GEM decreased at rates ranging from -0.9% to -3.3% per year at all sites where measurements began in the 1990s. Concentrations of total Hg in precipitation declined up to 3.7% yr^{-1}. Trends in RGM and TPM$_{2.5}$ were less clear due to shorter measurement periods and low concentrations, however, in spring at the high Arctic site (Alert) when RGM and TPM$_{2.5}$ concentrations were high, concentrations of both increased by 7%–10% per year.

Reprinted from *Atmosphere*. Cite as: Cole, A.S.; Steffen, A.; Eckley, C.S.; Narayan, J.; Pilote, M.; Tordon, R.; Graydon, J.A.; St. Louis, V.L.; Xu, X.; Branfireun, B.A. A Survey of Mercury in Air and Precipitation across Canada: Patterns and Trends. *Atmosphere* **2014**, *5*, 635–668.

1. Introduction

Mercury (Hg) is a pollutant of concern due to its health effects and elevated concentrations in aquatic food webs. Current and historical anthropogenic activities have mobilized Hg from stable mineral deposits to the atmosphere, oceans, soils and biota. The atmosphere transports Hg around the globe and deposits it to vegetation, soils and water, where under certain conditions it can be methylated to toxic, bio-accumulative methylmercury [1]. In the atmosphere, gaseous elemental mercury (GEM) dominates due to its estimated 6–12 month lifetime in the atmosphere [2–4];

73

shorter-lived reactive gaseous mercury (RGM) and total particulate mercury (TPM) in the atmosphere can either be emitted directly or created by oxidation of GEM and taken up on particles, depositing within hours to weeks [5]. Conversion of Hg between these three inorganic forms can take place on various timescales and media and influence deposition and surface emissions. As a result, levels of Hg in air and precipitation across Canada can vary on timescales from hours to decades. Short-term temporal changes in atmospheric concentrations of Hg, such as diel and seasonal patterns, arise from changes in meteorology (e.g., temperature, winds, sunlight, precipitation) that lead to changes in emissions, mixing, chemistry and deposition. Therefore, these patterns may yield information about the chemistry and air-surface exchange of Hg at a particular location, whereas long-term trends are reflective of multi-year changes in the Hg budget on local, regional and global scales [6].

Environment Canada's Canadian Atmospheric Mercury Measurement Network (CAMNet) operated between 7 and 14 sites across Canada measuring total gaseous mercury (TGM, the sum of GEM and RGM) from 1994 to 2007. Some of these sites, which are predominantly rural or remote, are currently operated under the Canadian Air and Precipitation Monitoring Network (CAPMoN). From 2002 onward, selected CAMNet sites began making measurements of speciated Hg (GEM, RGM and $TPM_{2.5}$). Canada has also had up to 18 precipitation monitoring sites, in conjunction with the U.S.-led Mercury Deposition Network (MDN), measuring total Hg in precipitation at some locations since 1996. Figure 1 shows the locations of all active and inactive sites across Canada where the data presented in this study were collected. While the map shows more than 30 sites widely spread across the country, measurements at some locations were collected for only a few months, as summarized in Tables 1–3, and not all sites made both air and precipitation measurements.

Here, concentrations of Hg in the atmosphere and in precipitation are profiled across Canada. The atmospheric data are reported as TGM when measured by a Tekran 2537 instrument alone, and GEM, RGM and $TPM_{2.5}$ when using the Tekran 2537/1130/1135 system. These data incorporate and update previous measurements reported for some sites [7–22] and include speciated Hg measurements at several newer sites for the first time. The data were collected and analyzed as part of a Canadian mercury science assessment [23]. Diel (Section 2.2) and seasonal (Section 2.3) patterns in atmospheric concentrations and precipitation fluxes, plus long-term trends at sites where Hg has been measured for several years (Section 2.4), are presented. Data collection and analysis methods are detailed in Section 3.

Table 1. Stations measuring total gaseous mercury (TGM) in Canada and mean daily concentrations over the entire measurement period.

Station	Network	Longitude (°W)	Latitude (°N)	Measurement Period for Data Included Here	Mean TGM ± SD (ng·m^{-3})
Little Fox Lake YK	NCP [a]	135.63	61.35	June 2007–November 2011	1.28 ± 0.17
Reifel Island BC	CAMNet	123.17	49.10	March 1999–Febuary 2004	1.67 ± 0.19
Saturna BC	CAPMoN	123.13	48.78	March 2009–December 2010	1.43 ± 0.20
Whistler BC	INCATPA [b]/CARA [c]	122.93	50.07	August 2008–November 2011	1.21 ± 0.20
Meadows AB	None	114.64	53.53	May 2005–December 2008	1.51 ± 0.21
Genesee AB	None	114.20	53.30	March 2004–December 2010	1.53 ± 0.25
Fort Chipewyan AB	CAMNet	111.12	58.78	June 2000–July 2001	1.36 ± 0.15
Esther AB	CAMNet	110.20	51.67	June 1998–April 2001	1.65 ± 0.15
Bratt's Lake SK	CAPMoN	104.71	50.20	May 2001–December 2010	1.44 ± 0.25
Flin Flon MB	CARA [c]	101.88	54.77	July 2008–June 2011	3.75 ± 2.22
Windsor ON	None	83.01	42.18	January 2007–December 2008	1.93 ± 0.80
Burnt Island ON	CAMNet	82.95	45.81	May 1998–December 2007	1.55 ± 0.22
Egbert ON	CAMNet	79.78	44.23	December 1996–December 2010	1.58 ± 0.29
Buoy ON	CAMNet	79.45	43.40	July–September 2005	1.71 ± 0.20
Kuujjuarapik PQ	CAMNet	77.73	55.30	August 1999–September 2009	1.68 ± 0.46
Point Petre ON	CAMNet	77.15	43.84	November 1996–December 2007	1.75 ± 0.33
St. Anicet PQ	CAMNet	74.28	45.12	August 1994–December 2009	1.60 ± 0.37
St. Andrews NB	CAMNet	67.08	45.09	January 1996–July 2007	1.38 ± 0.24
Kejimkujik NS	CAMNet	65.21	44.43	January 1996–December 2010	1.40 ± 0.31
Mingan PQ	CAMNet	64.17	50.27	January 1997–December 2000	1.57 ± 0.19
Southampton PE	CAMNet	62.58	46.39	January 2005–December 2006	1.23 ± 0.19
Alert NU	CAMNet	62.33	82.50	January 1995–December 2011	1.51 ± 0.37

[a] Northern Contaminants Program, Aboriginal Affairs and Northern Development Canada; [b] Intercontinental Atmospheric Transport of Anthropogenic Pollutants to the Arctic, Government of Canada Program for International Polar Year; [c] Clean Air Regulatory Agenda, Environment Canada.

Table 2. Stations measuring speciated mercury (GEM, RGM, and TPM$_{2.5}$) in Canada and mean values over the entire measurement period.

Station	Longitude (°W)	Latitude (°N)	Measurement Period for Data Included Here	Mean GEM (ng·m^{-3})	Mean RGM (pg·m^{-3})	Mean TPM$_{2.5}$ (pg·m^{-3})
Genesee AB	114.20	53.30	January–September 2009	1.40	5.0	4.5
Flin Flon MB	101.88	54.77	July 2010–May 2011	2.06	3.4	10.4
Churchill MB	94.07	58.75	March–August 2004	1.52	100.9	168.5
ELA ON	93.72	49.66	May 2005–December 2010	1.39	1.0	4.4
Mississauga ON	79.65	43.54	January–December 2009	1.40	3.7	6.5
Dorset ON	78.93	45.22	July 2008–March 2010	1.38	2.7	5.9
St Anicet PQ	74.28	45.12	January 2003–December 2010	1.52	3.0	17.5
Kejimkujik NS	65.21	44.43	January 2009–December 2010	1.34	0.5	4.2
Halifax NS	63.67	44.67	October 2009–December 2010	1.68	2.1	2.3
Alert NU	62.33	82.50	January 2002–December 2011	1.26	21.8	41.1

Table 3. Stations measuring Hg in precipitation in Canada and mean monthly concentration (volume-weighted means), precipitation, and deposition over the entire measurement period. MDN = Mercury Deposition Network; GSC = Geological Survey of Canada.

Station	Network	Long (°W)	Lat (°N)	Measurement Period for Data Presented Here	Mean Total Hg Conc. (ng·L^{-1})	Mean Monthly Precip (mm)	Mean Hg Dep. (μg·m^{-2}·month^{-1})
Reifel Island BC	MDN	123.17	49.10	April 2000–Febuary 2004	5.6	68	0.38
Saturna BC	MDN	123.13	48.78	September 2009–January 2011	4.5	91	0.41
Fort Vermillion AB	GSC	116.02	58.38	December 2006–January 2008	4.3	22	0.10
Genesee AB	MDN	114.20	53.30	July 2006–December 2009	12.8	32	0.44
Crossfield AB	GSC	114.00	51.29	May 2006–December 2007	9.3	23	0.22
Henry Kroeger AB	MDN	110.83	51.42	October 2004–December 2009	11.7	25	0.35
Esther AB	MDN	110.20	51.67	April 2000–May 2001	14.2	14	0.21
Bratts Lake SK	MDN	104.72	50.20	June 2001–December 2009	11.2	26	0.37
Flin Flon MB	Edmonton	101.88	54.77	September 2009–December 2010	59.9	30	1.80
Churchill MB	GSC	94.07	58.75	June 2006–December 2007	5.3	15	0.11
ELA ON	MDN	93.72	49.66	November 2009–January 2011	9.6	69	0.77
Burnt Island ON	MDN	82.95	45.81	November 2001–March 2003	9.2	61	0.56
Egbert ON	MDN	79.78	44.23	March 2000–January 2010	8.4	57	0.47
Dorset ON	MDN	78.93	45.22	January 1997–December 1998	9.7	56	0.55
Point Petre ON	MDN	77.15	43.84	November 2001–March 2003	9.1	58	0.54
Chapais PQ	MDN	74.98	49.82	December 2009–January 2011	6.4	71	0.46
St. Anicet PQ	MDN	74.03	45.20	April 1998–August 2007	7.9	70	0.56
St. Andrews NB	MDN	67.08	45.08	July 1996–December 2003	6.6	86	0.56
Kejimkujik NS	MDN	65.21	44.43	July 1996–January 2010	5.2	111	0.58
Mingan PQ	MDN	64.23	50.27	April 1998–August 2007	5.0	77	0.39
Stephenville NL	MDN	58.57	48.56	Febuary 2010–January 2011	5.6	97	0.54
Cormak NL	MDN	57.38	49.32	May 2000–January 2010	4.2	94	0.40

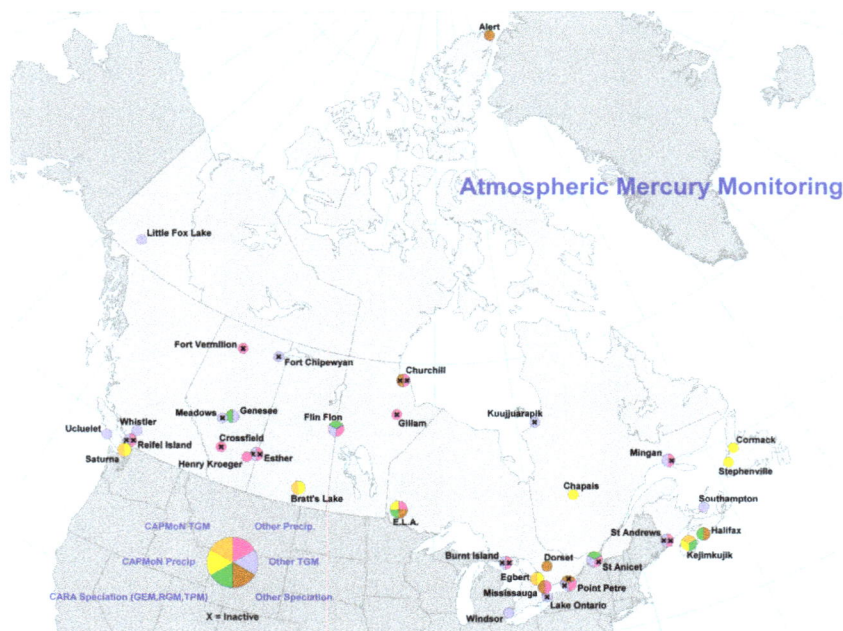

Figure 1. Map of current and past Canadian Hg monitoring stations.

2. Results and Discussion

2.1. Spatial Patterns of Hg in Canada

The distribution of average daily TGM concentrations at 22 measurement sites is summarized by the box and whisker plot in Figure 2. Mean TGM concentrations at all but one of the sites ranged from 1.21 to 1.93 ng·m^{-3} (Table 1), within or below northern hemispheric background levels [24,25]. The bulk of TGM is GEM, with an atmospheric lifetime on the order of a year. Therefore, it is expected that TGM concentrations will be fairly uniform at background sites that are far from emission sources, except at sites where other atmospheric processes dominate [9]. The mean TGM concentration of 3.75 ng·m^{-3} observed at Flin Flon, MB, is a result of local emissions [19].

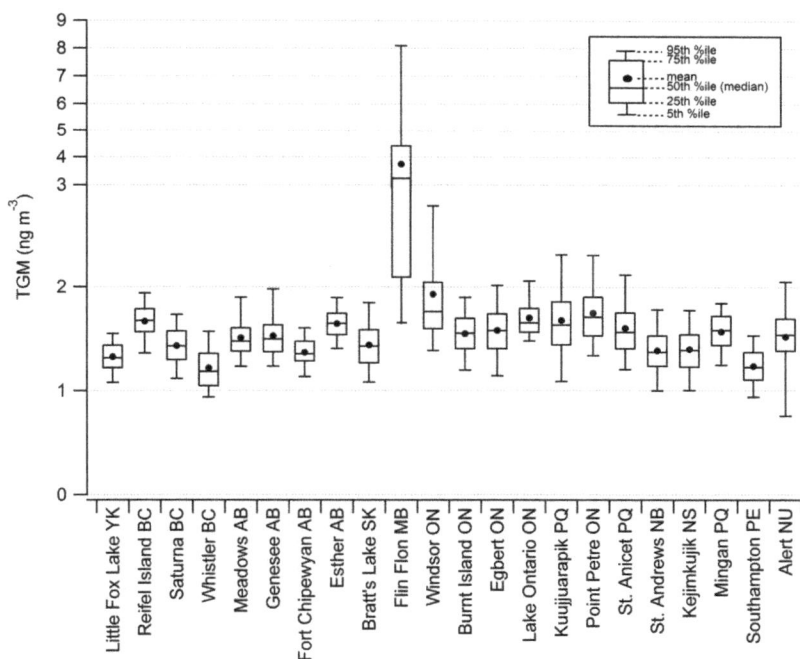

Figure 2. A comparison of total gaseous mercury (TGM) measurements, west to east. The top section is compressed to include Flin Flon, MB. Time periods covered by measurements are listed in Table 1.

Overall, TGM concentrations and distributions at these Canadian sites are similar to previously reported data [8,9] where available, with small decreases at some sites observed over the long term (discussed further in Section 2.4). The sites with the lowest TGM levels, ranging from 1.21–1.38 ng·m^{-3}, include two high-elevation sites near the west coast (Whistler and Little Fox Lake) and three maritime sites at Southampton, Kejimkujik and St. Andrews. While the latter two sites began measurements in the 1990s, the other three cover more recent periods and the lower concentrations may be partially due to decreases in TGM since the 1990s [25,26]. The highest TGM levels by far were observed at Flin Flon, due to local emissions from the copper smelter which was the largest point source of Hg in Canada before closure [19]. The sites that show the most variable TGM concentrations include Alert and Kuujjuarapik, which both experience episodes of rapid conversion of GEM to RGM and/or $TPM_{2.5}$ in the spring, known as atmospheric mercury depletion events (AMDEs) [27]. These AMDEs, as well as emissions from the snow and ocean in late spring and summer [28–31], lead to frequent and large fluctuations in TGM. Sites that were affected by local and regional pollution such as Windsor, Point Petre, Flin Flon and St. Anicet showed right-skewed distributions (*i.e.*, a long tail of higher concentrations) that were due to periodic episodes of high TGM concentrations. Similarly, the concentration data from the Lake Ontario Buoy, which is downwind of U.S. emission sources, and the Edmonton-area sites at Genesee and Meadows, which are downwind of local coal-fired power plants, also showed right-skewed distributions. TGM at Kuujjuarapik has a right-skewed distribution that is likely a result of the increased GEM concentrations in the spring and summer following the AMDE season. In contrast, TGM concentrations at Alert are left-skewed due to the periodic losses of GEM during AMDEs that have a greater effect on TGM levels in comparison to emissions. These seasonal cycles are discussed in more detail in Section 2.3.

Measurements of RGM, $TPM_{2.5}$ and GEM at sites measuring atmospheric Hg speciation concentrations are summarized in Table 2. A box and whisker plot of RGM and $TPM_{2.5}$ is shown in Figure 3. At Alert and Churchill, the two sites bordering sea ice, RGM and $TPM_{2.5}$ levels spike in the spring due to AMDEs, when GEM is oxidized to RGM periodically [27]. While the highest RGM and $TPM_{2.5}$ concentrations were measured at Churchill, it should be noted that the Alert probability distribution represents data from several years and includes low RGM measurements from fall and winter. Data from Churchill were only collected for one spring and summer and therefore are more weighted by the spring months when RGM and $TPM_{2.5}$ values are high. Both sites have RGM and $TPM_{2.5}$ distributions highly skewed towards higher concentrations (mean >median), as expected of a site that is observing a local "source;" in this case, formation of RGM by local or regional chemistry [27,32].

RGM is relatively short-lived with respect to deposition to particles or surfaces, with an estimated lifetime of a few hours to days [5]. Therefore, remote sites that are far from emission sources and are not affected by AMDEs, such as the ELA and Kejimkujik, report very low concentration levels of RGM. In fact, the RGM concentration at these locations were below the manufacturer's suggested detection limit of ~2 pg·m^{-3} more than 75% of the time. Further, even some sites that would be expected to be affected by local emission sources (such as Mississauga, St. Anicet and Halifax) have median RGM levels below this detection limit (Figure 3), though method detection limits determined by blank measurements at individual sites may be lower (see Experimental section). Genesee and Flin Flon sites had higher RGM concentrations due to local sources of Hg pollution (coal-fired power plants [17] and a copper smelter [19], respectively) but concentrations were not as high as those reported in the Arctic at Alert and Churchill. Concentrations of TPM$_{2.5}$ were usually, but not always, higher than RGM (Table 2; Figure 3) such that median values were above the 2 pg·m^{-3} threshold at most sites. Similar to RGM, mid-latitude sites had lower TPM$_{2.5}$ concentrations than the northern coastal sites, Alert and Churchill. Concentrations of RGM and TPM$_{2.5}$ at mid-latitude sites were comparable to those reported by Gay *et al.* [18] for remote sites across the United States, plus Kejimkujik, for 2009–2011.

Speciated Hg was measured at Flin Flon starting in July 2010, a few weeks after the shutdown of the copper smelter. Concentrations of RGM and TPM$_{2.5}$ were elevated above background levels (represented at the ELA site), likely due to residual effects from the smelter activities, and were similar to concentrations at other rural sites affected by regional pollution sources, such as St. Anicet and Genesee (Figure 3). The mean TGM concentration measured with the speciation system SSW of the smelter is higher than at other sites at 1.91 ng·m^{-3} but lower than measured at the TGM-only site SSE of the smelter during the same period (3.17 ng·m^{-3}), likely because it is usually upwind of the source [19].

It should also be noted that since RGM and TPM$_{2.5}$ are operationally defined, comparisons between different sites are problematic. For example, the specific compounds that make up RGM may vary between sites, and if those compounds are retained by the denuder column with different efficiencies [33], some of what is RGM may be measured as GEM. In addition, ozone has been shown to decrease the collection efficiency of RGM on the denuder [34], which may suppress observed RGM levels at more polluted sites. Finally, TPM$_{2.5}$ measures only those particles smaller than 2.5 μm, so persistent differences in the size of aerosol at different sites (e.g., due to differences in humidity or the age of particles) would also affect the relative amounts of TPM$_{2.5}$ measured at those sites.

Monthly volume-weighted mean total Hg concentrations in precipitation and monthly Hg wet deposition rates are shown in Figure 4 and Table 3. The site locations for the atmospheric measurements differ from the wet deposition sites with only 13 in common for both types of sampling. The range of values for mean monthly deposition of Hg was 0.10–0.77 μg·m^{-2}·month^{-1}, or 1.2–9.2 μg·m^{-2}·yr^{-1}, excluding Flin Flon. These values are similar to or lower than what was found for MDN sites across North America outside of the southeastern US, where deposition rates for

2005 were on the order of 10–20 $\mu g \cdot m^{-2} \cdot yr^{-1}$ [16]. As with TGM measurements, precipitation measurements at Flin Flon include time periods when the smelter was operational and thus the Hg deposition rates were 2–18 times higher than at other sites.

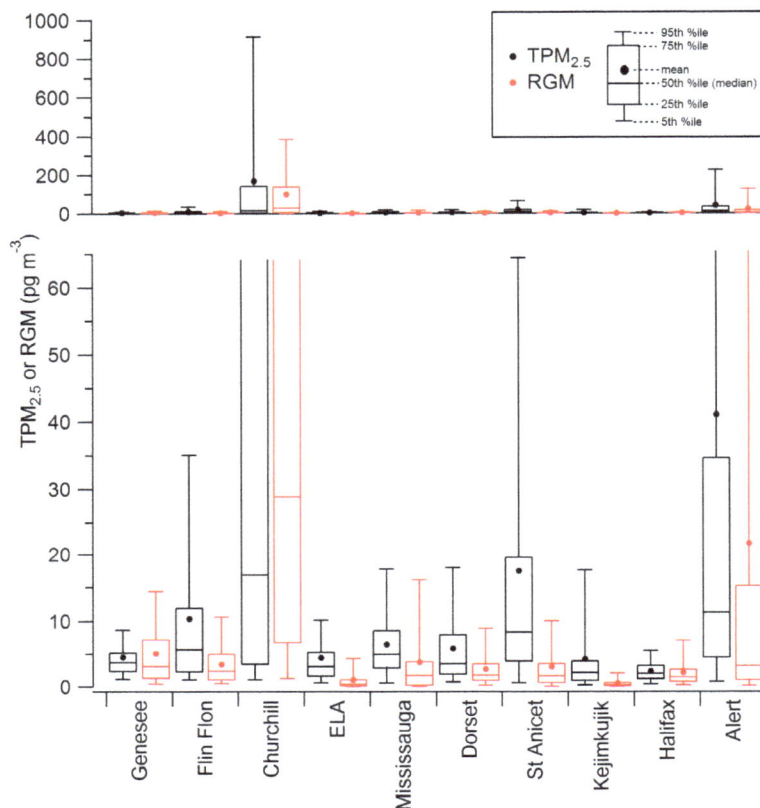

Figure 3. Comparison of reactive gaseous mercury (RGM) and total particulate mercury (TPM$_{2.5}$) at mercury speciation stations. The top plot shows the full range to include Alert, NU, and Churchill, MB. Time periods covered by measurements are listed in Table 2.

The lowest Hg wet deposition rates were observed at Fort Vermilion and Churchill, the most northerly of the precipitation monitoring sites, where concentrations of Hg in precipitation were also low. Sites with low precipitation volumes in Alberta and Saskatchewan had high concentrations of Hg in precipitation but low overall deposition. Low precipitation amounts can lead to low wet deposition and high concentrations of Hg as both TPM$_{2.5}$ and RGM can build up in the air or on cloud droplets before being deposited in rain or snow. At this time, there are insufficient co-located speciation measurements to confirm this hypothesis. In contrast, sites near the west and east coasts, where total precipitation volumes were high, had low concentrations of Hg in precipitation (due to dilution) but higher wet deposition than the drier remote sites. Finally, with the exception of the ELA, sites closer to industrial activity in Ontario, Québec and the northeastern United States had relatively high concentrations of Hg in rain and snow (similar to the drier western sites) and the highest wet deposition amounts as a group. The ELA showed the highest average wet deposition of all the sites aside from Flin Flon, however, it should be noted that fewer than two years of data are presented here from the ELA and 10 other sites, and our ability to discern general spatial patterns from these data is therefore limited due to inter-annual variability. An earlier long-term study at the ELA reported a volume-weighted mean concentration of total Hg of 5.7 ng·L^{-1} for 1992–2006 [35] rather than the 11.2 ng·L^{-1} reported here.

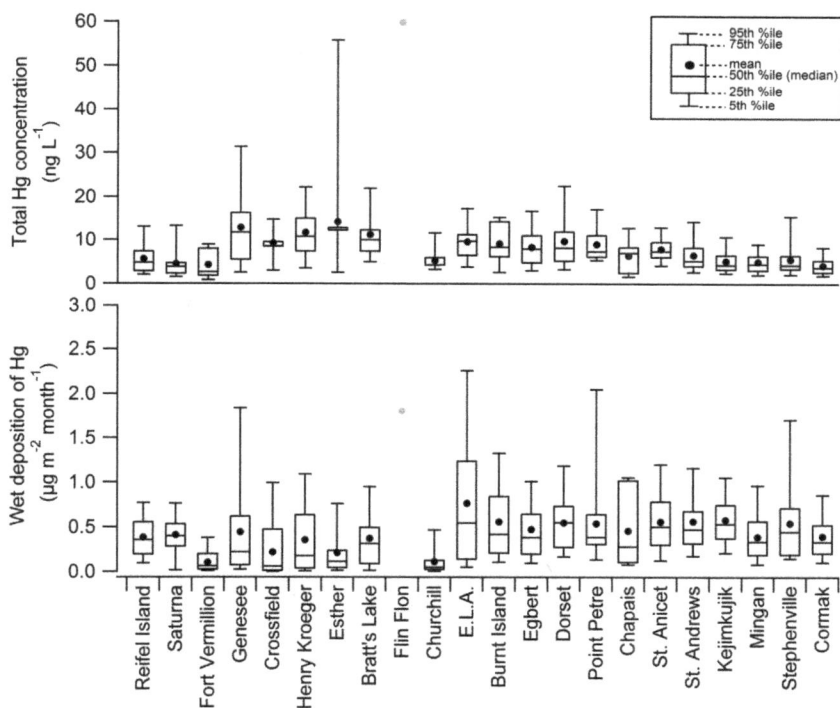

Figure 4. Comparison of total Hg concentrations in precipitation (top) and monthly wet deposition (bottom), with only the mean shown for Flin Flon. Time periods covered by measurements are listed in Table 3. The distributions are based on monthly volume-weighted means and monthly deposition amounts, which were first calculated from weekly samples (see Experimental).

2.2. Diel Patterns of Atmospheric Hg

Plots of diel patterns in TGM are shown Figure 5. In this and all following figures, Alert is assigned to Atlantic Standard Time (UTC-4) since the maximum solar radiation occurs at noon in that time zone. Figure 5 shows that 14 of the 22 sites, as well as Flin Flon [19], have a similar daily pattern of a pre-dawn minimum and afternoon maximum. The afternoon peak is broad because data from all seasons have been averaged in the overall profile, and peaks can vary in timing and amplitude by season, with the strongest cycles observed in spring and summer. The diel patterns reported here are consistent with previous reports [8,10]. Kellerhals *et al.* [8] attributed this pattern to two factors: (1) night time deposition of TGM from the shallow nocturnal boundary layer followed by mixing with more TGM-rich air aloft when the nocturnal inversion breaks down; and (2) surface emission of TGM during the daytime following photolytic or temperature-driven processes on soils or snow [31,36–38]. This pattern has also been observed at other non-urban sites in North America (e.g., [e.g. 39,40]). The TGM diel cycle at St. Anicet is slightly different, with a second peak around 22:00. Previous analysis of St. Anicet data identified the influence of nighttime regional anthropogenic and industrial sources at this location [41].

At Alert, TGM concentrations were quite stable with a variation of about 0.05 ng·m^{-3} or 3%, likely because it is in constant light or darkness for most of the year [13]. TGM peaked around 11:00 but dropped quickly after noon, reaching a minimum around 20:00 and slowly increasing throughout the night and morning. This pattern was strongest in the springtime with an amplitude of 15% in April.

The remaining six sites had atypical diel patterns. TGM concentrations at Little Fox Lake did not vary with time of day, while Southampton was the only site with a late morning minimum. While no meteorological data was available for Southampton, the diel pattern of TGM was almost exactly opposite the diel pattern in wind speed at the nearby weather station at St. Peter's, 6 km to the northeast. At Reifel Island, Saturna and on the Lake Ontario Buoy, TGM concentrations peaked in the early, mid-, and late morning, respectively, then dropped through the afternoon to an evening minimum. Windsor also saw minimum TGM concentrations in the evening, rising to a broad peak between 03:00 and 13:00. This pattern at Reifel Island was previously attributed to changes in wind direction over the day, with daytime sea breezes bringing clean air from the ocean and nighttime land breeze transporting the TGM-rich plume from Greater Vancouver past the site [8]. The other three sites are also influenced by sea or lake breezes and are near to metropolitan areas. Urban areas in North America that report similar diel cycles in TGM or GEM include Detroit, MI, with maximum GEM at 05:00 and minimum around 15:00 [42] and Birmingham, AL, with maximum GEM around 10:00 and minimum around 16:00 [40]. It has been postulated that the decrease in urban GEM concentrations through the day to an afternoon minimum is due to the presence of high surface concentrations of Hg from local sources that are increasingly diluted by cleaner air from above as the day warms and vertical mixing is enhanced [40]. However, this pattern is not seen at all urban sites [e.g. 43] and it remains unclear what drives the diel patterns at these four Canadian sites.

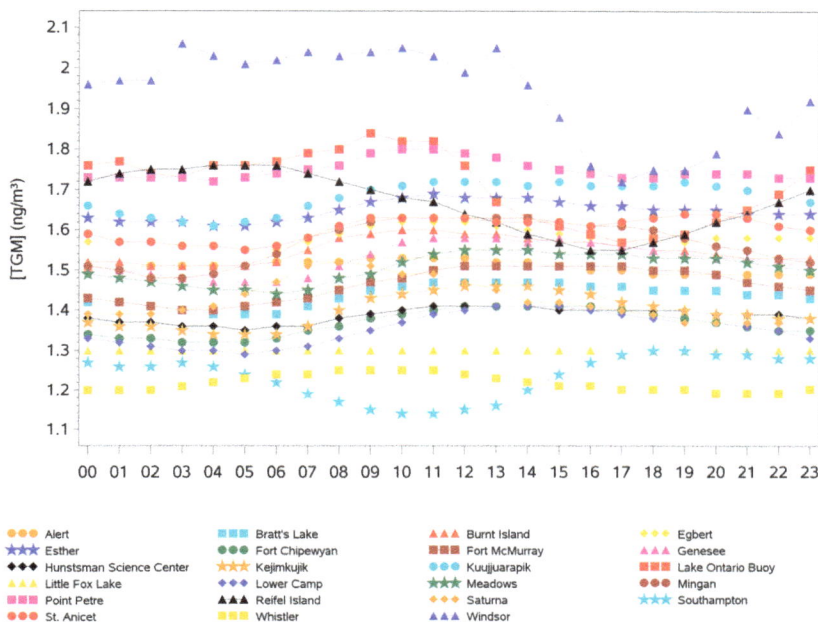

Figure 5. Diel cycles in TGM averaged over total measurement period at all sites. Hour 0 is from 12:00 to 1:00 a.m. local time.

Diel patterns of RGM, $TPM_{2.5}$ and GEM were also investigated for 10 different sites. These profiles are shown for every site except Flin Flon in Figure 6. RGM concentrations at all sites peaked in the afternoon, with amplitudes of 35%–180% of the mean. This suggests that short-lived RGM is formed through photo-oxidation of GEM in the atmosphere [44]. An afternoon maximum may also be due to higher RGM concentrations in the free troposphere (e.g., [45,46]) that mix down to the surface during the day. Higher concentrations aloft could arise from the slower nighttime dry deposition above the nocturnal boundary layer. These two mechanisms—*in situ* photo-oxidation and mixing with higher RGM levels above—could explain the consistent diel pattern across North American sites [39,40,42,43]. At certain sites, such as Genesee, local emission sources that release RGM above the nocturnal boundary layer, such that these releases are not mixed into the surface air until daytime [47], could contribute to the observed pattern as well.

The y-axis scales on both the Alert and Churchill plots in Figure 6 should be noted as the concentrations of RGM at these sites are significantly higher than at the others. This is a direct result of the springtime atmospheric chemistry that occurs at these locations. At Alert, the diel cycle of RGM is dominated by relatively large variability in the spring, when minimum and maximum concentrations can vary by up to 35 pg·m^{-3} (30% of the mean) with a broad peak in the afternoon. However, RGM concentrations vary widely at all times of day, and remain elevated throughout the night in the early spring when there is still a day/night cycle, suggesting that RGM has a lifetime of at least several hours, if not days, leading to elevated concentrations throughout the region. RGM levels at Alert could therefore be influenced by regionally elevated RGM and TPM$_{2.5}$ and depleted GEM [48], with an additional contribution from local photochemical production [32]. RGM concentrations at Churchill are similarly elevated throughout the day and night with a peak in the afternoon, again suggesting contributions from non-local as well as local chemistry.

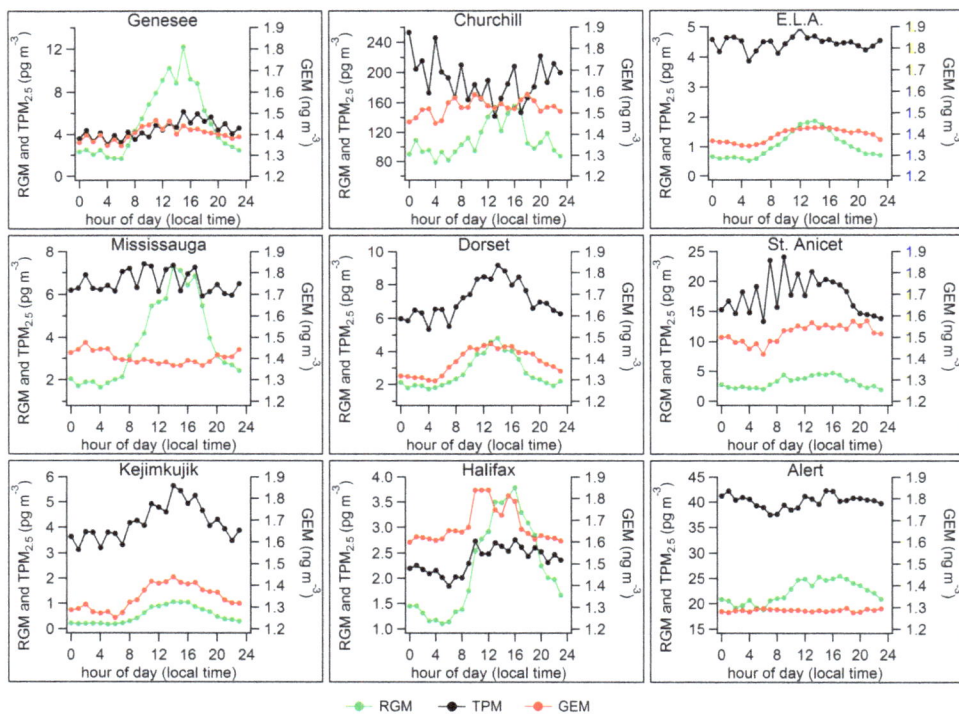

Figure 6. Diel cycles in concentrations of GEM, RGM and TPM$_{2.5}$ (mean concentrations for each hour) at nine sites. Axis scales for RGM and TPM$_{2.5}$ vary.

Diel patterns in TPM$_{2.5}$ concentrations are less consistent between sites than those observed for RGM. Particulate Hg is influenced by many factors, such as the amount of particulate available, the temperature (which can influence the partitioning between the gas and particle phases for semi-volatile compounds), transport of TPM from sources, and precipitation events that remove TPM from the air. At some sites, such as Dorset, St. Anicet, Kejimkujik and Genesee, TPM$_{2.5}$ is higher in the afternoon, though the time of maximum TPM$_{2.5}$ varies. At other sites such as ELA, Mississauga and Halifax, concentrations vary little across the day (<1 pg·m^{-3}). Only Churchill shows a minimum in TPM$_{2.5}$ levels during the daytime. TPM$_{2.5}$ patterns at other sites in North America are also inconsistent [10,39,40,42,49]. One factor that may influence the TPM$_{2.5}$ cycle is the vertical gradient of TPM$_{2.5}$ concentrations [40]. If surface TPM$_{2.5}$ concentrations are fairly low, and if there is dry deposition in the nocturnal boundary layer, TPM$_{2.5}$ may increase in the morning when vertical mixing brings more TPM$_{2.5}$ from aloft. This is consistent with an increase in concentrations in the morning and stable concentrations during the day, as at Halifax (though as noted, the differences are small). In contrast, if there are local emissions, TPM$_{2.5}$ levels may build up during the night and drop in the morning. This pattern was not seen at any of the sites presented here. In an alternative mechanism, increased RGM (from photo-oxidation of GEM) may increase TPM$_{2.5}$ during the day if some of the RGM is scavenged by particles. This would be consistent with patterns at Dorset, Kejimkujik, Genesee, and Alert. However, it should be noted that mean TPM$_{2.5}$ concentration at Alert are dominated by March and April peaks, while RGM concentrations are dominated by May values. Nevertheless, the photo-oxidation that produces high RGM in May likely produces high TPM$_{2.5}$ in March and April, when there is more particulate matter available and temperatures are colder [22], such that the diel cycles are similar. The diel cycle in TPM$_{2.5}$ at Churchill is opposite the diel cycle in RGM, suggesting that reactive Hg species generated during AMDEs around this site partition between the gas and particle phase depending on the temperature, with more adsorbed to particles during the cooler night hours. The erratic changes in all speciated Hg measurements from one hour to the next are a function of the short measurement period and the high variability in concentrations as AMDEs begin and end.

As expected—since GEM comprises the bulk of TGM—diel cycles in GEM are similar to those of TGM discussed above, with maximum amplitudes of 3%–9%. All but two sites (Mississauga and Alert) exhibited the typical non-urban diel cycle of minimum GEM at dawn and maximum GEM at mid-day or early afternoon. GEM in Mississauga followed the opposite cycle, with low values during the day and higher concentrations at night. This is more typical of urban sites, where local emissions build up TGM levels in the nocturnal boundary layer until they are diluted by cleaner air during the day when there is more mixing from aloft [12,40,42].

2.3. Seasonal Patterns in Atmospheric Hg and Hg in Precipitation

In Figure 7, monthly median TGM concentrations are plotted for the 11 sites with at least five years of data. Sites with shorter data coverage periods were excluded for clarity, but their cycles are discussed here. The data from the Lake Ontario Buoy reflect only three months of measurements and thus is not included in the overall seasonal discussion. At 12 of the 21 remaining sites (Reifel Island, Fort Chipewyan, Genesee, Esther, Burnt Island, St. Andrews, Kejimkujik, Egbert, Point Petre, Bratt's Lake and St. Anicet), maximum TGM concentrations are seen in late winter/early spring (February, March or April) and minima in the fall (September or October). This finding is consistent with previous results from Canada [8,9,11] and throughout the Northern Hemisphere [12,39,40,50]. This dominant pattern has previously been attributed to multiple factors, such as enhanced winter emissions from anthropogenic sources (primarily coal burning, as well as wood combustion), less vertical mixing in winter to clean out TGM that is emitted from surface sources [11,12], and increased oxidation and precipitation in summer [51,52]. However, modelling indicates that oxidation and precipitation are not significant factors in the TGM seasonality [23]. Rather, high revolatilization of Hg from snow significantly elevates GEM concentrations in winter in mid-latitudes. The evasion of Hg from snow and other surfaces increases with increasing solar radiation; this may cause the maximum in GEM to be in late winter /spring. In addition, several studies have shown that soil-air Hg emissions can be greatly enhanced as a result of warming temperatures [53] and increased soil moisture [54–57] that can occur following snowmelt. Terrestrial and oceanic emissions are at a maximum in summer. Therefore, the minimum in GEM concentrations generally occurs in fall when these various emission processes are least active. Five additional sites, not shown in Figure 7 (Saturna, Whistler, Little Fox Lake, Meadows and Mingan), had seasonal patterns similar to the dominant one discussed above, with slight differences in timing likely due to greater variability within these shorter measurement periods.

Alert showed the strongest seasonality with a minimum in the spring (April/May), maximum in the summer (July) and average concentrations in the fall and winter [13]. The spring depression is consistent with the AMDEs that occur at this site and the summer maximum is likely due to emission from various sources [13,29]. Kuujjuarapik also shows a summer maximum resulting from similar emissions as Alert, but the minimum is broader, extending from October to March. AMDEs in the spring at Kuujjuarapik are offset by rapid revolatilization of TGM after each event, resulting in no net depression of monthly TGM levels through the spring months.

Of the sites with less than five years of data, TGM concentrations at Southampton and Windsor, both with only two years of data, appear to have a bimodal seasonal pattern with low values of TGM in the spring and fall and high values in the summer and winter. The reason for this seasonality at Southampton is currently unclear for this relatively short and highly variable time series of data. Pollution episodes at Windsor are responsible for the summer maximum there. Similarly, Point Petrie was previously reported to have a summer maximum in TGM concentrations due to periodic elevated concentrations of TGM in the summer [8,11]; measurements of TGM that cover a longer time period have shown the spring maximum reported here. Finally, Flin Flon shows much higher TGM concentrations than any other site, with maxima in spring (April) and summer (July) and minima in the late fall/early winter [19].

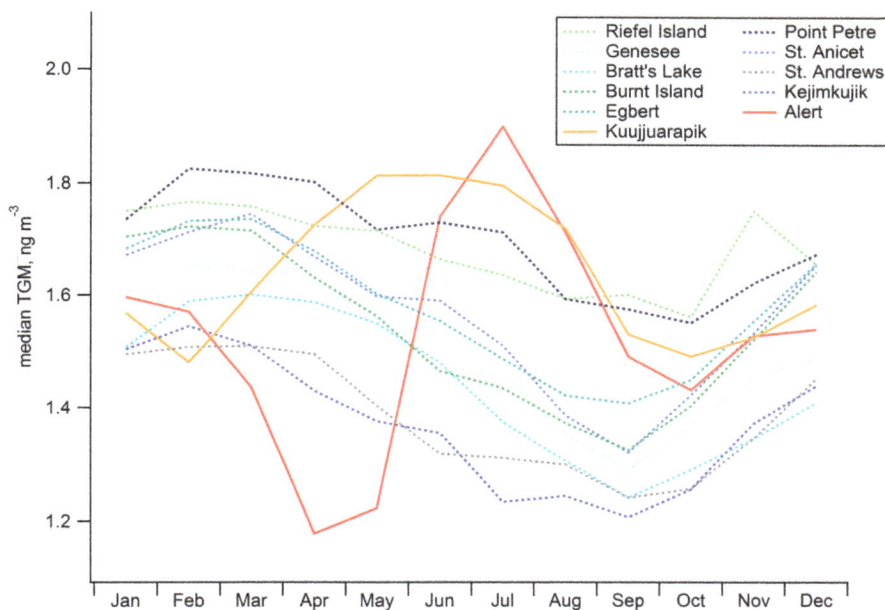

Figure 7. Seasonal cycles in total gaseous mercury (TGM) at the 11 sites with more than five years of data. Cycles from Alert and Kuujjuarapik differ from the dominant pattern and are therefore highlighted.

Seasonal variability in GEM, RGM and $TPM_{2.5}$ concentrations are shown in Figure 8. Of the 10 sites, three (Genesee, Flin Flon and Churchill) do not have measurements covering the entire year, so discussion of seasonal cycles is limited by the data coverage at these sites. Flin Flon "seasonal" patterns may not be typical since speciated Hg measurements began in July, shortly after shut down of the copper smelter, which is likely why the highest GEM, RGM and $TPM_{2.5}$ concentrations were measured in July. Results from Flin Flon are discussed in detail elsewhere [19].

As shown in Figure 8, GEM concentrations at the remaining nine sites follow similar patterns to TGM concentrations as expected (since TGM is largely GEM). Four of the sites (Genesee, St. Anicet, Kejimkujik, Alert) also have TGM instruments and the cycles in GEM agree with the TGM cycles discussed above. Measurements at Churchill cover March through August and were lowest in March and April and highest in June due to AMDE chemistry, similar to Alert. The remaining four sites had maximum GEM concentrations in March (Mississauga, Dorset, Halifax) or April (ELA) and minimum GEM concentrations in August (Dorset), September (ELA) or October (Mississauga, Halifax). These cycles are in agreement with the dominant TGM patterns across the country as discussed above.

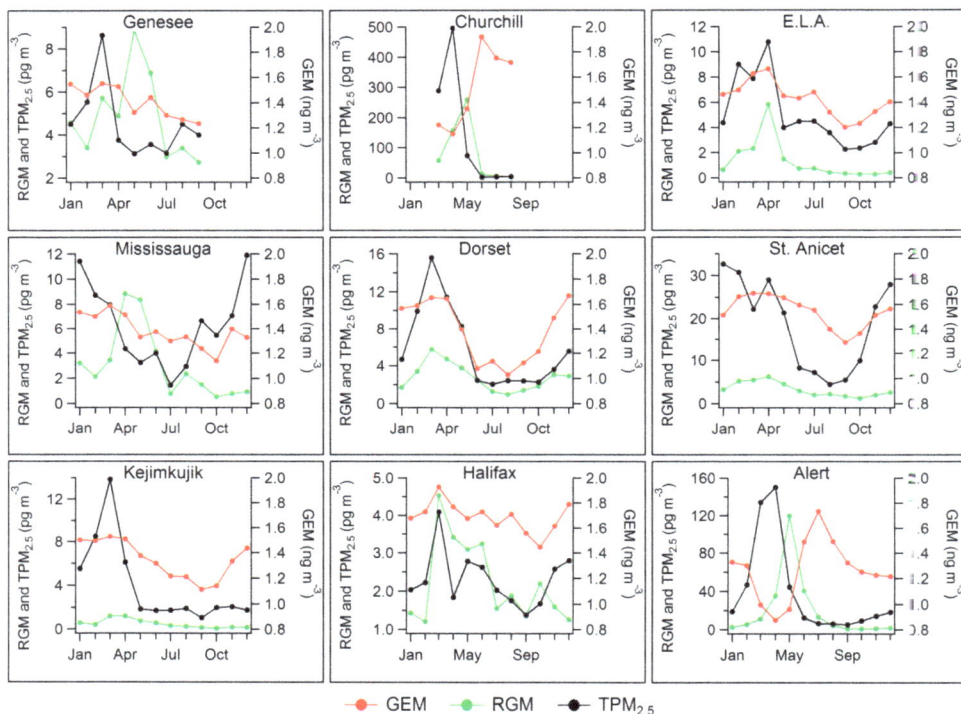

Figure 8. Seasonal cycles in speciated Hg (mean values for each month) at nine sites. Axis scales for RGM and TPM$_{2.5}$ vary.

RGM concentrations were highest in the spring (March, April, or May) at all sites. Minimum RGM concentrations occurred in September or October at most sites, with the exceptions of Dorset (August) and Halifax (December–February). At Alert and Churchill, RGM was elevated in springtime due to AMDE activity. While the photo-oxidation of GEM to produce RGM is expected to be highest in summer in other parts of the country due to more irradiation and oxidants, increased deposition of RGM may offset this increased source. Wet deposition of total Hg is highest in summer at most sites (see below) and dry deposition is expected to be much higher during summer and fall when vegetation is in full leaf [58]. Another factor may be a sampling artifact in the measurement of RGM, in which high concentrations of ozone in sampled air interfere with denuder absorption of RGM [34]. Since ozone concentrations are highest in summer, this artifact would suppress RGM concentrations the most during summer [59].

$TPM_{2.5}$ concentrations, like RGM, were high in March and April at the sites where AMDEs occur (Alert and Churchill) (Figure 8). Five non-AMDE sites also saw maximum $TPM_{2.5}$ concentrations in these months: Genesee, ELA, Dorset, Kejimkujik and Halifax. Mississauga and St. Anicet had elevated $TPM_{2.5}$ concentrations throughout the winter, peaking in December and January. These two sites are fairly close to urban centres, and may be affected by local and regional emissions that can be trapped close to the surface in the winter when mixing heights are low. $TPM_{2.5}$ may also be enhanced in cold months due to the gas-particle partitioning of semi-volatile Hg(II) species favouring the particle phase at low temperatures [60].

Finally, at many sites the peaks in GEM, RGM and $TPM_{2.5}$ coincide. This could point to common sources (such as coal combustion or other anthropogenic pollution at non-remote sites) or sinks (such as high summertime deposition) [11]. The interplay of various factors that may influence seasonal distributions of GEM, RGM and $TPM_{2.5}$ highlight the need for detailed modelling of these species in the atmosphere to explain these observations and determine their fate with respect to deposition.

The seasonal patterns of total Hg concentrations in rain and snow, precipitation amounts, and wet deposition of Hg at 22 sites were calculated by summing weekly samples over each calendar month. Monthly values were then averaged over all years of sampling. The seasonal pattern of Hg wet deposition showed maximum deposition in the summer (May–August) at 20 of 22 sites. The exceptions, Reifel Island and Dorset, had maxima in October and September, respectively. Twelve of the 22 sites reported here recorded less than three years of data, which increases the uncertainty in the seasonal cycling because there can be large inter-annual variability in precipitation amounts.

Seasonal cycles for four of the ten sites with at least three years of data are shown in Figure 9, illustrating representative patterns from those ten sites. Since Hg wet deposition flux depends on the total monthly precipitation amount and the concentration of total Hg in that precipitation, the seasonal cycles of both these parameters are shown along with the cycles of Hg wet deposition. For the sites not shown, seasonal patterns at Genesee and Henry Kroeger were similar to those at Bratt's Lake; patterns at St. Anicet, Mingan and Cormak resembled those at Egbert;

patterns at St. Andrews were similar to those at Kejimkujik. The summer maximum in Hg wet deposition is predominant. However, examining patterns of total Hg concentrations and precipitation amount reveal that at Bratt's Lake, the Hg wet deposition maximum corresponds to the maximum in precipitation amount, whereas at Kejimkujik the maximum Hg wet deposition occurs during the season of minimum precipitation but highest total Hg concentrations. At Egbert, a peak in both total Hg concentrations and precipitation volume inevitably results in a period of maximum wet deposition. The pattern at Reifel Island indicates that the amount of precipitation is the dominant factor in determining the wet deposition cycle at that site.

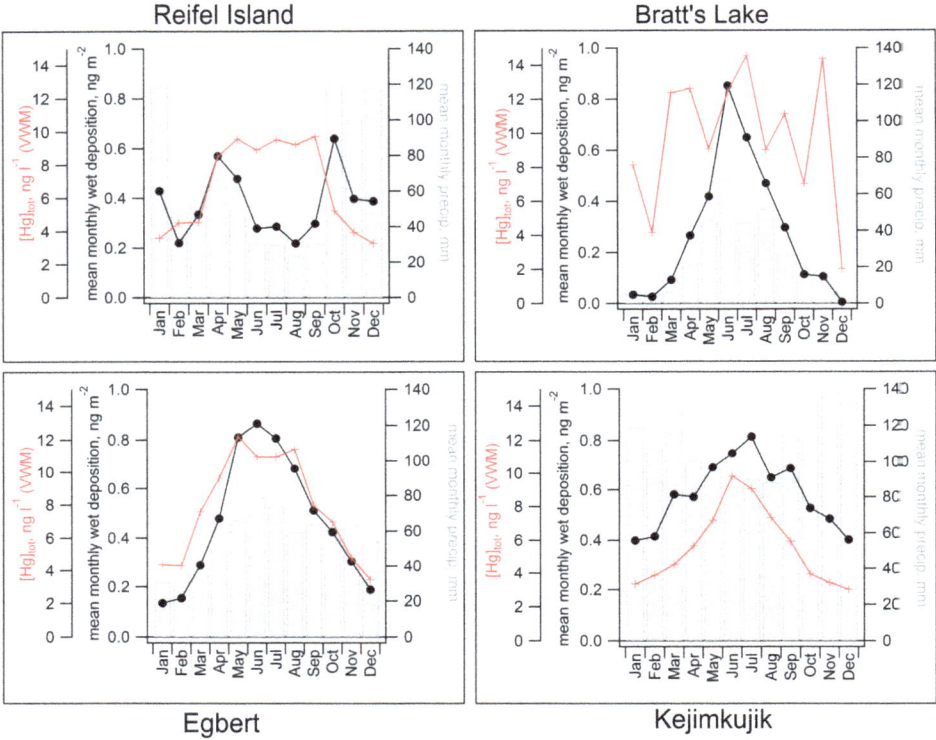

Figure 9. Seasonal cycles in monthly wet deposition (black), precipitation amounts (gray bars), and concentration of total Hg in precipitation (red) at four sites.

2.4. Long-Term Trends

Worldwide atmospheric measurements of TGM up to the early 2000s suggest that concentrations of atmospheric Hg increased from the 1970s to a peak in the 1980s and then plateaued in the late 1990s [24]. Similarly, measurements of air bubbles in a Greenland glacial firn indicated that GEM (the primary component of TGM) increased from the 1940s to the 1970s and reached a plateau around the mid-1990s [61]. In Canada, long-term trends in TGM [7,9,13,26], speciated atmospheric Hg [26] and Hg concentrations in precipitation [16,62] have been reported at some sites. The Canadian trends presented below are updated with the most comprehensive and recently available TGM, precipitation and Hg speciation data. The time period over which data is reported differs for each location. Linear trends were estimated for all available data from each site rather than limiting the analysis to only overlapping time periods. A minimum of five years of data were required to perform the Seasonal Kendall trend analysis, which is described in Section 3.

For TGM, the analysis used monthly median concentrations with the requirement that 75% of the month had valid data. Overall trend results for all sites with more than five years of data are listed in Table 4. Ten of the 11 sites experienced concentration decreases ranging from -0.9% to -3.3% per year, though time periods varied between five and 15 years. These decreases are comparable to a reported trend in background TGM concentration at Mace Head, Ireland of $-1.8\% \pm 0.1\%$ per year over the period 1996–2009 [63]. The six sites that best overlap with that time period (Egbert, Point Petre, St. Anicet, St. Andrews, Kejimkujik and Alert) recorded decreases of $0.8\%–1.8\%$ yr^{-1}. In contrast, TGM in the southern hemisphere declined at a faster rate of approximately -2.7% per year from 1996 to 2009, based on data from shipboard measurements and monitoring at Cape Point, South Africa [25]. Of the eleven sites, only Genesee, where measurements began in 2004, did not have a significant decrease in TGM over the measurement period.

Table 4. TGM trends (and 95% confidence limits) at sites with >5 years of measurements.

Site	Time Period	TGM Trend, $pg \cdot m^{-3} \cdot yr^{-1}$	TGM Trend, % yr^{-1}
Reifel Island	1999–2004	-55 (-70 to -40)	-3.3 (-4.2 to -2.4)
Genesee	2004–2010	-6 (-21 to $+1$) (ns)	-0.4 (-1.4 to $+0.1$) (ns)
Bratt's Lake	2001–2010	-37 (-48 to -23)	-2.5 (-3.4 to -1.6)
Burnt Island	1998–2007	-15 (-22 to -7)	-1.0 (-1.4 to -0.4)
Egbert	1996–2010	-20 (-27 to -16)	-1.3 (-1.7 to -1.0)
Kuujjuarapik	1999–2009	-40 (-55 to -23)	-2.4 (-3.4 to -1.4)
Point Petre	1996–2007	-29 (-38 to -20)	-1.7 (-2.2 to -1.2)
St. Anicet	1995–2009	-24 (-29 to -19)	-1.5 (-1.8 to -1.2)
St. Andrews	1996–2007	-30 (-42 to -20)	-2.2 (-3.1 to -1.5)
Kejimkujik	1996–2010	-14 (-20 to -6)	-1.0 (-1.4 to -0.5)
Alert	1995–2011	-14 (-18 to -10)	-0.9 (-1.1 to -0.6)

Seasonal trend analysis for TGM is exemplified for two sites—Reifel Island and Kejimikujik—in Figure 10. An overall decreasing trend of -0.055 $ng \cdot m^{-3} \cdot yr^{-1}$ or -55 $pg \cdot m^{-3} \cdot yr^{-1}$ was determined at Reifel Island from five years of data. To illustrate the result, the time series of monthly median TGM is shown in the bottom of Figure 10 where the red line indicates the slope. The results from Kejimkujik show that even in a data set that has greater seasonal variability, a longer time period narrows the uncertainty in the seasonal trends and the differences between months.

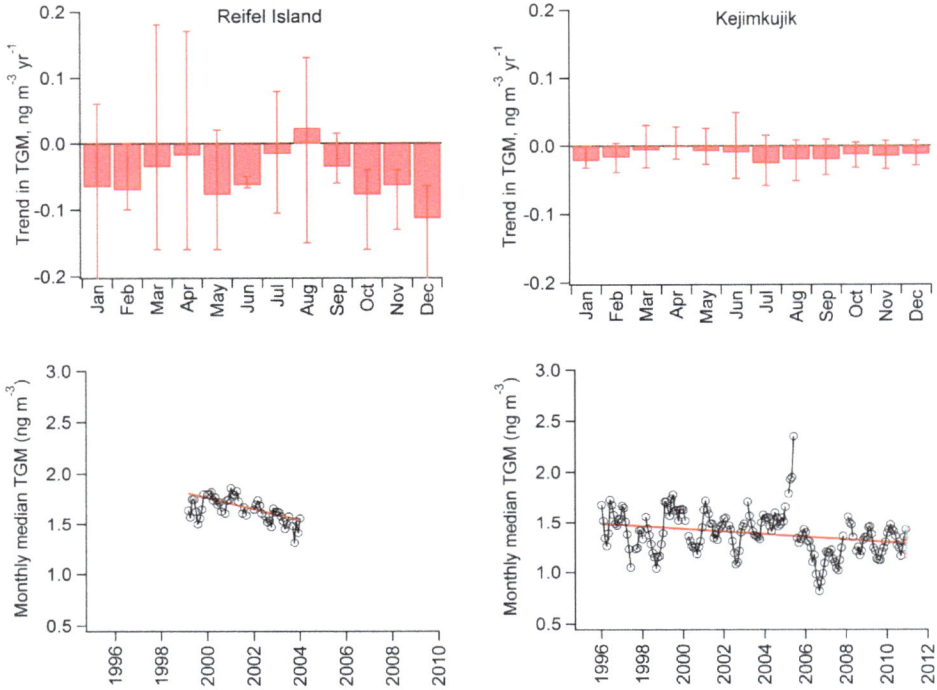

Figure 10. Long-term trends in TGM by season (**top**) and monthly median TGM with overall annual trend (**bottom**), at Reifel Island and Kejimkujik

An earlier analysis by Temme *et al.* [9] using a different statistical technique found significant declines in TGM at Burnt Island, Egbert, Point Petre, St. Anicet, St. Andrews and Kejimikujik and no significant change at Reifel Island, Bratt's Lake and Alert, based on measurements to 2005. The trends reported here with an additional 3–5 years of monitoring agree (within uncertainties) with the earlier trends at three sites (Burnt Island, Point Petre, St. Anicet) and show an increasingly negative trend at Reifel Island, Bratt's Lake, Egbert, St. Andrews, Kejimkujik and Alert.

Anthropogenic emissions of Hg from Canada and the United States, which decreased by approximately 60%–70% from 1995 to 2010, contribute roughly 15% or less of the TGM measured at Canadian sites, depending on the relative amounts of anthropogenic, natural and legacy (re-emissions of previously-deposited mercury) emissions [62,64,65]. Regional emission decreases may therefore account for approximately 0.5%–1% yr^{-1} in TGM concentration changes, at least in the eastern mid-latitudes where most of the sources are located. Since the bulk of TGM in Canada is from global and natural sources, the magnitude of the concentration decline is inconsistent with global emission estimates for the past 15 years that suggest that global anthropogenic Hg emissions have increased [66]. If these emission budgets are accurate, then either surface (natural and legacy) emissions have decreased or there has been increased deposition of Hg, either globally or near emission sources. The following sections discuss long-term trends in RGM and TPM$_{2.5}$ (which contribute to both wet and dry deposition fluxes) and trends in measured wet deposition of Hg.

Three sites in Canada—Alert, St. Anicet and ELA—have been collecting speciated Hg data for more than five years as shown in Figure 11. Daily mean concentrations of RGM, TPM$_{2.5}$ and GEM were used to calculate seasonal trends at these sites in months with sufficient valid data. The results of these calculations are shown in Figure 12, where a positive trend indicates a year-to-year increase in concentration for that month. While an overall trend was not reported for this data due to insufficient data in some months, a few interesting features in the seasonal trends were observed for RGM, TPM$_{2.5}$ and GEM, as discussed below.

For the Alert data, which showed significant amounts of RGM from February to July, the highest concentrations every year were seen in the month of May, with a median concentration of 110 pg·m^{-3}. Measurements from May over the ten years of monitoring have shown an RGM increase of 7.5 ± 3.5 pg·m^{-3}·yr^{-1}, or 6.8% per year. Concentrations of RGM in April and July also increased significantly, though the absolute changes were much smaller than in May due to lower levels of RGM. No significant trends were seen in the March and June RGM data. RGM measurements at St. Anicet and the ELA were below estimated detection limits for most of the year and thus trends are not reported for those sites. In general, seasonal trends were not calculated for months in which more than half of the measurements were below instrumental detection limits or for months that did not have measurements covering at least 75% of the days in that month for at least five years. It should be noted that further investigation into the detection limits of these instruments (discussed in Section 3) may validate more of the existing data and allow additional trends to be calculated in the future.

In contrast, measurements of TPM$_{2.5}$ are generally higher than RGM at these locations and therefore a few more valid monthly trends were reported in the seasonal trend analysis. Trends were variable and, for the lower latitude sites, found to be significant in both positive (increasing) and negative (decreasing) directions depending on the month. At the ELA, TPM$_{2.5}$ decreased in June, the only month with sufficient data coverage. At St. Anicet, TPM$_{2.5}$ decreased in July, increased in October and January, and did not show a significant trend in December. The magnitude of these trends ranged from −0.2 to +1.6 pg·m^{-3}·yr^{-1} or −3 to +12% yr^{-1}.

At Alert, $TPM_{2.5}$ showed a significant increase in the months where concentrations were the highest overall, namely March and April. The increase in March was $10 \pm 6 \text{ pg·m}^{-3}\text{·yr}^{-1}$ ($10\% \text{ yr}^{-1}$) and in April was roughly $7 \text{ pg·m}^{-3}\text{·yr}^{-1}$ or $7\% \text{ yr}^{-1}$, based on median concentrations of 99 and 98 pg·m^{-3}, respectively. Trends in the remainder of the months were either positive (May, July, August) or not significant. As with RGM, no negative trends in $TGM_{2.5}$ were observed at Alert.

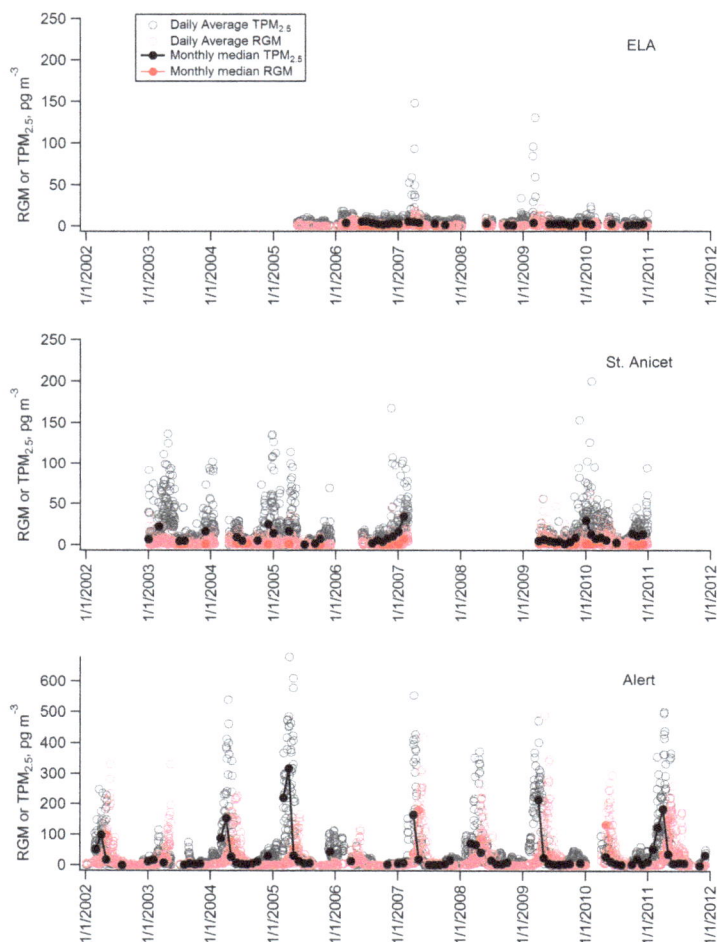

Figure 11. Reactive gaseous mercury (RGM) and total particulate mercury ($TPM_{2.5}$) at sites with more than five years of monitoring.

The overall decreasing trend of TGM reported from across the country is also evident in the GEM measurements at the ELA and in most months for St. Anicet and Alert. The latter two sites show more variability in the GEM than TGM trends at the same sites. This is likely due to a shorter measurement period for this speciation data compared to the TGM data.

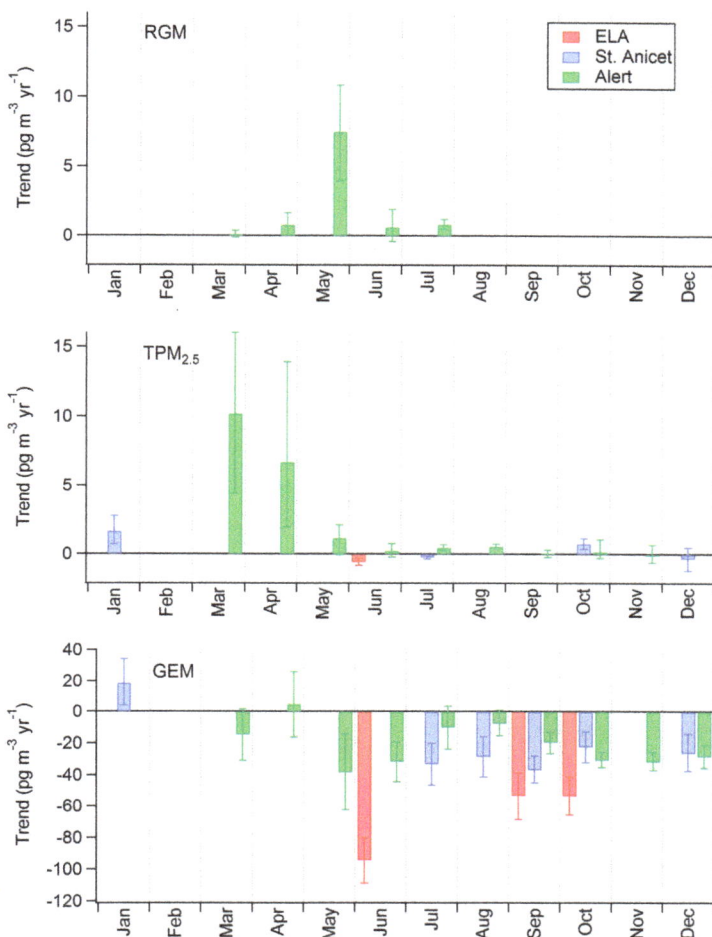

Figure 12. Seasonal trends in RGM, TPM$_{2.5}$ and GEM at the three sites with >5 years of speciation measurements. A positive trend indicates a year-over-year increase in concentrations for that month. Note scale change for GEM.

The springtime trends in speciated Hg at Alert are of interest because of their link to atmospheric photochemistry and ocean conditions at this time of year. An increase in RGM and TPM$_{2.5}$ during the AMDE season could indicate an overall increase in the amount of oxidation of GEM by halogen radicals. However, trends in GEM over the same period are not consistently decreasing (GEM decreased in May but did not change significantly in March or April) as would be expected from an increase in the oxidation rate. Also, these trends are still qualitatively uncertain due to the short time period and the fact that RGM and $TPM_{2.5}$ measurements are still not well-calibrated.

A long-term trend study from Resolute Bay, NU, was based on observations of filterable Hg (manual samples of Hg collected by passing air through particle filters which likely reflects the sum of $TPM_{2.5}$ and RGM combined) from 1974 to 2000 [67]. These authors reported a decrease of approximately 3% per year in total filterable Hg in summer and fall which is similar to the world-wide decrease in Hg emissions from anthropogenic activities between 1983 and 1995 and with reports on other atmospheric data [24]. Considerable variability was found in the data during the winter and early spring months suggesting some influence of AMDEs in the samples. These data precede the continuous measurements presented here and suggest that trends may be changing in recent decades.

To assess changes in the wet deposition of Hg, trends were calculated for the volume-weighted monthly mean concentration of total Hg in precipitation, the total monthly precipitation volume and the resulting monthly wet deposition Hg flux. Data from Egbert, St Anicet, St Andrews, Kejimkujik, Mingan and Cormak are reported because gaps in data from Bratt's Lake and Henry Kroeger limited the trend calculation using this technique.

The results are shown in Figure 13 and reveal that trends in Hg wet deposition at all sites are not significantly different from zero (95% confidence limits). Four sites report a significant decrease in the concentration of total Hg in precipitation (Egbert, St. Anicet, St. Andrews, and Kejimkujik). Mingan and Cormak had decreases that were not significantly different from zero. Overall deposition is a function of both Hg concentration and amount of precipitation, as seen at St. Anicet where a decrease in Hg concentration and an increase in precipitation amount resulted in a non-significant change in the wet deposition of Hg. Using a different statistical method, a similar result (no change in wet deposition flux of total Hg) was seen at ELA from 2001 to 2006 [35].

Overall time trends for Hg concentrations in precipitation are compared with those reported through the Mercury Deposition Network (MDN) in Table 5. The MDN results reported in 2009 are based on data up to 2005 [16]. The St. Andrews trend reported for the period 1996–2003 is small but decreasing and shows the strongest time overlap between trend analyses. However, some inconsistencies were found for the 2002 concentration between the 2009 publication and the MDN website and thus the trend value is flagged. At Egbert, the trend from the current analysis found a significant decrease of -2.1% yr^{-1}, which differs from the "no trend" reported earlier and is consistent with a recent report of a significant decrease between 2004 and 2010 [62]. At Cormak, the new ten-year trend is decreasing at a slower rate, -1.7% yr^{-1}, than the previous five-year trend, -4.4% yr^{-1}, and is not significant at the 95% confidence level. Finally, trends reported from St. Anicet,

Kejimkujik and Mingan have not changed significantly with the addition of more recent data. For these six sites, additional tests for spatial homogeneity, similar to the seasonal homogeneity tests, were performed to test if there was an overall trend across eastern Canada. Results showed that trends for wet deposition, Hg concentrations and precipitation amounts were homogeneous between the sites. As expected from individual site results, trends in wet deposition flux and precipitation amounts for the entire group were not significantly different from zero. Only Hg concentration in precipitation had a significant trend for the group, with an overall decrease estimated at -0.13 ± 0.04 ng·L^{-1}·yr^{-1} or $-2.1 \pm 0.6\%$ yr^{-1}.

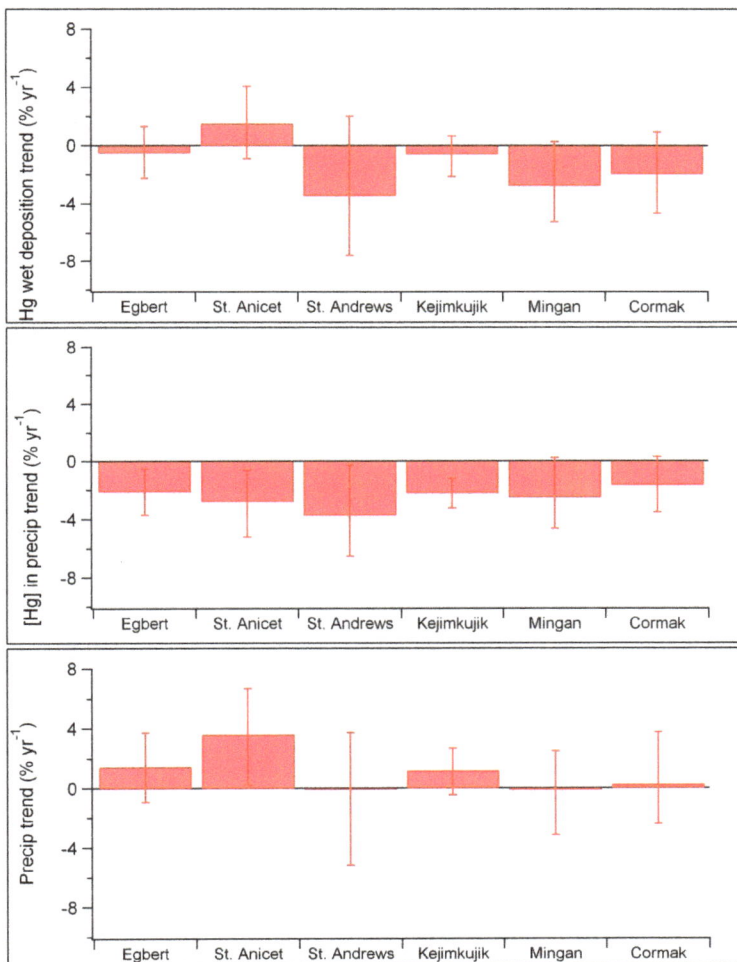

Figure 13. Overall trends in wet deposition, Hg concentration, and total precipitation at six sites with >5 years of Hg precipitation measurements.

At the two long-term sites in western Canada, Bratt's Lake and Henry Kroeger, low precipitation levels in the winter resulting in few data points restricted both the performance of a seasonal Kendall test and determination of an overall trend. However, a standard Mann-Kendall test for trend was performed over the entire time period and revealed no significant trend in Hg wet deposition or Hg precipitation concentrations at either site.

Table 5. Trends (with 95% confidence limits) in total Hg in precipitation at sites with >5 years of measurements, as well as results from Prestbo and Gay (PG) 2009 Concentrations are volume weighted means (ns = not significant).

Site	Time Period	[Hg] Trend, $ng \cdot L^{-1} \cdot yr^{-1}$	[Hg] Trend, % yr^{-1}	Time Period (PG 2009)	[Hg] Trend (PG 2009), % yr^{-1}
Egbert	2000–2010	−0.18 (−0.31 to −0.05)	−2.1 (−3.7 to −0.6)	2000–2005	ns
St. Anicet	1998–2007	−0.22 (−0.41 to −0.05)	−2.8 (−5.2 to −0.6)	1998–2005	−1.5
St. Andrews	1996–2003	−0.25 (−0.43 to −0.02)	−3.7 (−6.5 to −0.3)	1997–2003	−1.9 [**]
Kejimkujik	1996–2010	−0.12 (−0.17 to −0.06)	−2.2 (−3.3 to −1.2)	1997–2005	−2.0
Mingan	1998–2007	−0.13 (−0.23 to +0.01) [ns]	−2.5 (−4.6 to +0.2) [ns]	1998–2005	ns
Cormak	2000–2010	−0.07 (−0.15 to +0.01) [ns]	−1.7 (−3.5 to +0.3) [ns]	2000–2005	−4.4

** Possible error in data set.

Four of the long-term precipitation collection sites also measure TGM (Egbert, St. Anicet, St. Andrews and Kejimkujik). A comparison of the TGM concentrations in air and Hg concentrations in precipitation shows that they have both significantly decreased at all four sites. Quantitatively comparing the trends, trends in TGM concentration at Egbert, St. Anicet and St. Andrews agreed with trends in precipitation concentrations within the uncertainties, though measurement time periods were not identical. At Kejimkujik, the rate of decrease of Hg concentration in precipitation (-2.2% yr^{-1}) was significantly more negative than the decrease in TGM concentration (-1.0% yr^{-1}) over the same time period. Similar results were reported for earlier trend comparisons [9]. These differences suggest there may be changes in precipitation amounts or types, and/or in the distribution of GEM, RGM and $TPM_{2.5}$ (which are scavenged with differing efficiencies).

The wet deposition trends presented here for Canada as well as those for the United States [16,62] do not show a large-scale increase in wet deposition of Hg in eastern mid-latitude North America. Therefore, the additional decreases observed in TGM concentrations at these Canadian sites (on top of those due to regional emission controls) cannot be explained by increased wet deposition in this region as might be expected if changes in Hg speciation or precipitation were scavenging more Hg from the atmosphere. Increased deposition in other areas could have offset increasing global anthropogenic emissions; increased monitoring of wet deposition in under-sampled areas, such as ecologically sensitive northern regions, is needed to provide model constraints and track the response of the Arctic and sub-Arctic to changes in emissions and climate. Alternatively, significant decreases in natural emissions [25,68] or in re-emissions of recently deposited Hg could also result in a declining trend in air concentrations.

3. Methods

3.1. Data Collection

TGM measurements were made using automated Tekran® 2537 Hg vapour analyser [9,27,69]. The air is sampled at flow rates between 1.0 and 1.5 L·min^{-1} through a Teflon filter (47 mm diameter; 0.2 μm pore). The Hg in the sample air is pre-concentrated by amalgamation on two gold cartridges (5–30 min pre-concentration times) which alternate between collection and thermal desorption and detection for continuous monitoring. Elemental Hg is detected using Cold Vapour Atomic Fluorescence Spectrometry (CVAFS). The instruments are calibrated daily using an internal Hg permeation source and verified during routine site audits by manual injections of Hg from an external source. The data are quality controlled using the Environment Canada Research Data Management and Quality Control (RDMQ) system [70]. Limited studies suggest that TGM measurements consist of GEM and RGM [71,72]; however it is possible that under certain environmental conditions and sample inlet configurations, the RGM is removed.

Speciated Hg measurements (GEM, RGM and TPM$_{2.5}$) were made using Tekran® Mercury 1130, 1135 and 2537 speciation systems and are described in detail elsewhere [27,73]. Briefly, air is pulled into the analyzer through a Teflon® coated elutriator and impactor designed to remove particles >2.5 μm at flow rates of 10.0 L·min^{-1}. The sample air flows over a KCl coated quartz denuder to trap the RGM in the 1130 unit and then passes over a quartz particulate filter to trap TPM$_{2.5}$ in the 1135 unit. The remaining GEM is carried into the 2537 analyser (at a flow rate of 1 L·min^{-1}) for analysis. RGM and TPM$_{2.5}$ are accumulated at a high flow rate for 1 to 3 h while the GEM is collected every 5 minutes. RGM and TPM$_{2.5}$ are subsequently thermally desorbed and pyrolysed to GEM and then analysed by the 2537, interrupting the sampling. Exact identification of RGM and TPM$_{2.5}$ fractions are still not known and thus are operationally defined. Therefore, in lieu of RGM and TPM$_{2.5}$ standards, rigorous procedures during and after sample collection/analysis have been put in place to ensure standard methods are used at all site locations [70].

The analytical detection limit of GEM in the Tekran 2537 is <0.1 ng/m^3 in 7.5 L, or <0.75 pg. Therefore, detection limits for RGM and TPM$_{2.5}$ in the 2537 alone would be <0.4–1.2 pg·m^{-3} (depending on the collection time), though the manufacturer estimated 2 pg·m^{-3} as the more reasonable limit. Detection limits ranging from 1 to 3 pg·m^{-3} for RGM and TPM$_{2.5}$ were determined at particular sites with multiple years of measurements using three times the standard deviation of the blanks (zero-air samples) during each desorption cycle. Discussions regarding determination of the instrument detection limits are currently ongoing within the research community. For individual RGM and $TPM_{2.5}$ measurements the RDMQ process validates/invalidates measurements taking into consideration the magnitude of the species concentrations relative to the blank values for each desorption.

Precipitation samples were collected according to the Mercury Deposition Network (MDN) SOP using an automatic precipitation sampler (Modified Aerochem Metrics Model 301 or N-Con Systems Co. Automatic Precipitation Sampler). The sampler was open to the atmosphere only during precipitation events that were detected using a sensor. The sampling system/train consists of a borosilicate glass funnel connected to an acid-cleaned, 3-mm capillary tube. The other end of the capillary tube is expanded to form a sphere with a small hole that allows water to drain into a 2-liter borosilicate glass bottle that is pre-charged with dilute hydrochloric acid to prevent microbial activity and volatilization [16]. The sample collection system housing is insulated and temperature controlled. Field operators collect samples once a week using clean techniques and replace the entire sampling train with fresh equipment that that has been rigorously cleaned. All sample analysis and glassware cleaning is performed by the Mercury (Hg) Analytical Laboratory (HAL) at Frontier Global Sciences. The weekly precipitation amount is recorded for deposition determination using a precision precipitation gauge (Belfort Universal Precipitation Gauge 5–780 or NOAH IV, ETI Instrument Systems, Inc).

Precipitation samples were analysed at the laboratory using the EPA method 1631 to measure total Hg. Bromine chloride (BrCl) in hydrochloric acid (HCl) is added to the sample bottle to oxidize all forms of Hg to Hg^{2+}. Stannous chloride ($SnCl_2$) is added to reduce Hg^{2+} to $Hg°$ and the sample is purged with ultra-pure nitrogen onto gold-coated silica traps. A dual gold trap amalgamation technique is used to concentrate and focus the Hg. The Hg is removed from the gold traps via thermal desorption and is analysed using CVAFS and quantified by peak height. Field blanks, system blanks and laboratory blanks are all routinely analysed as part of quality assurance procedures. The laboratory performs weekly standard reference material and spike recovery tests. Sample Hg concentrations are blank corrected and the method of detection limit is approximately $0.1 \ ng·L^{-1}$ based on three standard deviations of the laboratory blanks. All data is reviewed and validated by the Mercury Analytical Laboratory.

3.2. Data Analysis

Final data obtained from the National Atmospheric Chemistry database [74] or provided by principal investigators were first averaged to give daily values for TGM and speciated Hg concentrations and monthly values for wet deposition and volume-weighted mean concentrations in precipitation, with 75% data coverage required within the day or month for acceptance of the daily TGM or monthly precipitation average. Due to differing sampling intervals for speciation data, a minimum of three samples in a day was required to accept daily averages for GEM, RGM and TPM$_{2.5}$. Weekly precipitation samples that covered a change in the calendar month were partially attributed to each of the two months, weighting the sample in each monthly average (or sum) according to the number of days from that week that belonged to the month in question. Atmospheric speciation measurements that sampled over midnight were similarly treated. Daily/monthly averages were then used to calculate summary statistics for each site (see supplementary data file). Diel patterns were calculated for TGM using hourly averages, again requiring 75% coverage within the hour. Since speciated Hg concentrations used 1–3 h collection intervals, the start and end times of the collection period were rounded to the nearest hour and each hour within the adjusted start and end times was assigned the concentration value for that sample. No concentration was assigned during the analysis period.

Long-term trends were calculated using the seasonal Kendall test for trend and the related Sen's slope calculation [75]. This method is an extension of the non-parametric Mann-Kendall test for trend in which data from the 12 months are treated as 12 separate data sets. The Mann-Kendall test is recommended for data sets in which there are missing points, which applies to many of the data sets used here, and when the data are not normally distributed, which is true for some sites (e.g., Alert). Non-parametric tests are also more robust for long-term trend analysis (compared to a parametric linear regression) since they are less sensitive to outliers at the beginning and end of the data set. For each month, the presence of a trend is confirmed or rejected by the Mann-Kendall test and a slope is estimated using Sen's nonparametric estimator of slope. An overall annual trend can then be estimated from the monthly trend statistics; however, this estimate is less reliable if the monthly trends are not sufficiently homogeneous. Thus, to ensure reliability of the data, a test for seasonal homogeneity was performed as well [76]. If seasonal trends were homogeneous, the results were used to determine an overall trend for the entire period. If they were not homogeneous, or when there were insufficient data to calculate a trend in certain months, only trends for individual months were reported.

4. Conclusions

A synthesis of all available Hg measurements from Canadian air and precipitation monitoring sites is presented, including diel and seasonal cycles and long-term trends. These results provide a wealth of data on long-term patterns in Hg species for validation of models of the transformation and deposition of atmospheric Hg. They also provide a baseline for evaluating the response of Hg concentrations and deposition to emission reduction strategies both nationally and under the Minamata Convention, an international agreement on Hg signed by 99 countries to date [77]. With broad geographic coverage including remote sites with few local or regional sources, Canada's observation network is well positioned to evaluate future Hg trends in the Northern Hemisphere.

TGM concentrations, with the exception of Flin Flon, were in the range of 1.2–1.9 ng·m^{-3} across the country, similar to northern hemispheric background levels. RGM and TPM$_{2.5}$ ranged from a few pg·m^{-3} over much of the country to hundreds of pg·m^{-3} during spring AMDEs at Alert and Churchill. Wet deposition amounts averaged between 0.1 and 0.8 μg·m^{-2}·month^{-1}, except at Flin Flon. Spatial patterns of Hg measurements showed the influence of local and regional sources and sinks. For example, high air concentrations of Hg species and higher Hg deposition were seen in the vicinity of the Flin Flon smelter and in the populated Great Lakes region, while AMDE chemistry in the spring and oceanic emissions in the summer influenced seasonal patterns in GEM, RGM and TPM$_{2.5}$ in Arctic and sub-Arctic locations.

At all sites in this survey, RGM levels were highest in the mid-afternoon. This pattern is consistent with photochemical formation of RGM and a relatively short lifetime (on the order of hours), though elevated levels of RGM persisted through the night in Churchill and Alert during the AMDE season, suggesting the lifetime may be longer depending on conditions. Other atmospheric Hg species had more diverse diel patterns. At remote sites, TGM and GEM were generally lowest in the early morning, as deposition in the shallow nocturnal boundary layer depleted TGM relative to the free troposphere above. Sites influenced by urban emissions and/or on- and off-shore breezes did not follow this pattern.

Seasonal patterns in TGM and GEM concentrations were similar at most mid-latitude sites, with maxima in the late winter or early spring and minima in the fall, perhaps due to surface evasion. Sites influenced by AMDEs were the exception, with highest concentrations in the summer and lower in the spring (Alert) or winter (Kuujjuarapik). RGM levels peaked in the spring, in contrast to the summer maximum that might be expected due to its photochemical source. Increased wet and dry deposition, or analytical issues, may play a role. TPM$_{2.5}$ levels were also highest in spring at remote sites and in winter at more polluted locations. Wet deposition of total Hg was almost always highest in summer, except at Reifel Island, BC, where the dry summer months resulted in much lower wet deposition.

Concentrations of both TGM and Hg in precipitation have declined at most sites since the mid-1990s, while Hg wet deposition trends were not significant due to variability in precipitation amounts. The TGM trends extend the number of global sites reporting decreases over the last two decades [7,25,26,63]. Increases in springtime RGM and TPM$_{2.5}$ previously noted at Alert for 2002–2009 [26] are further supported by the additional two years of data included here.

Moving forward, in addition to continued monitoring, identification of the compounds making up RGM and TPM$_{2.5}$ and the development of calibration standards continue to be needed. Knowledge of these exact chemical species would also lead to improved understanding of the chemistry and wet and dry deposition rates of RGM and TPM$_{2.5}$ in different air masses. Wet deposition measurements in Arctic and sub-Arctic locations would assist modellers in constraining the atmospheric Hg budget in this vast area, as would additional direct measurements of dry deposition across the country.

Acknowledgments: Funding for CAMNet and CAPMoN was provided by Environment Canada (EC). Additional funding was provided by the Northern Contaminants Program (Aboriginal Affairs and Northern Development Canada), the Geological Survey of Canada, the Natural Sciences and Engineering Research Council, the Government of Canada Program for International Polar Year, EC's Global Atmospheric Watch program and Whistler program, and Manitoba Hydro. Additional data was contributed by the Meteorological Service of Canada Air Quality Unit in Edmonton and by Geoffrey Stupple. Thanks to the numerous former and current scientists and technicians for CAMNet and CAPMoN, particularly Steve Beauchamp, Cathy Banic, Keith Puckett for the creation of CAMNet; to Patrick Lee for support at other EC sites; to Greg Skelton and Tina Scherz for many years of data QC; to Mike Shaw for the site map; and to Heather Morrison and Pierrette Blanchard for coordinating this project for the Canadian Mercury Science Assessment.

Author Contributions: Amanda S. Cole created several figures and wrote the majority of the text, along with Alexandra Steffen. Julie Narayan generated statistics for a large part of the data as well as additional figures. Chris S. Eckley assembled and quality controlled several data sets and wrote parts of the experimental section. Rob Tordon collected data from St. Andrews, Kejimkujik, Cormak and Halifax and contributed to writing the experimental section. Data were collected and editorial input provided by Martin Pilote (St. Anicet, Mingan and Kuujjuarapik), Jennifer A. Graydon and Vincent L. St. Louis (ELA), Xiaohong Xu (Windsor) and Brian A. Branfireun (Mississauga and Dorset).

Conflicts of Interest: The authors declare no conflict of interest.

References

1. Selin, N.E. Global biogeochemical cycling of mercury: A review. *Ann. Rev. Environ. Resour.* **2009**, *34*, 43–63.
2. Corbitt, E.S.; Jacob, D.J.; Holmes, C.D.; Streets, D.G.; Sunderland, E.M. Global source-receptor relationships for mercury deposition under present-day and 2050 emissions scenarios. *Environ. Sci. Technol.* **2011**, *45*, 10477–10484.
3. Holmes, C.D.; Jacob, D.J.; Corbitt, E.S.; Mao, J.; Yang, X.; Talbot, R.; Slemr, F. Global atmospheric model for mercury including oxidation by bromine atoms. *Atmos. Chem. Phys.* **2010**, *10*, 12037–12057.
4. Lindberg, S.; Bullock, R.; Ebinghaus, R.; Engstrom, D.R.; Feng, X.; Fitzgerald, W.F.; Pirrone, N.; Prestbo, E.; Seigneur, C. A synthesis of progress and uncertainties in attributing the sources of mercury in deposition. *Ambio* **2007**, *36*, 19–32.
5. Schroeder, W.H.; Munthe, J. Atmospheric mercury—An overview. *Atmos. Environ.* **1998**, *32*, 809–822.

6. Hynes, A.J.; Donohoue, D.L.; Goodsite, M.E.; Hedgecock, I.M. Our current understanding of major chemical and physical processes affecting mercury dynamics in the atmosphere and at the air-water/terrestrial interfaces. In *Mercury Fate and Transport in the Global Atmosphere*; Pirrone, N., Mason, R.P., Eds.; Springer: New York, NY, USA, 2009; pp. 427–457.

7. Cole, A.S.; Steffen, A. Trends in long-term gaseous mercury observations in the Arctic and effects of temperature and other atmospheric conditions. *Atmos. Chem. Phys.* **2010**, *10*, 4661–4672.

8. Kellerhals, M.; Beauchamp, S.; Belzer, W.; Blanchard, P.; Froude, F.; Harvey, B.; McDonald, K.; Pilote, M.; Poissant, L.; Puckett, K.; *et al.* Temporal and spatial variability of total gaseous mercury in Canada: Results from the Canadian Atmospheric Mercury Measurement Network (CAMNet). *Atmos. Environ.* **2003**, *37*, 1003–1011.

9. Temme, C.; Blanchard, P.; Steffen, A.; Beauchamp, S.T.; Poissant, L.; Tordon, R.J.; Wiens, B. Trend, seasonal and multivariate analysis study of total gaseous mercury data from the Canadian Atmospheric Mercury Measurement Network (CAMNet). *Atmos. Environ.* **2007**, *41*, 5423–5441.

10. Poissant, L.; Pilote, M.; Beauvais, C.; Constant, P.; Zhang, H.H. A year of continuous measurements of three atmospheric mercury species (GEM, RGM and Hg_p) in southern Quebec, Canada. *Atmos. Environ.* **2005**, *39*, 1275–1287.

11. Blanchard, P.; Froude, F.A.; Martin, J.B.; Dryfhout-Clark, H.; Woods, J.T. Four years of continuous total gaseous mercury (TGM) measurements at sites in Ontario, Canada. *Atmos. Environ.* **2002**, *36*, 3735–3743.

12. Kim, K.-H.; Ebinghaus, R.; Schroeder, W.H.; Blanchard, P.; Kock, H.H.; Steffen, A.; Froude, F.A.; Kim, M.-Y.; Hong, S.; Kim, J.-H. Atmospheric mercury concentrations from several observatory sites in the Northern Hemisphere. *J. Atmos. Chem.* **2005**, *50*, 1–24.

13. Steffen, A.; Schroeder, W.H.; Macdonald, R.; Poissant, L.; Konoplev, A. Mercury in the Arctic atmosphere: an analysis of eight years of measurements of GEM at Alert (Canada) and a comparison with observations at Amderma (Russia) and Kuujjuarapik (Canada). *Sci. Total Environ.* **2005**, *342*, 185–198.

14. Xu, X.; Akhtar, U.S. Identification of potential regional sources of atmospheric total gaseous mercury in Windsor, Ontario, Canada using hybrid receptor modeling. *Atmos. Chem. Phys.* **2010**, *10*, 7073–7083.

15. Sanei, H.; Outridge, P.M.; Goodarzi, F.; Wang, F.; Armstrong, D.; Warren, K.; Fishback, L. Wet deposition mercury fluxes in the Canadian sub-Arctic and southern Alberta, measured using an automated precipitation collector adapted to cold regions. *Atmos. Environ.* **2010**, *44*, 1672–1681.

16. Prestbo, E.M.; Gay, D.A. Wet deposition of mercury in the U.S. and Canada, 1996–2005: Results and analysis of the NADP mercury deposition network (MDN). *Atmos. Environ.* **2009**, *43*, 4223–4233.

17. Mazur, M.; Mintz, R.; Lapalme, M.; Wiens, B.J. Ambient air total gaseous mercury concentrations in the vicinity of coal-fired power plants in Alberta, Canada. *Sci. Total Environ.* **2009**, *408*, 373–381.

18. Gay, D.A.; Schmeltz, D.; Prestbo, E.; Olson, M.L.; Sharac, T.; Tordon, R. The Atmospheric Mercury Network: measurement and initial examination of an ongoing atmospheric mercury record across North America. *Atmos. Chem. Phys.* **2013**, *13*, 11339–11349.

19. Eckley, C.S.; Parsons, M.T.; Mintz, R.; Lapalme, M.; Mazur, M.; Tordon, R.; Elleman, R.; Graydon, J.A.; Blanchard, P.; St Louis, V.L. Impact of closing Canada's largest point-source of mercury emissions on local atmospheric mercury concentrations. *Environ. Sci. Technol.* **2013**, *47*, 10339–10348.

20. Kirk, J.L.; St. Louis, V.L.; Sharp, M.J. Rapid reduction and reemission of mercury deposited into snow packs during atmospheric mercury depletion events at Churchill, Manitoba, Canada. *Environ. Sci. Technol.* **2006**, *40*, 7590–7596.

21. Stupple, G.W.; Branfireun, B.A. Canopy mercury accumulation in urban and rural forests in southern Ontario. *Environ. Pollut.* **2014**. in preparation.

22. Steffen, A.; Bottenheim, J.; Cole, A.; Ebinghaus, R.; Lawson, G.; Leaitch, W.R. Atmospheric mercury speciation and mercury in snow over time at Alert, Canada. *Atmos. Chem. Phys.* **2014**, *14*, 2219–2231.

23. Steffen, A.; Cole, A.S.; Ariya, P.; Banic, C.; Dastoor, A.; Durnford, D.; Eckley, C.; Graydon, J.A.; Mintz, R.; Pilote, M.; *et al.* Chapter 4: Atmospheric processes, transport, levels and trends. In *Canadian Mercury Science Assessment*; Environment Canada: Toronto, ON, Canada, 2014; in press.

24. Slemr, F.; Brunke, E.; Ebinghaus, R.; Temme, C.; Munthe, J.; Wängberg, I.; Schroeder, W.H.; Steffen, A.; Berg, T. Worldwide trend of atmospheric mercury since 1977. *Geophys. Res. Lett.* **2003**, *30*, 23–21.

25. Slemr, F.; Brunke, E.G.; Ebinghaus, R.; Kuss, J. Worldwide trend of atmospheric mercury since 1995. *Atmos. Chem. Phys.* **2011**, *11*, 4779–4787.

26. Cole, A.; Steffen, A.; Aspmo Pfaffhuber, K.; Berg, T.; Pilote, M.; Tordon, R.; Hung, H. Ten-year trends of atmospheric mercury in the high Arctic compared to Canadian sub-Arctic and mid-latitude sites. *Atmos. Chem. Phys.* **2013**, *13*, 1535–1545.

27. Steffen, A.; Douglas, T.; Amyot, M.; Ariya, P.; Aspmo, K.; Berg, T.; Bottenheim, J.; Brooks, S.; Cobbett, F.D.; Dastoor, A.; *et al.* A synthesis of atmospheric mercury depletion event chemistry in the atmosphere and snow. *Atmos. Chem. Phys.* **2008**, *8*, 1445–1482.

28. Durnford, D.; Dastoor, A.; Ryzhkov, A.; Poissant, L.; Pilote, M.; Figueras-Nieto, D. How relevant is the deposition of mercury onto snowpacks?—Part 2: A modeling study. *Atmos. Chem. Phys.* **2012**, *12*, 9251–9274.

29. Fisher, J.A.; Jacob, D.J.; Soerensen, A.L.; Amos, H.M.; Steffen, A.; Sunderland, E.M. Riverine source of Arctic Ocean mercury inferred from atmospheric observations. *Nat. Geosci.* **2012**, *5*, 499–504.

30. Hirdman, D.; Aspmo, K.; Burkhart, J.F.; Eckhardt, S.; Sodemann, H.; Stohl, A. Transport of mercury in the Arctic atmosphere: Evidence for a springtime net sink and summer-time source. *Geophys. Res. Lett.* **2009**.

31. Lalonde, J.D.; Poulain, A.J.; Amyot, M. The role of mercury redox reactions in snow on snow-to-air mercury transfer. *Environ. Sci. Technol.* **2002**, *36*, 174–178.

32. Lindberg, S.E.; Brooks, S.; Lin, C.-J.; Scott, K.J.; Landis, M.S.; Stevens, R.K.; Goodsite, M.; Richter, A. Dynamic oxidation of gaseous mercury in the Arctic troposphere at polar sunrise. *Environ. Sci. Technol.* **2002**, *36*, 1245–1256.

33. Huang, J.; Miller, M.B.; Weiss-Penzias, P.; Gustin, M.S. Comparison of gaseous oxidized Hg measured by KCl-coated denuders, and nylon and cation exchange membranes. *Environ. Sci. Technol.* **2013**, *47*, 7307–7316.

34. Lyman, S.N.; Jaffe, D.A.; Gustin, M.S. Release of mercury halides from KCl denuders in the presence of ozone. *Atmos. Chem. Phys.* **2010**, *10*, 8197–8204.

35. Graydon, J.A.; St. Louis, V.; Hintlemann, H.; Lindberg, S.E.; Sandilands, K.A.; Rudd, J.W.M.; Kelly, C.A.; Hall, B.; Mowat, L.D. Long-term wet and dry deposition of total and methyl mercury in the remote boreal ecoregion of Canada. *Environ. Sci. Technol.* **2008**, *42*, 8345–8351.

36. Poulain, A.J.; Garcia, E.; Amyot, J.D.; Campbell, P.G.C.; Raofie, F.; Ariya, P.A. Biological and chemical redox transformations of mercury in fresh and salt waters of the high Arctic during spring and summer. *Environ. Sci. Technol.* **2007**, *41*, 1883–1888.

37. Poulain, A.J.; Lalonde, J.D.; Amyot, J.D.; Shead, J.A.; Raofie, F.; Ariya, P.A. Redox transformations of mercury in an Arctic snowpack at springtime. *Atmos. Environ.* **2004**, *38*, 6763–6774.

38. Poissant, L.; Casimir, A. Water-air and soil-air exchange rate of total gaseous mercury measured at background sites. *Atmos. Environ.* **1998**, *32*, 883–893.

39. Choi, H.-D.; Holsen, T.M.; Hopke, P.K. Atmospheric mercury (Hg) in the Adirondacks: Concentrations and sources. *Environ. Sci. Technol.* **2008**, *42*, 5644–5653.

40. Nair, U.S.; Wu, Y.; Walters, J.; Jansen, J.; Edgerton, E.S. Diurnal and seasonal variation of mercury species at coastal-suburban, urban, and rural sites in the southeastern United States. *Atmos. Environ.* **2012**, *47*, 499–508.

41. Poissant, L. Total gaseous mercury in Quebec (Canada) in 1998. *Sci. Total Environ.* **2000**, *259*, 191–201.

42. Liu, B.; Keeler, G.J.; Dvonch, J.T.; Barres, J.A.; Lynam, M.M.; Markis, F.J.; Morgan, J.T. Temporal variability of mercury speciation in urban air. *Atmos. Environ.* **2007**, *41*, 1911–1923.

43. Engle, M.A.; Tate, M.T.; Krabbenhoft, D.P.; Schauer, J.J.; Kolker, A.; Shanley, J.B.; Bothner, M.H. Comparison of atmospheric mercury speciation and deposition at nine sites across central and eastern North America. *J. Geophys. Res.: Atmos.* **2010**.

44. Engle, M.A.; Tate, M.T.; Krabbenhoft, D.P.; Kolker, A.; Olson, M.L.; Edgerton, E.S.; DeWild, J.F.; McPherson, A.K. Characterization and cycling of atmospheric mercury along the central US Gulf Coast. *Appl. Geochem.* **2008**, *23*, 419–437.

45. Fain, X.; Obrist, D.; Hallar, A.G.; McCubbin, I.; Rahn, T. High levels of reactive gaseous mercury observed at a high elevation research laboratory in the Rocky Mountains. *Atmos. Chem. Phys.* **2009**, *9*, 8049–8060.

46. Swartzendruber, P.C.; Jaffe, D.A.; Prestbo, E.; Weiss-Penzias, P.; Selin, N.E.; Park, R.; Jacob, D.J.; Strode, S.; Jaegle, L. Observations of reactive gaseous mercury in the free troposphere at the Mount Bachelor Observatory. *J. Geophys. Res.: Atmos.* **2006**.

47. Poissant, L.; Pilote, M.; Xu, X.; Zhang, H.; Beauvais, C. Atmospheric mercury speciation and deposition in the Bay St. Francois wetlands. *J. Geophys. Res.: Atmos.* **2004**.

48. Bottenheim, J.; Chan, H.M. A trajectory study into the origin of spring time Arctic boundary layer ozone depletion. *J. Geophys. Res.: Atmos.* **2006**.

49. Weiss-Penzias, P.; Gustin, M.S.; Lyman, S.N. Observations of speciated atmospheric mercury at three sites in Nevada: Evidence for a free tropospheric source of reactive gaseous mercury. *J. Geophys. Res.: Atmos.* **2009**.

50. Iverfeldt, A. Occurrence and turnover of atmospheric mercury over the Nordic countries. *Water Air Soil Pollut.* **1991**, *56*, 251–265.

51. Kock, H.H.; Bieber, E.; Ebinghaus, R.; Spain, T.G.; Thees, B. Comparison of long-term trends and seasonal variations of atmospheric mercury concentrations at the two European coastal monitoring stations Mace Head, Ireland, and Zingst, Germany. *Atmos. Environ.* **2005**, *39*, 7549–7556.

52. Slemr, F.; Schell, H.E. Trends in atmospheric mercury concentrations at the summit of the Wank Mountain, southern Germany. *Atmos. Environ.* **1998**, *32*, 845–853.

53. Corbett-Hains, H.; Walters, N.E.; Van Heyst, B.J. Evaluating the effects of sub-zero temperature cycling on mercury flux from soils. *Atmos. Environ.* **2012**, *63*, 102–108.

54. Briggs, C.; Gustin, M.S. Building upon the conceptual model for soil mercury flux: Evidence of a link between moisture evaporation and Hg evasion. *Water Air Soil Pollut.* **2013**, *224*, 1–13.

55. Gustin, M.S.; Stamenkovic, J. Effect of watering and soil moisture on mercury emissions from soils. *Biogeochemistry* **2005**, *76*, 215–232.

56. Lin, C.J.; Gustin, M.S.; Singhasuk, P.; Eckley, C.; Miller, M. Empirical models for estimating mercury flux from soils. *Environ. Sci. Technol.* **2010**, *44*, 8522–8528.

57. Song, X.; Van Heyst, B. Volatilization of mercury from soils in response to simulated precipitation. *Atmos. Environ.* **2005**, *39*, 7494–7505.

58. Zhang, L.; Wright, L.P.; Blanchard, P. A review of current knowledge concerning dry deposition of atmospheric mercury. *Atmos. Environ.* **2009**, *43*, 5853–5864.

59. Baker, K.R.; Bash, J.O. Regional scale photochemical model evaluation of total mercury wet deposition and speciated ambient mercury. *Atmos. Environ.* **2012**, *49*, 151–162.

60. Amos, H.M.; Jacob, D.J.; Holmes, C.D.; Fisher, J.D.; Wang, Q.; Corbitt, E.S.; Galarneau, E.; Rutter, A.P.; Gustin, M.S.; Steffen, A.; *et al.* Gas-particle partitioning of atmospheric Hg(II) and its effect on global mercury deposition. *Atmos. Chem. Phys.* **2012**, *12*, 591–603.

61. Fain, X.; Ferrari, C.P.; Dommergue, A.; Albert, M.R.; Battle, M.; Severinhaus, J.; Arnaud, L.; Barnola, J.-M.; Cairns, W.; Barbante, C.; *et al.* Polar firn air reveals large-scale impact of anthropogenic mercury emissions during the 1970s. *Proc. Natl. Acad. Sci. USA* **2009**, *106*, 16114–16119.

62. Zhang, Y.; Jaeglé, L. Decreases in mercury wet deposition over the United States during 2004–2010: Roles of domestic and global background emission reductions. *Atmosphere* **2013**, *4*, 113–131.

63. Ebinghaus, R.; Jennings, S.G.; Kock, H.H.; Derwant, R.G.; Manning, A.J.; Spain, T.G. Decreasing trends in total gaseous mercury in baseline air at Mace Head, Ireland from 1996–2009. *Atmos. Environ.* **2011**, *159*, 1577–1583.

64. Durnford, D.; Dastoor, A.; Figueras-Nieto, D.; Ryjkov, A. Long range transport of mercury to the Arctic and across Canada. *Atmos. Chem. Phys.* **2010**, *10*, 6063–6086.

65. Strode, S.A.; Jaegle, L.; Jaffe, D.A.; Swartzendruber, P.C.; Selin, N.E.; Holmes, C.D.; Yantosca, R.M. Trans-Pacific transport of mercury. *J. Geophys. Res.: Atmos.* **2008**.

66. Streets, D.G.; Devane, M.K.; Lu, Z.; Sunderland, E.M.; Jacob, D.J. All-time releases of mercury to the atmosphere from human activities. *Environ. Sci. Technol.* **2011**, *45*, 10485–10491.

67. Li, C.; Cornett, J.; Willie, S.; Lam, J. Mercury in Arctic air: The long-term trend. *Sci. Total Environ.* **2009**, *407*, 2756–2759.

68. Soerensen, A.L.; Jacob, D.J.; Streets, D.G.; Witt, M.L.I.; Ebinghaus, R.; Mason, R.P.; Andersson, M.; Sunderland, E.M. Multi-decadal decline of mercury in the North Atlantic atmosphere explained by changing subsurface seawater concentrations. *Geophys. Res. Lett.* **2012**.

69. Poissant, L. Field observation of total gaseous mercury behaviour: Interactions with ozone concentration and water vapour mixing ratio in air at a rural site. *Water Air Soil Pollut.* **1997**, *97*, 341–353.

70. Steffen, A.; Scherz, T.; Olson, M.L.; Gay, D.A.; Blanchard, P. A comparison of data quality control protocols for atmospheric mercury speciation measurements. *J. Environ. Monit.* **2012**, *14*, 752–765.

71. Slemr, F.; Ebinghaus, R.; Brenninkmeijer, M.; Hermann, M.; Kock, H.H.; Martinsson, B.G.; Schuck, T.; Sprung, D.; Van Velthoven, P.; Zahn, A.; *et al.* Gaseous mercury distribution in the upper troposphere and lower stratosphere observed onboard the CARIBIC passenger aircraft. *Atmos. Chem. Phys.* **2009**, *9*, 1957–1969.

72. Temme, C.; Einax, J.W.; Ebinghaus, R.; Schroeder, W.H. Measurements of atmospheric mercury species at a coastal site in the Antarctic and over the South Atlantic Ocean during polar summer. *Environ. Sci. Technol.* **2003**, *37*, 22–31.

73. Landis, M.; Stevens, R.K.; Schaedlich, F.; Prestbo, E.M. Development and characterization of an annular denuder methodology for the measurement of divalent inorganic reactive gaseous mercury in ambient air. *Environ. Sci. Technol.* **2002**, *36*, 3000–3009.

74. The Canadian National Atmospheric Chemistry (NAtChem) Database and Analysis System. Available online: http://www.ec.gc.ca/natchem/ (accessed on 29 August 2014).

75. Gilbert, R.O. *Statistical Methods for Environmental Pollution Monitoring*; Van Nostrand Reinhold Company: New York, NY, USA, 1987.

76. van Belle, G.; Hughes, J.P. Nonparametric tests for trend in water quality. *Water Resour. Res.* **1984**, *20*, 127–136.

77. Minamata Convention on Mercury. Available online: http://www.mercuryconvention.org/Convention/tabid/3426/Default.aspx (accessed on 27 June 2014).

Tracing Sources of Total Gaseous Mercury to Yongheung Island off the Coast of Korea

Gang S. Lee, Pyung R. Kim, Young J. Han, Thomas M. Holsen and Seung H. Lee

Abstract: In this study, total gaseous mercury (TGM) concentrations were measured on Yongheung Island off the coast of Korea between mainland Korea and Eastern China in 2013. The purpose of this study was to qualitatively evaluate the impact of local mainland Korean sources and regional Chinese sources on local TGM concentrations using multiple tools including the relationship with other pollutants, meteorological data, conditional probability function, backward trajectories, and potential source contribution function (PSCF) receptor modeling. Among the five sampling campaigns, two sampling periods were affected by both mainland Korean and regional sources, one was caused by mainland vehicle emissions, another one was significantly impacted by regional sources, and, in the remaining period, Hg volatilization from oceans was determined to be a significant source and responsible for the increase in TGM concentration. PSCF identified potential source areas located in metropolitan areas, western coal-fired power plant locations, and the southeastern industrial area of Korea as well as the Liaoning province, the largest Hg emitting province in China. In general, TGM concentrations generally showed morning peaks (07:00~12:00) and was significantly correlated with solar radiation during all sampling periods.

Reprinted from *Atmosphere*. Cite as: Lee, G.S.; Kim, P.R.; Han, Y.J.; Holsen, T.M.; Lee, S.H. Tracing Sources of Total Gaseous Mercury to Yongheung Island off the Coast of Korea. *Atmosphere* **2014**, *5*, 273–291.

1. Introduction

Mercury (Hg) is a toxic heavy metal of concern throughout the Northern Hemisphere. It is emitted from both anthropogenic and natural sources into the atmosphere mostly as inorganic forms. Atmospheric Hg does not generally constitute a direct public health risk at the level of exposure usually found [1]. However, once Hg is deposited into aquatic systems, it can be transformed into methylmercury (MeHg) which is very toxic and readily bioaccumulates through the food web and can affect the health of humans and wildlife. For MeHg in aquatic systems, it has been suggested that atmospheric deposition of inorganic Hg is an important source in many previous studies [2–5]; therefore, there is a need for research on the behavior of atmospheric Hg.

Anthropogenic sources of atmospheric Hg include combustion of coal and other fuels, mining activities, non-metal smelters, and waste incinerators, which emit

1960 ton/yr globally [6]. Atmospheric mercury exists mainly in three operationally defined inorganic forms: gaseous elemental mercury (GEM, Hg^0), gaseous oxidized mercury (GOM, Hg^{2+}), and particulate bound mercury (PBM, Hg(p)). GOM is highly soluble in water and readily deposits to surfaces. Thus, it has short residence time of 1–2 days [1,7]. The residence time of PBM is dependent on the size of particles, but, generally, it has been assumed to be a few days [8,9]. The predominant form of Hg in ambient air is GEM due to its low solubility in water, small deposition velocity, and relatively low reaction rates; therefore, it can be transported long distances and is often considered as a global transboundary pollutant (residence time = 0.5–1 yr) [7,10].

The region of largest anthropogenic Hg emissions is East and Southeast Asia, contributing 39.7% (396–1690 ton) of the total anthropogenic emissions according to an estimation in 2008 [6]. China accounts for three-quarters of these emissions, or about one third of the global total [6]. In addition, Hg emissions in China have dramatically increased since 1990, primarily because coal burning for power generation and for industrial purposes continues to increase while Hg emissions from Europe and North American have decreased. There are several studies predicting the contribution of Asian sources to Hg levels on other continents. Signeur *et al.* [11] estimated that the contribution of Asian anthropogenic emissions to the total Hg deposition over the continental United States ranged from 5% to 36% using a CTM (Chemical Transport Model). In addition, Jaffe *et al.* [12] observed a significant increase in Hg concentrations with prevailing westerly winds from continental Asia. Obrist [13] also measured enhanced mercury concentrations in the Colorado Rocky Mountains in the United States due to long-range transport from Asia with westerly winds.

In Korea, the history of atmospheric mercury measurements is relatively shorter than in the USA and Canada. Total gaseous mercury (TGM) was first measured by Kim and Kim [14] starting in the late 1980s in Seoul, Korea. They reported high concentrations of 14.4 ± 9.8 $ng \cdot m^{-3}$. Recent studies showed much lower TGM concentrations ranging from 2.1 $ng \cdot m^{-3}$ to 3.9 $ng \cdot m^{-3}$ [15–19] due to the wider use of air pollution controls and more stringent regulations. Although Hg emissions in Korea have generally decreased since 1990, Hg levels in Korea may be still seriously affected by Chinese emissions because Korea is situated just west of China, the biggest Hg emitter in the world. This study was designed to identify the contribution of both Chinese emissions and local emissions on atmospheric Hg concentrations in Korea. The sampling site was the most western island in Korea, located in between eastern China and Korea, so that, depending on wind patterns, the effect of Chinese and local Hg emissions could be evaluated.

2. Results and Discussion

2.1. Sampling Description

The biggest local Hg source on Yongheung Island (Figure 1) is the Yongheung coal power plant (YCPP) (0.11 ton Hg·yr^{-1} [20]) located approximately 4.5 km southwest of the sampling site. To the east of the sampling site, there are multiple mainland Hg sources in the industrial area of Incheon including the steel industry (1–57 kg·yr^{-1}) and waste incinerator (0–3 kg·yr^{-1}) (Figure 1). In addition, the Hg concentrations at this site can be impacted by Chinese emissions through long-range transport when there are prevailing westerly winds. There are also other local sources in western and northern areas (waste incinerator) and southern areas (coal-fired power plant) within 50 km of the sampling site. The total TGM emissions rate from all anthropogenic Hg sources in Korea is 32.2 ton·yr^{-1} in 2005 [21].

2.2. General TGM Patterns

The average TGM concentration was 2.87 ± 1.07 ng·m^{-3} during the sampling periods (Figure 2). The seasonal TGM concentration was the highest in winter (January, February) (3.60 ± 0.97 ng·m^{-3}), followed by in spring (April, May) (2.43 ± 0.83 ng·m^{-3}) and in summer (August) (2.29 ± 0.85 ng·m^{-3}) (Tukey HSD test, *p-value* < 0.001) (Table 1, Figure 2). Higher TGM concentrations in winter are generally observed in the Northern Hemisphere at least in part due to increased emissions and by the distinctive meteorological conditions including reduced mixing layer heights [22,23]. On the other hand, generally reduced TGM concentrations were seen during summer.

Table 1. Summarized seasonal total gaseous mercury (TGM) concentrations for five sampling periods.

Sampling	Date	Season	TGM (ng·m^{-3})	Wind Speed (m·s^{-1})	Temperature (°C)	Solar Radiation (W m^{-2})
1st Period	2013.01.17–2013.01.21	Winter	3.49 ± 0.81	1.45 ± 1.05	1.00 ± 2.71	68.94 ± 136.49
2nd Period	2013.02.25–2013.03.01		3.67 ± 0.91	1.55 ± 1.41	2.59 ± 2.33	40.55 ± 81.21
3rd Period	2013.04.08–2013.04.13	Spring	2.09 ± 0.40	6.98 ± 2.56	5.58 ± 1.01	238.99 ± 324.75
4th Period	2013.05.20–2013.05.25		2.80 ± 1.00	1.56 ± 1.08	13.60 ± 2.93	252.30 ± 318.84
5th Period	2013.08.19–2013.08.24	Summer	2.29 ± 0.85	1.03 ± 1.15	25.74 ± 1.91	180.55 ±274.40

These concentrations were considerably lower than those measured at urban sites in China and Taiwan, but in a similar range as urban sites in Korea and higher than in suburban and background areas of USA and Canada (Table 2). Considering that background TGM concentrations in locations not impacted by local emission source are typically 1.4–2.0 ng·m^{-3} [24], TGM concentrations measured at this study were indeed affected by local and regional anthropogenic sources.

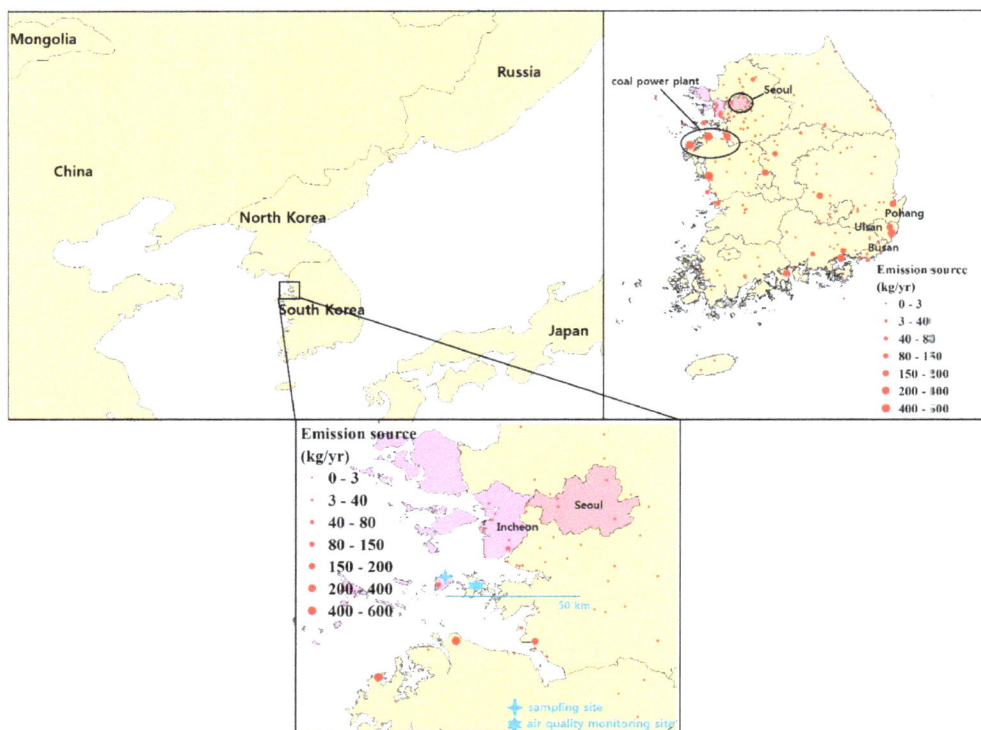

Figure 1. The Hg sampling site (with a cross mark) in this study and the national air quality monitoring site (with a star mark) for other atmospheric pollutants. The upper right panel indicates the anthropogenic TGM emission sources in Korea.

During the sampling period, westerly and southwesterly winds were predominant (Figure 3); however, high TGM concentrations were associated with all wind directions (Figure 3). Low TGM concentrations (<2 ng·m^{-3}) were not observed with easterly winds, but the frequency of easterly winds was too low to indicate that mainland Korean sources were more important than southwesterly located YCP? or regional Chinese sources. The relative importance of local and regional sources will be discussed in more detail in the next section.

Diel variations in TGM concentration generally showed morning peaks (07:00–12:00) during all sampling periods (Figure 4). In urban areas, TGM concentrations are typically higher during the nighttime than during the daytime [25,26] due to a combination of GEM loss by daytime oxidation, increased use of household heating systems and decreased mixing height at night. In this study, higher nighttime TGM concentrations were not found in any sampling period (Figure 4). The sampling site is located in a remote island having a population of only 5815; therefore, Hg emissions from nighttime household heating systems were presumed to be not significant. Although the biggest anthropogenic source on this island (YCPP) operates continuously all year round, actual electricity usage generally fluctuates with a larger rate during the daytime than in the night (Korea Electric Power Generation, http://cyber.kepco.co.kr), which also might contribute to the higher daytime TGM concentration measured in this study.

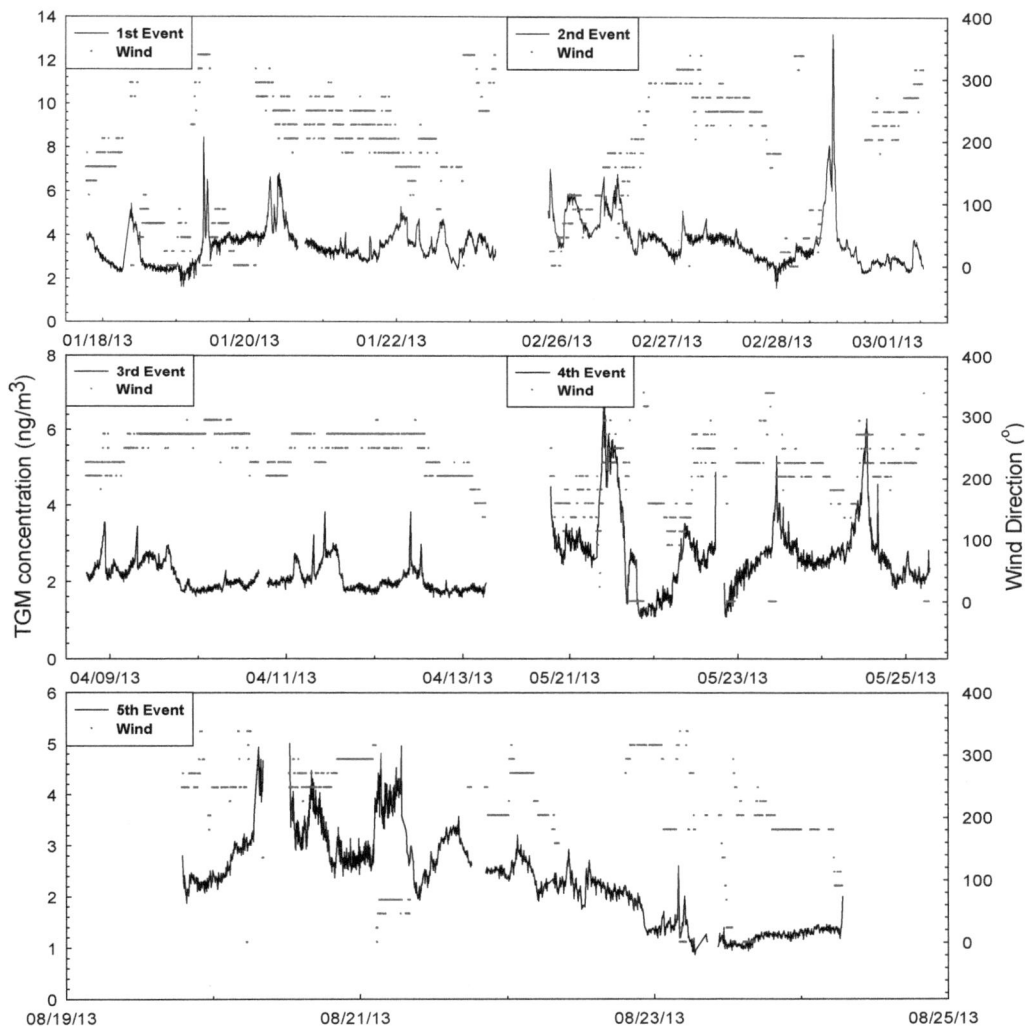

Figure 2. Measured TGM concentrations (left y-axis) with wind direction (right y-axis) for five sampling periods.

Some previous studies in rural and remote areas found that a maximum TGM concentration in mid-morning and a minimum at night was due to Hg emissions from natural surfaces including oceans and soils during times of increased solar radiation [27–29]. There are many studies showing that emissions of GEM from water and soil surfaces increase with increasing solar radiation [30–32]. Since the sampling site is located next to the beach, the emissions of Hg from the ocean surface could possibly elevate atmospheric TGM concentrations. Solar radiation was positively correlated with TGM concentration for all sampling periods (Table 3), suggesting the possibility of measurable volatilization from natural sources. TGM was also negatively correlated with wind speed and atmospheric temperature, and positively correlated with relative humidity for the whole sampling period although the relationship was not consistent for every individual sampling period (Table 3).

Table 2. Comparison of TGM concentrations with other studies.

Country	Site	Year	Remarks	TGM $(ng \cdot m^{-3})$	Reference
Korea	Seoul	2005–2006	Urban	3.4 ± 2.1	33
	Seoul	2005–2006	Urban	3.2 ± 2.1	23
	Jeju*	2006–2007	Island	3.9 ± 1.7	18
	Chuncheon	2006–2009	Rural	2.1 ± 1.5	25
	This study	2013	Island	2.9 ± 1.1	
China	Pearl River	2008	Background	2.9	34
	Guiyang	2009	Urban	9.7 ± 10.2	35
	Beijing	1998	Urban	7.9 ± 34.9	36
Taiwan	Central Taiwan	2010–2011	Urban	6.1 ± 3.7	37
U.S.A.	Reno, Nevada	2007–2009	Suburban	2.0 ± 0.7	38
	Great Smoky Mt.	2007	Background	1.65	22
	Detroit	2002	Urban	3.1	39
Canada	Nova Scotia	2010–2011	Rural	1.4 ± 0.2	24

Table 3. Correlation coefficients of TGM with meteorological data. Correlation coefficients with an asterisk indicate statistically significant relationships at $\alpha = 0.05$.

	Wind speed	Temp	RH	Solar
1st Period	0.065* (0.012)	0.178* (<0.001)	0.153* (<0.001)	0.106* (<0.001)
2nd Period	−0.189* (<0.001)	0.067 (0.053)	−0.544* (<0.001)	0.209* (<0.001)
3rd Period	0.307* (<0.001)	0.367* (<0.001)	0.254* (<0.001)	0.368* (<0.001)
4th Period	0.137* (<0.001)	−0.153* (<0.001)	0.377* (<0.001)	0.680* (<0.001)
5th Period	0.032 (0.275)	0.011 (0.716)	0.035 (0.255)	0.325* (<0.001)
Total	−0.242* (<0.001)	−0.345* (<0.001)	0.148* (<0.001)	0.103* (<0.001)

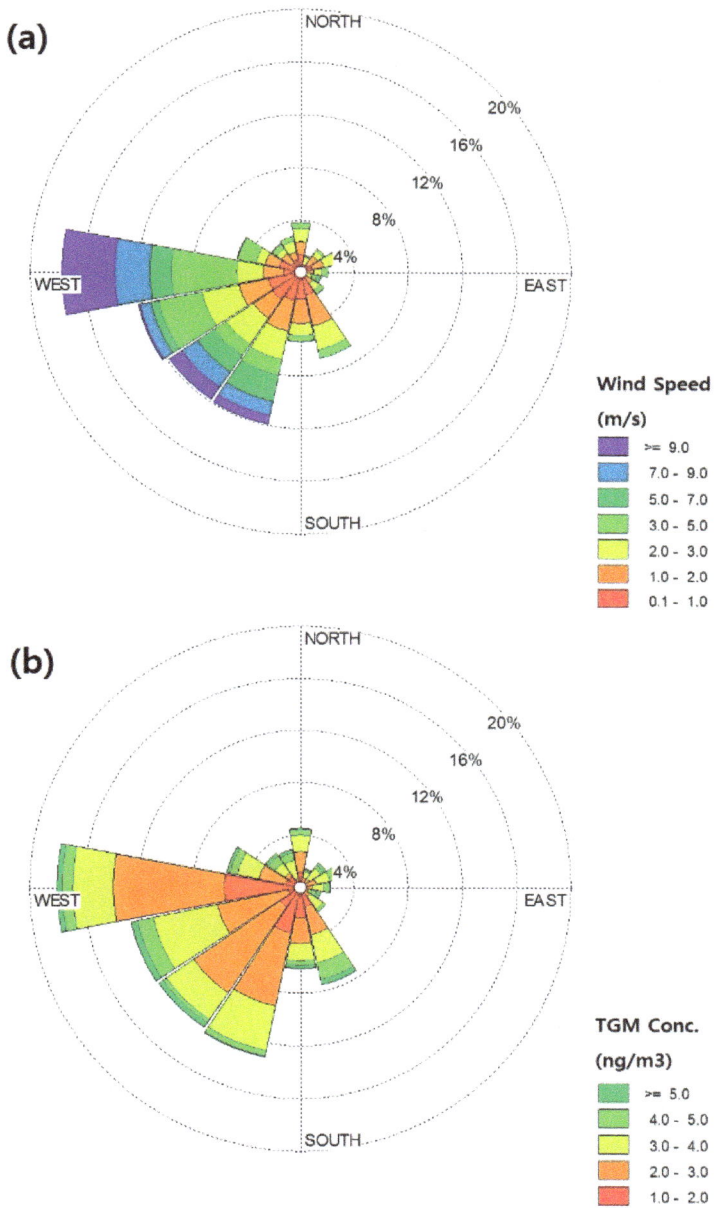

Figure 3. Wind rose (**a**) and pollution rose (**b**) for the entire sampling period.

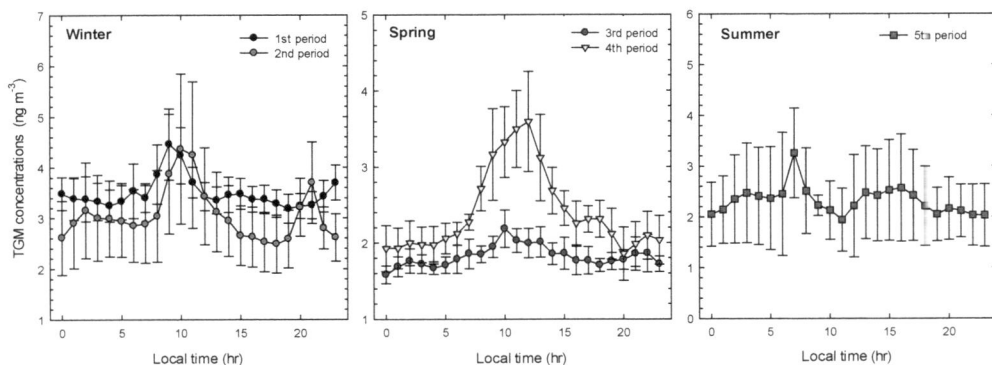

Figure 4. Diel pattern of mean TGM concentration for each sampling period. The error bar indicates one standard deviation.

2.3. Impact of Local vs. Regional Sources

2.3.1. First Sampling Period

For the first sampling period (17–21 January 2013), TGM concentration was well correlated with SO_2 and CO, but not with NO_2 (Table 4). The atmospheric residence time of NO_2 (1 day) is shorter than SO_2 (3 days) or CO (60 days) [40]. The major source of NO_2 in Korea is vehicle emission which contributes about 55% of the total NO_2 emissions [41]. These results suggest that vehicles were not a major source of TGM for the first sampling period, and the good correlation of TGM with SO_2 and CO suggests that coal combustion was important for enhancing TGM concentrations. CPF shows that the high TGM concentration occurred with NW and SE directions, indicating the possible effects of both Chinese and mainland Korean sources (Figure 5). Figure 6 also indicates that some back-trajectories for the top 10% TGM concentration samples for this period originated from mainland of Korea but most traveled through industrial areas of China (Shandong and Henan Provinces). Back-trajectories for the bottom 10% TGM concentration samples all passed through northern China originating in Mongolia (Figure 6). In total, these results suggest that both Korean and Chinese sources affected TGM concentration at the sampling site for this sampling period and likely sources include coal combustion.

116

Table 4. Pearson correlation coefficient (R) with four representative air pollutants for TGM and their corresponding concentrations for each sampling period. Correlation coefficients with an asterisk indicate statistically significant relationships at $\alpha = 0.05$.

	SO$_2$		NO$_2$		CO		O$_3$	
	R (p-value)	Conc. (ppb)	R (p-value)	Conc. (ppb)	R (p-value)	Conc. (ppm)	R (p-value)	Conc. (ppb)
1st Period	0.474* (<0.001)	5.76 ± 4.24	0.155 (0.078)	26.71 ± 12.38	0.465* (<0.001)	0.52 ± 0.17	−0.135 (0.129)	21.71 ± 13.54
2nd Period	0.552* (<0.001)	7.31 ± 3.02	0.558* (<0.001)	28.84 ± 19.10	0.362* (0.001)	0.65 ± 0.19	−0.532* (<0.001)	28.33 ± 14.67
3rd Period	0.402* (<0.001)	4.83 ± 1.68	0.304* (0.001)	8.61 ± 3.33	0.543* (<0.001)	0.35 ± 0.08	−0.136 (0.162)	46.76 ± 6.04
4th Period	0.286* (0.003)	5.67 ± 2.48	0.040 (0.684)	17.18 ± 5.91	0.438* (<0.001)	0.48 ± 0.11	0.224* (0.020)	36.15 ± 19.40
5th Period	0.161 (0.103)	4.65 ± 1.59	0.254* (0.009)	10.20 ± 3.76	0.484* (<0.001)	0.50 ± 0.12	0.241* (0.014)	40.93 ± 29.58
Total	0.441* (<0.001)	5.58 ± 2.98	0.560* (<0.001)	18.22 ± 13.11	0.560* (<0.001)	0.50 ± 0.16	−0.242* (<0.001)	34.53 ± 20.40

2.3.2. Second Sampling Period

Among the five sampling periods, the average TGM concentration was the highest for the second period and the coefficient of variation (the standard deviation divided by the arithmetic mean) was also high (Table 1, Figure 2), suggesting a higher influence of local sources compared to regional background contributions, since contributions from background sources are generally less variable than contributions from local sources [24,26,42–44]. During this period, TGM was statistically well correlated with SO$_2$, NO$_2$, and CO, with the Pearson correlation coefficient highest with NO$_2$ (Table 4). The correlation coefficient between NO$_2$ and CO was the highest during this period (0.47) relative to 0.21–0.37 for the other periods. Since the biggest source of NO$_2$ and CO in Korea is vehicle exhaust emissions [41,45], these findings suggest that this source was one of the causes of the elevated TGM concentrations, along with coal-fired power plants. This conclusion is supported by the diel variation in TGM concentrations which had two major peaks during rush hour for only the second sampling period (Figure 4). The relatively lower correlation coefficient with CO than with NO$_2$ and SO$_2$ also suggests that the long-range transport from regional sources were not as important as local sources. For this period, a large number of back-trajectories passed through western mainland Korea (Figure 6). CPF also indicated that TGM concentration increased with easterly and southerly winds (Figure 5). Therefore, for the second sampling period, there is a high probability that mainland Korean vehicle emissions were the most important cause of enhanced TGM concentrations.

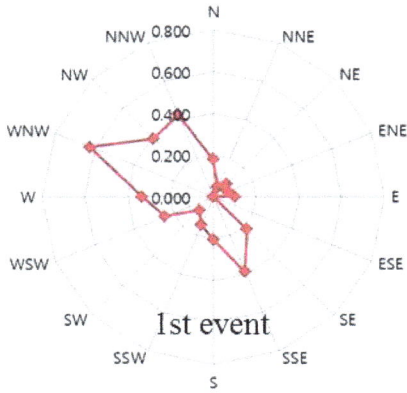

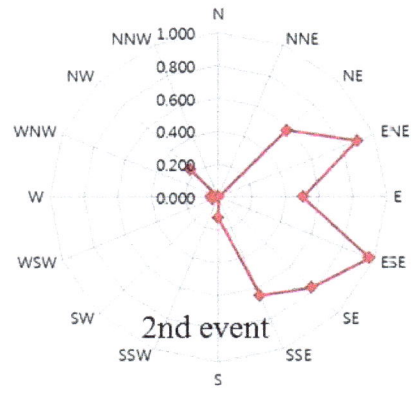

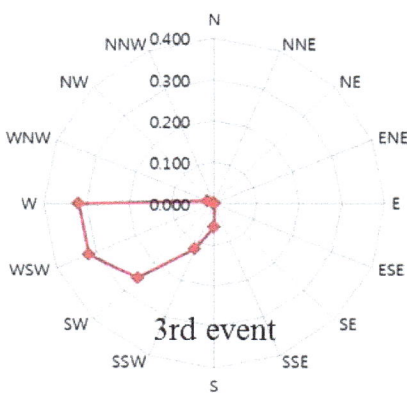

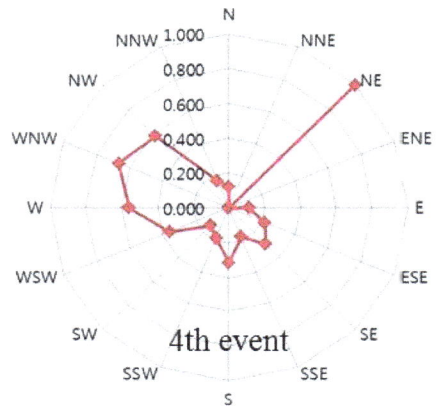

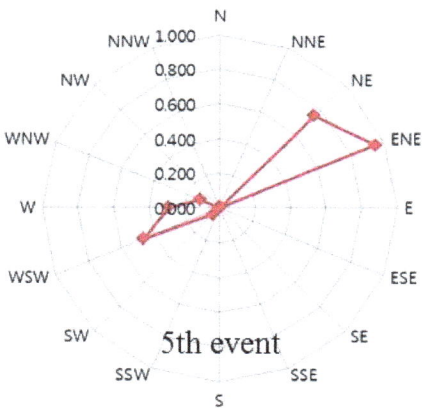

Figure 5. CPF (Conditional Probability Function) plot for each sampling period.

2.3.3. Third Sampling Period

Among the five sampling periods, the lowest concentrations of TGM, NO_2, and CO and the second lowest of SO_2 were measured during the third sampling period (Table 4). NO_2, representing local sources, had an average concentration of only 47% of all five sampling periods. For this period, the winds were consistently from the west (Figure 2) and TGM had the strongest correlation with CO; therefore, fossil fuel combustion in China was a likely major source. High TGM concentrations were associated with westerly winds in CPF analysis (Figure 5), and most of the back-trajectories originated from Northeastern China and did not pass through mainland Korea (Figure 6). A minimal effect from local sources was also indicated by the lowest coefficient of variation of 0.19 among the five sampling periods. Another notable fact is that a high wind speed (6.98 ± 2.56 m·s^{-1}) was observed; significantly larger than the average wind speed for other sampling periods (1.5 ± 1.4 m·s^{-1}) (Table 1). Generally, high wind speeds cause effective horizontal dilution in ambient air, leading to low TGM concentration; over the complete sampling period there was a statistically significant negative correlation between TGM concentration and wind speed (p-value < 0.001) (Table 3). To conclude, it is likely that Chinese emissions along with high wind speed and without local source contributions were responsible for the relatively low TGM concentrations observed during the third sampling period.

2.3.4. Fourth Sampling Period

During the fourth sampling period, TGM had a statistically significant correlation with SO_2 and CO, but not with NO_2, similar to the first sampling period. However, the back-trajectories did not pass through mainland Korean sources or industrial areas of China. Most trajectories stayed over the Yellow Sea for long periods of time (Figure 6). It is known that gaseous elemental mercury (GEM) is emitted from the natural surfaces, with the ocean being the largest natural source, contributing approximately 33% of total Hg emissions from both anthropogenic and natural sources [6]. According to Han *et al.* [46], the Atlantic Ocean was suggested to be a significant source of TGM measured in NY State, USA using back-trajectory based models. If the ocean was a significant source of TGM, concentrations should increase as solar radiation increases because Hg emissions are controlled primarily by a photo-reduction process initiated by solar radiation [47–49]. For the fourth sampling period, the correlation coefficient between TGM and solar radiation was the highest (r = 0.680) among all five sampling periods (Table 3), and TGM also was positively correlated with O_3 (Table 4) which is produced by photochemical reactions. In addition, the diel pattern of TGM showed one major peak at noon (Figure 4) while the main peak appeared at 7–10 am for all other sampling periods. In addition, the increment from the nighttime minimum (at 20:00) to the daytime maximum (at 12:00) was the largest among the five sampling periods (Figure 4). These results altogether suggest that the volatilization of Hg from the ocean was an important source for this sampling period.

Since the YCPP was located approximately 4.5 km southwest of the sampling site TGM concentrations were expected to be enhanced during southwesterly winds if the impact of YCPP was significant. During this period, high TGM concentrations were regularly associated with wind directions between 180 and 315° (Figures 2 and 5). The large coefficient of variation (0.36) for TGM concentrations suggests that TGM concentrations were influenced by local sources and local chemistry to a greater extent than regional sources during this period [50,51]. Therefore, in the fourth sampling period, both natural sources and the local YCPP source were important.

2.3.5. Fifth Sampling Period

During the fifth sampling period TGM had the strongest correlations with CO, followed by NO_2 (Table 4), suggesting that both long-range transport and local sources concurrently affected TGM concentrations. The correlation coefficient between TGM and NO_2 was weak, suggesting that vehicle emissions were not significant; in addition, no rush hour peaks for TGM were observed (Figure 4). During this period, the top 10% concentration samples were associated with distinctly different back-trajectories than the lowest 10% concentration samples. High concentration samples passed through Northern China, North Korea, and metropolitan areas of Korea (Figure 6). For the lowest 10% samples, all back-trajectories originated from the Yellow Sea (Figure 6). CPF also shows a relationship between winds from NE and high TGM concentrations (Figure 5). Unlike the ocean associated trajectories from period 4, which were associated with high concentrations, heavy rain occurred (precipitation depth = 75 mm) during the fifth sampling period (on 23 August 2013). Since the evasion of Hg from the water surface is primarily due to photoreduction initiated by solar radiation, heavy rain possibly inhibited the emission of Hg from the ocean surface and consequently caused low TGM concentrations in the ambient air [52] (Figure 2). Solar radiation measured during the fifth sampling period was much lower than that during the fourth sampling period (Table 1).

2.4. Potential Source Contribution Function

In order to identify the possible source areas associated with elevated TGM concentrations, potential source contribution function (PSCF) was used. Hourly back-trajectories were calculated for each hourly observation of TGM concentration. The number of back-trajectories was 540 for each arrival height of 200 m and 500 m. The criterion value used to distinguish between high and low concentrations was set to the top 25% value (4.29 $ng \cdot m^{-3}$) to identify the larger sources. To reduce the uncertainty in a grid cell with a small number of endpoints, an arbitrary weight function W_{ij} was applied when the number of the end points in a particular cell was less than three times the average number of end points (N_{ave}) for all cells [35,46,53,54].

$$W_{ij} = \begin{pmatrix} 1.0 \ N_{ij} > 3N_{ave} \\ 0.70 \ 3N_{ave} > N_{ij} > 1.5N_{ave} \\ 0.40 \ 1.5N_{ave} > N_{ij} > N_{ave} \\ 0.20 \ N_{ave} > N_{ij} \end{pmatrix} \tag{1}$$

PSCF combines both measurement data at the sampling site and meteorological data; therefore, sources outside the region of back-trajectory pathways are not identified. To indicate which areas would be included in the PSCF modeling, the total residence time of back-trajectories (total number of endpoints in each grid) is shown in Figure 6. The prevailing winds were generally from the east, and the west.

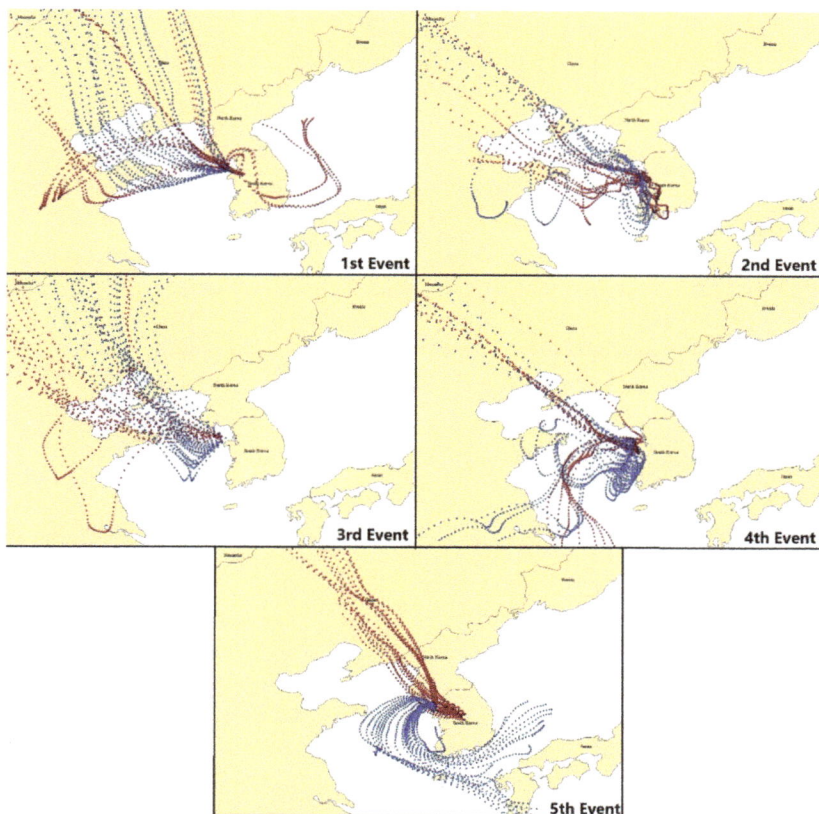

Figure 6. Back-trajectories during each sampling period. Red and blue points indicate top 10% and bottom 10% of TGM concentrations, respectively.

121

Potential source areas identified by the PSCF modeling include Liaoning province, the biggest Hg emitting province in China, which was linked to non-ferrous smelters [55] (Figure 7). The Yellow Sea and East Sea (Sea of Japan) were also identified as possible sources; however, it is not certain whether these areas were actual natural sources or appeared due to trailing effects. If the East Sea (Sea of Japan) area was caused by the trailing effect, the actual source is probably the southeastern industrial area of Korea, which was identified as a local mainland Korean source in Figure 7.

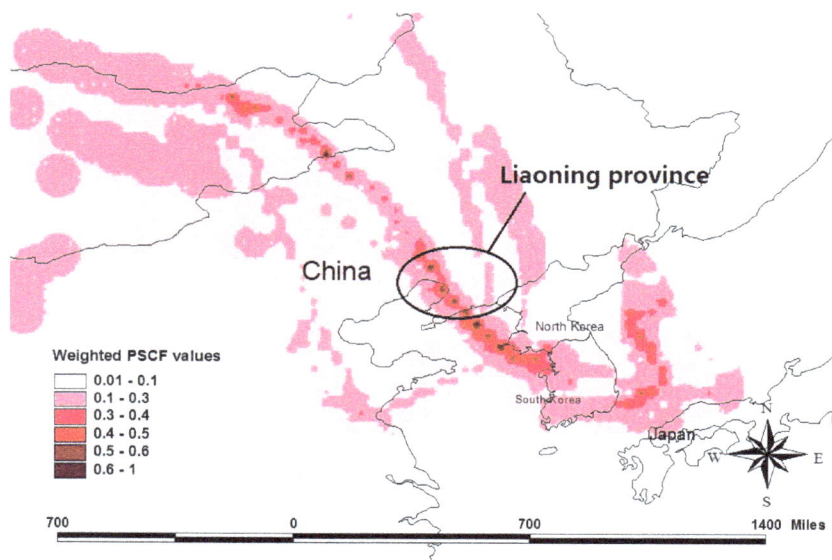

Figure 7. PSCF result for tracing regional sources. A grid cell size of 0.5° by 0.5° was used.

3. Experimental Section

3.1. Sampling and Analysis

The sampling site was located on the roof of three-story building one Yongheung Island, Korea (lat: 37.15, lon: 126.28). Yongheung Island is a small island located about 15~20 km west from the mainland of Korea, having a population of 5815 according to a 2013 census. Tourism has been this Island's main source of revenue. The Yongheung Coal-fired Power Plant (YCPP) was constructed in 2004, emitting about 0.11 ton·yr^{-1} of Hg. The sampling site was located approximately 4.8 km northeast from the YCPP. At this site, TGM (GEM+GOM) was measured during five intensive periods in winter, spring, and summer in 2013 using a Tekran

2537B (Table 1). Outdoor air at a flow rate of 1.5 L·min^{-1} was transported through a 3-m-long heated sampling line (1/4″ OD Teflon) into the analyzer.

The Tekran 2537B underwent automated daily calibrations using an internal permeation source. Manual injections were also used to evaluate these automated calibrations before each sampling campaign using a saturated mercury vapor standard. The relative percent difference between manual injections and automated calibration was less than 2%. The method detection limit was calculated as three times the standard deviation obtained after injecting 1 pg of the mercury vapor seven times (0.04 ng·m^{-3}). The recovery rate was obtained by directly injecting Hg vapor into the sampling line between the sample inlet and the Tekran 2537B in a zero-air stream. It was between 85% and 110% (96 ± 3%). All Teflon products were acid-cleaned following EPA method 1631E before use.

Meteorological data including temperature, wind direction, wind speed, relative humidity, solar radiation, and precipitation depth were also measured every 5 min at the sampling site using a meteorological tower (DAVIS Inc weather station, Vintage Pro2^{TM}). During the sampling period, the atmospheric temperature and wind speed ranged from −6.2 to 31.0 °C (9.6 ± 9.4°C) and 0.0 to 13.0 m·s^{-1} (2.7 ± 2.8 m·s^{-1}).

3.2. Other Atmospheric Pollutants

The concentrations of NO_2, SO_2, CO, and O_3 were obtained from the nearest national air quality monitoring station located approximately 8 km east from the sampling site. NO_2, SO_2, CO and O_3 were measured by a chemiluminescent method, pulse UV florescence method, non-dispersive infrared method, and UV photometric method, respectively (http://www.airkorea.or.kr/). These concentrations were compared with those measured at another national air quality monitoring station located approximately 24 km west from the Hg sampling site, and there were no statistical difference (p-value < 0.001), indicating that the spatial distribution of these pollutants were relatively uniform across the area.

3.3. Backward Trajectory

The three-day backward trajectories were calculated using the NOAA HYSPLIT 4.7 with GDAS (Global Data Assimilation System) meteorological data. The GDAS archive has 3-hourly, global, 1 degree latitude longitude datasets of the pressure surface. In this study, hourly back-trajectories were calculated for each hourly averaged sample concentrations, and the arrival heights of 200 m and 500 m were used to describe the local and the regional transport meteorological pattern.

3.4. Conditional Probability Function

The conditional probability function (CPF), which is the conditional probability that a given concentration from a given wind direction will exceed a predetermined threshold criterion, was calculated as the following equation.

$$CPF_{\Delta\theta} = \frac{m_{\Delta\theta}}{n_{\Delta\theta}} \tag{2}$$

where $m_{\Delta\theta}$ is the number of occurrences from wind sector $\Delta\theta$ where the TGM concentration is in the upper 25th percentile, and $n_{\Delta\theta}$ is the total number of occurrence from this wind sector.

3.5. Potential Source Contribution Function

The PSCF model counts each trajectory segment endpoint that terminates within given grid cell. The probability of an event at the receptor site is related to the number of endpoints in that cell relative to the total number of endpoints for all of the sampling dates [56,57]. If N is the total number of trajectory endpoints over the study period and if n is the number of endpoints of trajectories falling in a given ijth cell, the probability of this event ($P[A_{ij}]$) is calculated by n_{ij}/N. Also, if m_{ij} is the number of endpoints associated with concentration than a criterion value in ijth cell, the probability of this high concentration event, B_{ij}, is given by $P[B_{ij}]$ of m_{ij}/N. ThePSCF value in a given grid ijth cell is then calculated using the following equation.

$$\text{PSCF value} = P[B_{ij}]/P[A_{ij}] = m_{ij}/n_{ij} \tag{3}$$

Grid cells containing sources enhancing the TGM concentration measured at the receptor site are recognized as possible source areas in PSCF. The criterion value used was the top 25% of TGM concentration, which is 4.29 ng·m^{-3}. The cell size of 0.5° by 0.5° was used for tracing sources. More detailed descriptions of the PSCF model are in previous publications [46,58].

4. Conclusions

TGM concentrations were measured in the farthest western island of Korea in between eastern China and mainland Korea in 2013 in order to identify important TGM sources to this location. In general, westerly and southwesterly winds were predominant during the sampling periods; however, the TGM concentration was not directionally dependent. TGM concentrations showed a distinct diel variation with higher values during the daytime than during the nighttime. Concentrations were generally positively correlated with solar radiation, indicating that volatilization of natural surfaces were significant.

Multiple tools including the relationship with other atmospheric pollutants, TGM diel variation, backward trajectories, and meteorological data were used to identify the relative impact of mainland Korean and regional Chinese sources on TGM concentrations at the sampling site for each of five sampling periods. For two periods (January and August 2013), TGM was found to be enhanced by both mainland Korean and regional Chinese coal-fired sources based on the good correlations with SO_2 and CO and back-trajectories that passed through both mainland Korea and industrial areas of China. One period (February 2013) appeared to be affected by mainland vehicle emissions because TGM was significantly correlated with SO_2, NO_2, and CO and had two major peaks during rush hours. Also, a large number of back-trajectories originated from mainland Korea, suggesting that mainland Korean sources were important. During the April period, low concentrations and low coefficients of variation for TGM and other atmospheric pollutants associated with back-trajectories passing through China suggest that regional Chinese sources affected the TGM concentration. The very high wind speed observed in this sampling period was also an important factor for reducing atmospheric TGM concentrations. For the remaining sampling period (May 2013), a significant number of trajectories remained above the Yellow Sea (Eastern China Sea) and TGM concentration was significantly correlated with solar radiation, suggesting that TGM by volatilization from the ocean was important.

PSCF was also used to locate possible source areas. For mainland Korea sources, metropolitan areas (Seoul), the western coal-fired power plants area, and the southeast industrial area were identified as significant source areas. In addition, Liaoning province, the biggest Hg emitting province in China was found to be associated with increased TGM concentrations at the receptor site. These results highlight the need for international cooperation between Korea and China to reduce atmospheric Hg concentrations in Korea.

Acknowledgments: This work was supported by Basic Science Research Program through the National Research Foundation of Korea (NRF) funded by the Ministry of Education, Science, and Technology (2012R1A1A2042150).

Author Contributions: The work presented here was carried out in collaboration between all authors. Gang S. Lee analyzed data and wrote the paper. Pyung R. Kim performed the experiments and interpreted the results. Young J. Han defined the research theme, interpreted the results, and wrote the paper. Thomas M. Holsen also interpreted the results and approved the final paper. Seung H. Lee contributed to trajectory-based modeling.

Conflicts of Interest: The authors declare no conflict of interest.

References

1. Driscoll, C.T.; Han, Y.J.; Chen, C.Y.; Evers, D.C.; Lambert, K.F.; Holsen, T.M.; Kamman, N.C.; Munson, R.K. Mercury contamination in forest and freshwater ecosystems in the Northeastern United States. *Appl. Environ. Microbiol.* **2007**, *57*, 17–28.

2. Buehler, S.S.; Hites, R.A. The Great Lakes integrated atmospheric deposition network. *Environ. Sci. Technol.* **2002**, *36*, 354A–359A.

3. Landis, M.; Vette, A.F.; Keeler, G.J. Atmospheric mercury in the Lake Michigan Basin: Influence of the Chicago/Gary Urban Area. *Environ. Sci. Technol.* **2002**, *36*, 4508–4517.

4. Munthe, J. Recovery of mercury-contaminated fisheries. *Ambio* **2007**, *36*, 33–44.

5. Lyman, S.N.; Gustin, M.S.; Prestbo, E.M.; Kilner, P.I.; Edgerton, E.; Hartsell, B. Testing and application of surrogate surfaces for understanding potential gaseous oxidized mercury dry deposition. *Environ. Sci. Technol.* **2009**, *43*, 6235–6241.

6. UNEP. *The Global Atmospheric Mercury Assessment*; UNEP Chemicals Branch: Geneva, Switzerland, 2013.

7. Schroeder, W.H.; Munthe, J. Atmospheric mercury—An overview. *Atmos. Environ.* **1998**, *32*, 809–822.

8. Zhang, L.; Gong, S.; Padro, J.; Barrie, L. A size-segregated particle dry deposition scheme for and atmospheric aerosol module. *Atmos. Environ.* **2001**, *35*, 549–560.

9. Fang, G.C.; Zhang, L.; Huang, C.S. Measurement of size-fractioned concentration and bulk dry deposition of atmospheric particulate bound mercury. *Atmos. Environ.* **2012**, *61*, 371–377.

10. Slemr, F.; Schuster, G.; Seiler, W. Distribution, speciation and budget of atmospheric mercury. *J. Atmos. Chem.* **1985**, *3*, 407–434.

11. Seigneur, C; Vijayaraghavan, K.; Lohman, K.; Karamchandani, P.; Scott, C. Global source attribution for mercury deposition in the United States. *Environ. Sci. Technol.* **2004**, *38*, 555–569.

12. Jaffe, D.; Prestbo, E.; Swartzendruder, P.; Weiss-Penzias, P.; Kato, S.; Takami, A.; Hatakeyama, S.; Kaiji, Y. Export of atmospheric mercury from Asia. *Atmos. Environ.* **2005**, *39*, 3029–3038.

13. Obrist, D.; Gannet, H.A.; McCubbin, I.; Stephens, B.B.; Rahn, T. Atmospheric mercury concentrations at Storm Peak Laboratory in the Rocky Mountains: Evidence for long-range transport from Asia, boundary layer contributions, and plant mercury uptake. *Atmos. Environ.* **2008**, *42*, 7579–7589.

14. Kim, K.H.; Kim, M.Y. The effects of anthropogenic sources on temporal distribution characteristics of total gaseous mercury in Korea. *Atmos. Environ.* **2000**, *34*, 3337–3347.

15. Kim, K.H.; Kim, M.Y.; Kim, J.; Lee, G.W. The concentrations and fluxes of total gaseous mercury in a western coastal area of Korea during late March 2001. *Atmos. Environ.* **2002**, *34*, 3413–3427.

16. Shon, Z.H.; Kim, K.H.; Song, S.K.; Kim, M.Y.; Lee, J.S. Environmental fate of gaseous elemental mercury at an urban monitoring site based on long-term measurements in Korea (1997–2005). *Atmos. Environ.* **2008**, *42*, 142–155.

17. Gan, S.Y.; Choi, E.M.; Seo, Y.S.; Yi, S.M.; Han, Y.J. Characteristic of atmospheric speciated mercury measured in Seoul, Chuncheon and Ganghwado. *Korean Soc. Atmos. Environ.* **2009**, *5*, 213–214.

18. Nguyen, T.H.; Kim, M.Y.; Kim, K.H. The influence of long-range transport on atmospheric mercury on Jeju Island, Korea. *Sci. Total Environ.* **2010**, *408*, 1295–1307.

19. Kim, K.H.; Shon, Z.H.; Nguyen, H.T.; Jung, K.; Park, C.G.; Bae, G.N. The effect of man made source processes on the behavior of total gaseous mercury in air: A comparison between four urban monitoring sites in Seoul Korea. *Sci. Total Environ.* **2011**, *409*, 3801–3811.

20. Seo, Y.C. Personal Communication. Yonsei University: Seoul, Korea, 2007.

21. *AMAP/UNEP Technical Background Report on the Global Anthropogenic Mercury Assessment*; Arctic Monitoring and Assessment Programme/UNEP Chemicals Branch: Geneva, Switzerland, 2008; p. 159.

22. Kim, S.H.; Han, Y.J.; Holsen, T.M.; Yi, S.M. Characteristic of atmospheric speciated mercury concentrations (TGM, Hg(II) and Hg(p)) in Seoul, Korea. *Atmos. Environ.* **2009**, *43*, 3267–3274.

23. Cheng, I.; Zhang, L.; Mao, H.; Blanchard, P.; Tordon, R.; John, D. Seasonal and diurnal patterns of speciated atmospheric mercury at a coastal-rural and coastal-urban site. *Atmos. Environ.* **2014**, *82*, 193–205.

24. Valente, R.; Shea, C.; Lynn, H.K.; Tanner, R. Atmospheric mercury in the Great Smoky Mountains compared to regional and global levels. *Atmos. Environ.* **2007**, *41*, 1861–1873.

25. Han, Y.J.; Kim, J.E.; Kim, P.R.; Kim, W.J.; Yi, S.M.; Seo, Y.S.; Kim, S.H. General trends of Atmospheric mercury concentrations in urban and rural areas in Korea and characteristics of high-concentration events. *Atmos. Environ.* **2014**. submitted.

26. Kim, K.H.; Yoon, H.O.; Richard, J.C.B.; Jeon, E.C.; Sohn, J.R.; Jung, K.; Park, C.C.; Kim, I.S. Simultaneous monitoring of total gaseous mercury at four urban monitoring stations in Seoul, Korea. *Atmos. Res.* **2013**, *132–133*, 199–208.

27. Feng, X.; Yan, H.; Wang, S.; Qiu, G.; Tang, S.; Shang, L.; Dai, Q.; Hou, Y. Seasonal variation of gaseous mercury exchange rate between air and water surface over Baihua reservoir, Guizhou, China. *Atmos. Environ.* **2004**, *38*, 4721–4732.

28. Poissant, L.; Pilote, M.; Beauvais, C.; Constant, P.; Zhang, H.H. A year of continuous measurements of three atmospheric mercury species (GEM, RGM and Hg$_p$) in southern Quebec, Canada. *Atmos. Environ.* **2005**, *39*, 1275–1287.

29. Choi, H.D.; Holsen, T.M.; Hopke, P.K. Atmospheric mercury in the Adirondacks: Concentrations and sources. *Environ. Sci. Technol.* **2008**, *42*, 5644–5653.

30. Amyot, M.; Auclair, J.C.; Poissant, L. In situ high temporal resolution analysis of elemental mercury in natural waters. *Anal. Chimica Acta* **2001**, *447*, 153–159.

31. O'Driscoll, N.J.; Siciliano, S.D.; Lean, D.R.S. Continuous analysis of dissolved gaseous mercury in freshwater lakes. *Sci. Total Environ.* **2003**, *304*, 285–294.

32. Choi, H.D.; Holsen, T.M. Gaseous mercury emission from the forest floor of the Adirondacks. *Environ. Pollut.* **2009**, *157*, 592–600.

33. Choi, E.M.; Kim, S.H.; Holsen, T.M.; Yi, S.M. Total gaseous concentration in mercury in Seoul, Korea: Local sources compared to long-range transport from China and Japan. *Environ. Pollut.* **2009**, *157*, 816–822.

34. Li, Z.; Xia, C; Wang, X.; Xiang, Y.; Xie, Z. Total gaseous mercury in Pearl River Delta region, China during 2008 winter period. *Atmos. Environ.* **2011**, *45*, 834–838.

35. Fu, X.; Feng, X.; Qiu, G.; Shang, L.; Zhang, H. Speciated atmospheric mercury and its potential source in Guiyang, China. *Atmos. Environ.* **2011**, *45*, 4205–4212.

36. Liu, S.; Nadim, F.; Perkins, C.; Carley, R.J.; Hoag, G.E.; Lin, Y.; Chen, L. Atmospheric mercury monitoring survey in Beijing, China. *Chemosphere* **2002**, *48*, 97–107.

37. Huang, J.; Liu, C.K.; Huang, C.S.; Fang, G.C. Atmospheric mercury pollution at an urban site in central Taiwan: Mercury emission sources at ground level. *Chemosphere* **2012**, *87*, 579–585.

38. Lyman, S.N.; Gustin, M.S. Determinants of atmospheric mercury concentrations in Reno, Nevada, U.S.A. *Sci. Total Environ.* **2009**, *408*, 431–438.

39. Lyman, M.M.; Keeler, G.J. Source-receptor relationships for atmospheric mercury in urban Detroit, Michigan. *Atmos. Environ.* **2006**, *40*, 3144–3155.

40. Hidy, G.M. *Atmospheric Sulfur and Nitrogen Oxides: Eastern North America Source-Receptor Relationships*; Academic Press Inc.: San Diego, CA, USA, 1994; p. 447.

41. Pandey, S.K.; Kim, K.H.; Chung, S.Y.; Cho, S.J.; Kim, M.Y.; Shon, Z.H. Long term study of NOx behavior at urban roadside and background locations in Seoul, Korea. *Atmos. Environ.* **2008**, *42*, 607–622.

42. Lee, D.S.; Dollard, G.J.; Pepler, S. Gas-phase mercury in the atmosphere of the United Kingdom. *Atmos. Environ.* **1998**, *32*, 855–864.

43. Han, Y.-J.; Holsen, T.M.; Hopke, P.K.; Cheong, J.-P.; Kim, H.; Yi, S.-M. Identification of source locations for atmospheric dry deposition of heavy metals during yellow-sand events in Seoul, Korea in 1998 using hybrid receptor models. *Atmos. Environ.* **2004**, *38*, 5353–5361.

44. Chen, L.; Liu, M.; Fan, R.; Ma, S.; Xu, Z.; Ren, M.; He, Q. Mercury speciation and emission from municipal solid waste incinerators in the Pearl River Delta, South China. *Sci. Total Environ.* **2013**, *447*, 396–402.

45. Kim, K.H.; Shon, Z.H. Nationwide shift in CO concentration levels in urban areas of Korea after 2000. *J. Hazard Mater.* **2011**, *188*, 235–246.

46. Han, Y.J.; Holsen, T.M.; Hopke, P.K. Estimation of source location of total gaseous mercury measured in New York State using trajectory-based models. *Atmos. Environ.* **2007**, *41*, 6033–6047.

47. Zhang, H.; Lindberg, S.E. Sunlight and iron (III)-induced photochemical production of dissolved gaseous mercury in freshwater. *Environ. Sci. Technol.* **2001**, *5*, 928–935.

48. Siciliano, S.D.; O'Driscoll, N.J.; Lean, D.R.S. Microbial reduction and oxidation of mercury in freshwater lakes. *Environ. Sci. Technol.* **2002**, *36*, 3064–3067.

49. Ahn, M.C.; Kim, B.C.; Holsen, T.M.; Yi, S.M.; Han, Y.J. Factor influencing concentrations of dissolved gaseous mercury (DGM) and total mercury (TM) in an artificial reservoir. *Environ. Pollut.* **2010**, *158*, 347–355.

50. Han, Y.J.; Holsen, T.M.; Lai, S.O.; Hopke, P.K.; Yi, S.M.; Liu, W.; Pangano, J.; Falanga, L.; Milligan, M.; Andolina, C. Atmospheric gaseous mercury concentration in New York State: relationships with meteorological data and other pollutants. *Atmos. Environ.* **2004**, *38*, 6431–6446.

51. Chen, L.; Liu, M.; Xu, Z.; Fan, R.; Tao, J.; Chen, D.; Zhang, D.; Xie, D.; Sun, J. Variation trends and influencing factors of total gaseous mercury in the Pearl River Delta-A highly industrialised region in South China influenced by seasonal monsoons. *Atmos. Environ.* **2013**, *77*, 757–766.

52. Fu, X.; Feng, X.; Wan, Q.; Meng, B.; Yan, H.; Guo, Y. Probing Hg evasion from surface waters of two Chinese hyper/meso-eutrophic reservoirs. *Sci. Total Environ.* **2010**, *408*, 5887–5896.

53. Polissar, A.V.; Hopke, P.K.; Harris, J.M. Source regions for atmospheric aerosol measured at Barrow, Alaska. *Environ. Sci. Technol.* **2001**, *35*, 4214–4226.

54. Polissar, A.V.; Hopke, P.K.; Poirot, R.L. Atmospheric aerosol over Vermont: Chemical composition and sources. *Environ. Sci. Technol.* **2001**, *35*, 4604–4621.

55. Fu, X.; Feng, X.; Sommar, J.; Wang, S. A review of studies on atmospheric mercury in China. *Sci. Total Environ.* **2012**, *421–422*, 73–81.

56. Ashbaugh, L.L.; Malm, W.C.; Sadeh, W.D. A residence time probability analysis of sulfur concentrations at Grand Canyon National Park. *Atmos. Environ.* **1985**, *19*, 1263–1270.

57. Malm, W.C.; Johnson, C.E.; Bresch, J.F. Application of principal component analysis for purposes of identifying source-receptor relationships in receptor methods for source apportionment. In *Receptor Methods for Source Apportionment*; Air Pollution Control Association: Pittsburgh, PA, USA, 1986; pp. 127–148.

58. Han, Y.J. Mercury in New York State: Concentrations and Source Identification Using Hybrid Receptor Modeling. Ph.D. Thesis, Clarkson University, Potsdam, NY, USA, 2003.

Mercury Speciation at a Coastal Site in the Northern Gulf of Mexico: Results from the Grand Bay Intensive Studies in Summer 2010 and Spring 2011

Xinrong Ren, Paul Kelley, Mark Cohen, Fong Ngan, Richard Artz, Jake Walker,
Steve Brooks, Christopher Moore, Phil Swartzendruber, Dieter Bauer,
James Remeika, Anthony Hynes, Jack Dibb, John Rolison,
Nishanth Krishnamurthy, William M. Landing, Arsineh Hecobian,
Jeffery Shook and L. Greg Huey

Abstract: During two intensive studies in summer 2010 and spring 2011, measurements of mercury species including gaseous elemental mercury (GEM), gaseous oxidized mercury (GOM), and particulate-bound mercury (PBM), trace chemical species including O_3, SO_2, CO, NO, NO_Y, and black carbon, and meteorological parameters were made at an Atmospheric Mercury Network (AMNet) site at the Grand Bay National Estuarine Research Reserve (NERR) in Moss Point, Mississippi. Surface measurements indicate that the mean mercury concentrations were 1.42 ± 0.12 ng·m^{-3} for GEM, 5.4 ± 10.2 pg·m^{-3} for GOM, and 3.1 ± 1.9 pg m^{-3} for PBM during the summer 2010 intensive and 1.53 ± 0.11 ng·m^{-3} for GEM, 5.3 ± 10.2 pg·m^{-3} for GOM, and 5.7 ± 6.2 pg·m^{-3} for PBM during the spring 2011 intensive. Elevated daytime GOM levels (>20 pg·m^{-3}) were observed on a few days in each study and were usually associated with either elevated O_3 (>50 ppbv), BrO, and solar radiation or elevated SO_2 ($>$a few ppbv) but lower O_3 ($\sim$20–40 ppbv). This behavior suggests two potential sources of GOM: photochemical oxidation of GEM and direct emissions of GOM from nearby local sources. Lack of correlation between GOM and Beryllium-7 (^{7}Be) suggests little influence on surface GOM from downward mixing of GOM from the upper troposphere. These data were analyzed using the HYSPLIT back trajectory model and principal component analysis in order to develop source-receptor relationships for mercury species in this coastal environment. Trajectory frequency analysis shows that high GOM events were generally associated with high frequencies of the trajectories passing through the areas with high mercury emissions, while low GOM levels were largely associated the trajectories passing through relatively clean areas. Principal component analysis also reveals two main factors: direct emission and photochemical processes that were clustered with high GOM and PBM. This study indicates that the receptor site, which is located in a coastal environment of the Gulf of Mexico, experienced impacts from mercury sources that are both local and regional in nature.

Reprinted from *Atmosphere*. Cite as: Ren, X.; Luke, W.T.; Kelley, P.; Cohen, M.; Ngan, F.; Artz, R.; Walker, J.; Brooks, S.; Moore, C.; Swartzendruber, P.; Bauer, D.; Remeika, J.; Hynes, A.; Dibb, J.; Rolison, J.; Krishnamurthy, N.; Landing, W.M.; Hecobian, A.; Shook, J.; Huey, L.G. Mercury Speciation at a Coastal Site in the Northern Gulf of Mexico: Results from the Grand Bay Intensive Studies in Summer 2010 and Spring 2011. *Atmosphere* **2014**, *5*, 230–251.

1. Introduction

Mercury (Hg) is a ubiquitous and toxic pollutant in the environment. It exists in several distinct chemical and physical forms that dictate to a large degree, the ultimate impact of Hg on the environment. The main pathway of the release of mercury to the environment is through the atmosphere. The release of mercury compounds to the atmosphere, followed by their transport and deposition, often constitutes the main pathway for the global dispersion of mercury and the dominant loading mechanism of new mercury to water bodies and watersheds [1–4]. Human activities, such as smelting and coal burning, have significantly increased mercury levels in the atmosphere, surface soils, fresh waters, and oceans [2,5,6]. Mercury deposits to watersheds and receiving water bodies where it can be converted to methylmercury, a highly toxic form, and, thus, enters the food chain through bioaccumulation [7]. Methylmercury can adversely affect the nervous system, particularly those of fetuses and young children [8]. Human exposure to mercury is primarily from the consumption of contaminated fish and other aquatic organisms [5,8,9].

The mercury in the atmosphere arises from a variety of sources, both anthropogenic and natural [5,10,11]. Direct emissions of gaseous elemental mercury (GEM) and two operationally-defined forms of mercury, gaseous oxidized mercury (GOM), and particulate-bound mercury (PBM), from anthropogenic sources account for the bulk of mercury injected into the atmosphere [11]. GEM is also evaded from soils and the ocean surface; chemical reactions in the atmosphere transform natural and anthropogenic GEM into GOM and PBM species. Thus, it is important to understand where mercury emissions originate, how and where mercury is transported and deposited, and what changes will occur due to emission controls so that policy-makers and regulators can deal effectively with mercury emissions.

Studies have shown that the Gulf of Mexico region is plagued by persistently high total mercury in precipitation [12]. Data from the National Atmospheric Deposition Program's Mercury Deposition Network (MDN) indicate that mercury concentrations in precipitation in the Gulf of Mexico region are some of the highest in the United States [13]. Meanwhile, fish consumption in coastal areas is typically much higher than the national average, and every state along the Gulf of Mexico has widespread fish consumption advisories for mercury. The reasons why mercury deposition in the Gulf of Mexico region is especially high are not entirely clear.

Previous monitoring of atmospheric Hg in the Gulf of Mexico area shows a strong diel pattern for GOM with peaks in the afternoon [12,14]. The elevated GOM levels were attributed to the photochemical oxidation of GEM by atmospheric oxidants, with enhancement of GOM from local emissions [12]. An early morning enhancement of GOM at an urban site in Birmingham, Alabama was also observed and it was attributed to boundary layer processes such as the erosion of the nocturnal inversion and subsequent vertical mixing [14]. Long-term monitoring and intensive studies of mercury speciation are needed to characterize these processes and to assess both regional and global atmospheric budget and cycling of mercury.

In this paper, we present an analysis of two atmospheric mercury intensive studies at a coastal site in the northern Gulf of Mexico (see Figure 1 for the location of the site) in summer 2010 and in spring 2011. The main purpose of this study is to understand processes important to explain the variations in the observed mercury data. Back trajectory simulations and principal component analysis were conducted to try to examine the Hg source-receptor relationships in this costal environment.

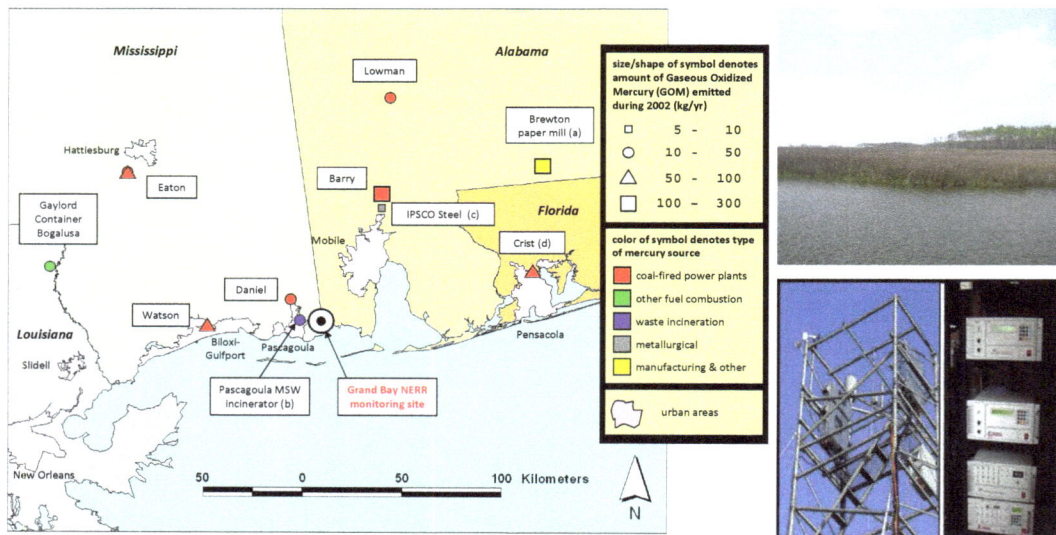

Figure 1. (**Left**) location of the Grand Bay NERR monitoring station, along with large point sources of gaseous oxidized mercury (GOM) in the region, based on the US EPA's 2005 National Emissions Inventory (NEI). (**Right top**) site view from the Grand Bay NERR atmospheric mercury measurement tower. (**Right bottom**) the measurement tower at Grand Bay NERR, and two sets of Tekran mercury speciation units in a climate-controlled shelter adjacent to the tower.

2. Results and Discussion

2.1. Surface Observations

A statistical summary of the measurements at the monitoring site are listed in Table 1 for both intensive studies in summer 2010 and spring 2011. We chose these two particular seasons because of reactive photochemistry in summer and enhanced GOM levels in spring from the historical mercury observations at this site.

2.1.1. Intensive Study in Summer 2010

Meteorological conditions during the 2010 intensive study were intensely hot and humid but largely free of precipitation (Figure 2). In the beginning of the study period, a high-pressure system was dominant in the Gulf of Mexico. A high-pressure system with stagnant conditions can results in reactive photochemical oxidation at the site while a cold front passage can bring polluted air to the site, all of which can significantly affect the mercury observations at the site. On 2 August, as the high-pressure system relaxed a weak front was approaching the coast of Mississippi. Daytime peak temperatures ranged from 30 °C to 36 °C and nighttime minima were about 25 °C, while relative humidity ranged from 45% to 60% during mid-day and typically greater than 80%–90% overnight. Winds were commonly from northerly directions overnight and in the early morning, usually shifting to southerly-southwesterly during the day. Mid-day winds remained largely from the north from 29 July to 9 August, and were generally easterly-southeasterly toward the end of the intensive (10–12 August) as a tropical depression moved into the area, eventually passing to the south and west of the site. Wind speeds were light to moderate, with mid-day wind speeds ranging from 3 m·s^{-1} to 5 m·s^{-1}. Precipitation events were recorded on 3 August, 7 August, 8 August, 11 August, and 12 August.

Table 1. Statistical summary of hourly measurements during the two intensives in summer 2010 (the first number in each cell) and spring 2011 (the second number in each cell).

Parameter	Mean ± Std	Median	Maximum	Minimum
[GEM] (ng·m^{-3})	$1.42 \pm 0.12, 1.53 \pm 0.11$	1.44, 1.53	1.70, 3.12	1.06, 1.07
[GOM] (ng·m^{-3})	$5.4 \pm 10.2, 5.3 \pm 10.2$	1.83, 0.9	70.8, 68.7	0.0, 0.0
[PBM] (ng·m^{-3})	$3.1 \pm 1.9, 5.7 \pm 6.2$	2.7, 3.2	8.8, 37.0	0.0, 0.0
Temperature (°C)	$29.4 \pm 3.0, 21.9 \pm 4.0$	29.2, 23.4	36.3, 27.4	24.3, 8.9
Relative Humidity (%)	$75.1 \pm 14.3, 73.9 \pm 17.1$	77.7, 79.5	97.2, 96.9	41.0, 23.6
Rain (mm, hour)	$0.12 \pm 0.85, 0.023 \pm 0.38$	0, 0	10.2, 9.0	0, 0
Solar Radiation (W·m^{-2})	$258 \pm 328, 266 \pm 332$	34, 45	1037, 983	0, 0
Wind Speed (m·s^{-1})	$2.2 \pm 1.4, 4.7 \pm 2.3$	1.7, 4.8	6.5, 10.8	0.05, 0.02
[O3] (ppbv)	$34.5 \pm 16.5, 38.4 \pm 12.5$	32.6, 36.6	91.0, 71.9	3.1, 9.5
[NO] (ppbv)	$0.24 \pm ,0.43, 0.16 \pm 0.32$	0.08, 0.08	3.07, 3.06	0.028, 0.029
[NOy] (ppb)	$4.13 \pm 2.63, 1.91 \pm 2.36$	3.64, 1.03	18.2, 18.4	0.30, 0.22
[CO] (ppbv)	$155 \pm 35, 139 \pm 26$	156, 141	267, 321	72, 86
[Black Carbon] (μg·m^{-3})	$0.40 \pm 0.23, 0.28 \pm 0.17$	0.39, 0.24	1.59, 1.35	0.03, 0.05

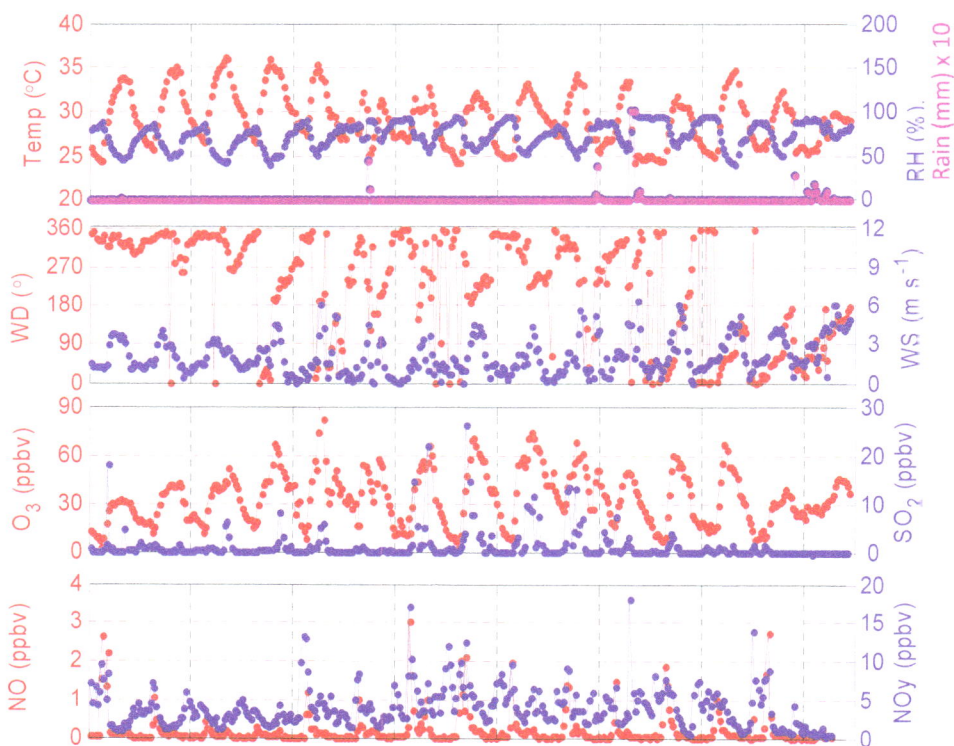

Figure 2. *Cont.*

134

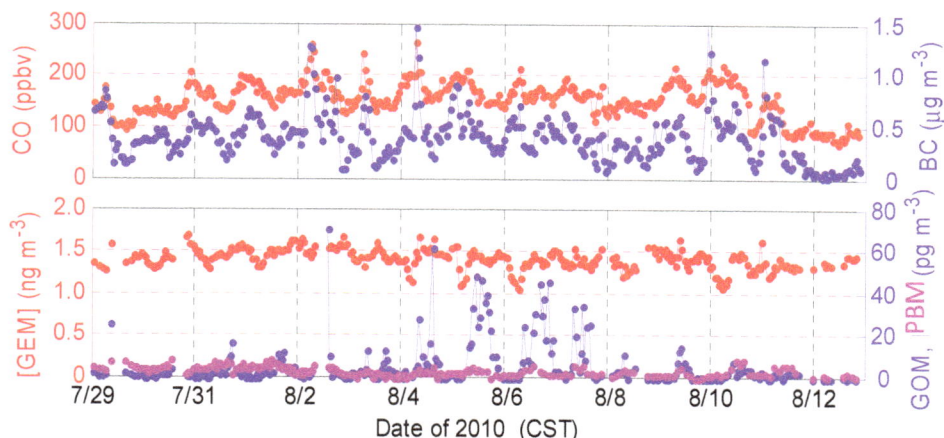

Figure 2. Measurements of meteorological parameters, trace chemical pollutants, and mercury species during the Grand Bay Intensive in summer 2010.

Concentrations of reactive nitrogen compounds (NO, NO_Y) typically peaked during the morning rush hour, when northerly flow transported mobile and stationary source emissions from upwind urban areas to the site. An interstate highway (I-10) and a state highway (Hwy 90) are located ~5 km to the north of the site. NO concentrations were typically less than 2 ppbv in the morning, decreasing to less than 0.2 ppbv during mid-day and to below the detection limit overnight. NO_Y concentrations decreased from about 6–10 ppbv during the morning rush hour (with occasional excursions to greater than 15 ppbv) to less than 5 ppbv at mid-day. Concentrations of CO and black carbon displayed similar behavior, with largest concentrations in the morning (150–250 ppbv for CO, 0.5–1.5 $\mu g \cdot m^{-3}$ for black carbon) and lower levels at mid-day (100–150 ppbv for CO, 0.2–0.4 $\mu g \cdot m^{-3}$ for black carbon). Under the clean easterly-southeasterly flow ahead of the tropical depression on 10–11 August, CO levels dropped to less than 100 ppbv, while NO_Y ranged from 1 ppbv to1.5 ppbv and black carbon levels were typically less than 0.1 $\mu g \cdot m^{-3}$.

Concentrations of SO_2 exhibited a very different behavior. These values were typically, but not always, low during the morning rush hour. Occasionally, the site was likely fumigated by specific upwind industrial sources (e.g., energy generating units or EGUs) to the west and southwest. In the absence of such plume impactions, however, SO_2 in northerly flow was rather low. SO_2 concentrations more often peaked during mid-day, when winds shifted from northerly to southwesterly. Such emissions may have come from the Chevron petroleum refinery plant located approximately 10 km to the southwest of the site.

Concentrations of GEM typically ranged from 1 $ng \cdot m^{-3}$ to 1.7 $ng \cdot m^{-3}$ at standard conditions (*i.e.*, 0 °C and 1 atm) with an average of 1.42 $ng \cdot m^{-3}$ and a standard deviation of 0.12 ng·m^{-3} during the intensive and occasionally reached very low concentrations (approximately 1 ng·m^{-3}) in the early morning hours. GEM exhibited little dependence on wind direction, and no discernible diurnal pattern, which is consistent with observations at a few other sites in the region [14]. GOM concentrations were below the detection limit at night, only a few pg·m^{-3} in the morning hours, and peaked typically around 10–20 pg·m^{-3} by midday. The mean GOM concentration for the entire 2010 intensive was 5.4 pg·m^{-3} with a standard deviation of 10.2 pg·m^{-3}. Much higher mid-day peaks in the range of 20–60 pg·m^{-3} were observed on 2, 4, 5, 6, and 7 August. Highest concentrations were observed in winds from 190 to 330 degrees (south/southwest to north/northwest. Concentrations of particulate-bound mercury (PBM) were low, ranging from 4 pg·m^{-3} to 8 pg·m^{-3} at mid-day with a mean concentration of 3.1 pg·m^{-3} and a standard deviation of 1.9 pg·m^{-3} for the entire 2010 intensive.

Interestingly, on 4–6 August 2010, a decrease of GEM (from ~1.4 to 1.1–1.2 ng·m^{-3}) was observed in the morning followed by GOM peaks in the afternoon. Similar behavior was observed on a few days during the spring intensive study in 2011, as discussed in Section 2.1.2.

2.1.2. Intensive in Spring 2011

During the 2011 intensive study, meteorological conditions were typical for spring with mild temperatures and infrequent precipitation during most of the study. Frontal activity was confined largely to the north of the Grand Bay region so the site area was dominated by southerly flows (Figure 3). Under southerly flow, temperatures ranged from 20 °C to 25 °C with little variations from day to day or over the diurnal cycle. On 16 April, 28 April, and 3 May, cold fronts passed through the region, bringing continental cold air masses to the monitoring station. Those days experienced typical post-frontal conditions—dry air, low night-time temperatures, and light northeasterly winds in the morning with southerly sea breezes in the afternoon. Wind speeds ranged from 0 m·s^{-1} to 10 m·s^{-1}. Precipitation events were observed on 16 April, 26 April, and 3 May. Temperature dropped to as low as 10 °C after the cold front passage on 3 May.

The average GEM concentration during the 2011 intensive was 1.53 ng·m^{-3} with a standard deviation of 0.11 ng·m^{-3}, which is slightly higher than values measured during the summer 2010 intensive and can be explained by the seasonal variations as observed by some other studies in the region [12,14]. GEM exhibited little variation with no distinct dependence on wind direction. The mean concentration of GOM was 5.3 pg·m^{-3} with a standard deviation of 10.2 pg·m^{-3}. Elevated GOM levels were observed on 17 April, 29 April and 4–9 May, with peak values in a range of 30–70 pg·m^{-3} (Figure 3). The mean concentration of PBM was 5.7 pg·m^{-3} with a standard deviation of 6.2 pg·m^{-3} for the entire spring 2011 intensive.

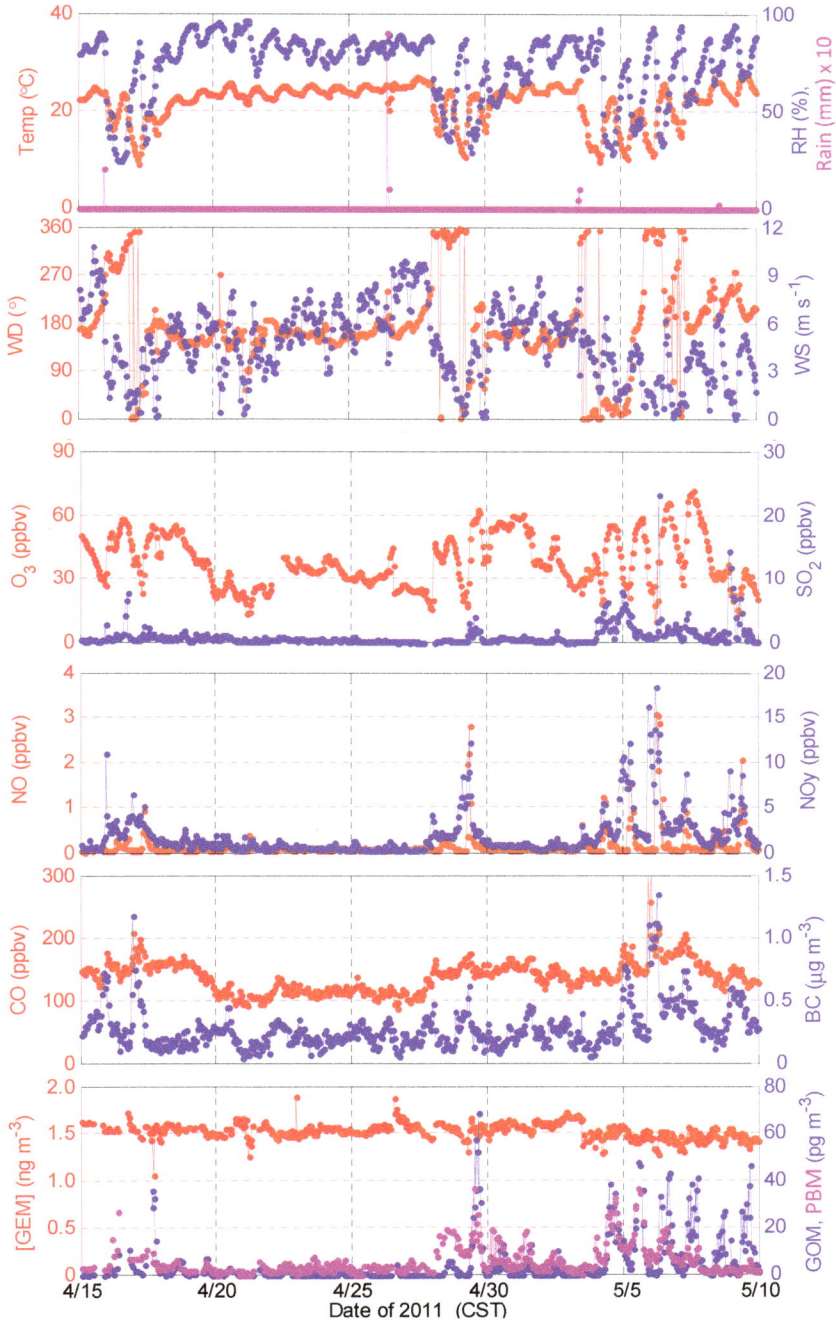

Figure 3. Measurements of meteorological parameters, trace chemical pollutants, and mercury species during the Grand Bay Intensive in spring 2011.

One interesting point to note is the period after 3 May when GEM dropped suddenly from ~1.6 to 1.4 ng·m^{-3} and remained lower while GOM and PBM were high for the next few days. At the same time, NOy, CO, BC, SO$_2$, and NO were also enhanced during this period. It seems that both enhanced transport of pollution and more active photochemistry were observed during this period. We note that on 3 May, the wind shifted from south to north at the same time when GEM concentration dropped and it rained for a few hours. It is most likely that an air mass change behind the cold front is responsible for the change in GEM concentration.

Similar to our observations in summer 2010, a decrease of GEM in the morning followed by a GOM peak in the afternoon was observed on 17 April, 29 April, and, to a lesser extent, on 4–7 May. The reasons for this GEM decrease are not clear and require more investigation. Similar GEM decrease events have been observed in middle latitude regions [12,15,16]. For example, [16] found nearly 100% depletion of GEM, suggesting that the residence time of GEM can be as short as hours to days under those conditions [16]. The GEM depletion events in the two intensive studies were usually associated with high relative humidity. Similar to what was observed by [16], the GEM decreases we observed are not accompanied by simultaneous depletion of ozone, which distinguishes them from the halogen driven atmospheric mercury depletion events (AMDEs) observed in polar regions [17,18] and other areas [19]. We also found that the decrease of GEM we observed occurred typically in the morning before sunrise when relative humidity was typically the highest of the day, which is consistent with observations of [12] and [14], but different from the afternoon events observed by [16]. We suspect that heterogeneous processes might be responsible for the decrease of GEM we observed and additional measurements of possible mercury oxidants are hence called for to reveal the chemical mechanism to assess its importance on larger scales. Similar GEM depletion events have also been observed at a suburban site in the mid-Atlantic US [20].

Concentrations of reactive nitrogen compounds (NO and NO$_Y$), CO, SO$_2$, and black carbon were generally low from 18 April to 27 April and from 30 April to 3 May, when southerly winds dominated and brought clean marine air masses to the site. During frontal passage periods when the wind shifted from southerly to northerly, transport of mobile and stationary source emissions from upwind urban areas to the site occurred, with hourly averaged NO$_Y$ concentrations from a few ppbv up to 18 ppbv, and SO$_2$ concentrations from a few ppbv up to 23 ppbv, and elevated CO concentration up to 320 ppbv.

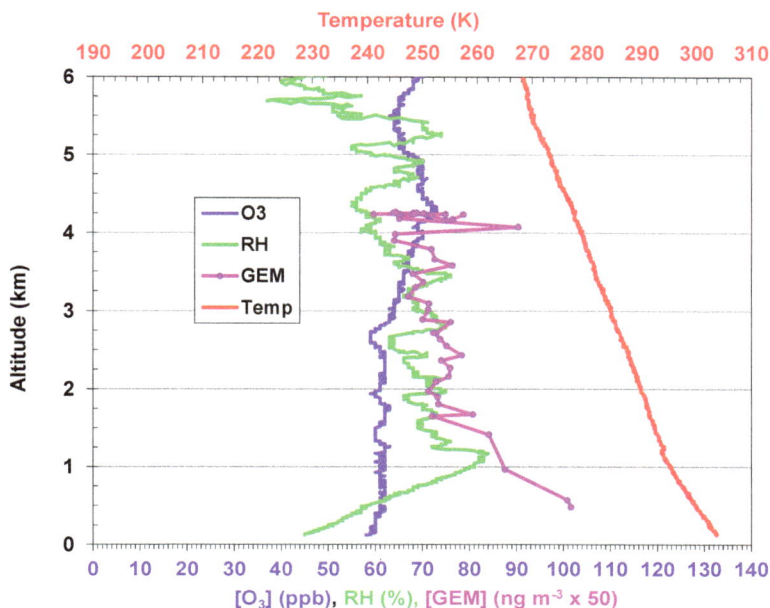

Figure 4. (**Left**) Time series of GEM and altitude during the flight on 6 August 2010. (**Right**) vertical profiles of aircraft GEM concentration and ozonesonde data, including ozone, relative humidity, and temperature. The ozonesonde was launched at 10:55 CST on 6 August 2010.

2.2. Aircraft and Ozonesonde Measurements

Vertical profiles of GEM, GOM, ozone, SO_2, and condensation nuclei (CN) were measured during four flights in the summer 2010 study and 11 flights in the spring 2011 study. Details of these measurements will be presented elsewhere (Hynes *et al.*, manuscript in preparation). Figure 4 shows an example of the aircraft GEM measurement as well as ozonesonde data on 6 August 2010. In the free troposphere between 2 km and 4 km, GEM concentrations were relatively constant, ranging between 1.3 ng·m^{-3} and 1.5 ng·m^{-3}. GEM concentrations increased from ~1.5 to 2 ng·m^{-3} when the airplane flew into the mixing layer with its top height of about 1.5 km. A high-pressure system with stagnant conditions (low wind speeds of 0–10 m·s^{-1} from surface to ~10 km) was dominant in the area on this day after a weak cold front passage on 2 August. The ozonesonde was launched at 10:55 Central Standard Time (CST) on this day and it indicated the mixing layer height was about 1.3 km. A slightly higher mixing layer was observed by the aircraft about an hour later during its descent prior to landing, indicating the mixing layer was still rising in height and did not reach its maximum until around 13:00–14:00 CST in this area.

139

2.3.1. Correlation among GOM, O_3, SO_2, and BrO

Results from the spring 2011 intensive study show that elevated GOM values during the day were usually associated with two different sets of chemical and physical conditions, as shown by two groups of data points in Figure 5. For the data points in Group #1, GOM concentrations are positively correlated with O_3 in air masses associated with high O_3 levels (>50 ppbv) and high solar radiation (>500 $W \cdot m^{-2}$). Lower O_3 levels (20–40 ppbv) but elevated SO_2 levels (> a few ppbv) were observed for the data points in Group #2. Similar correlations between GOM and O_3/SO_2 have been observed in two previous studies in the Gulf of Mexico region [12,14]. This is consistent with two possible processes that could lead to elevated GOM levels: (1) photochemical conversion of GEM under conditions when ozone levels are high during the midday, and (2) direct emissions of GOM from local emission sources in which high SO2 levels were present.

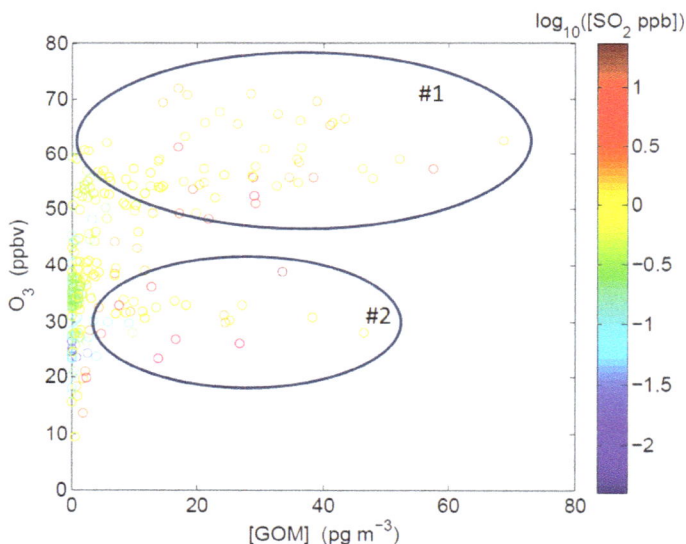

Figure 5. Hourly averaged ozone *versus* GOM color-coded with $\log_{10}([SO_2])$ during daytime in the spring 2011 intensive.

The diurnal variations of GOM, ozone, and SO_2 can also be used to differentiate between direct emissions (with narrow plumes of SO_2 and GOM fumigating the site, leading to short-term spikes) and photochemistry (with longer term increases during midday), for example, on 5 August 2010 around 9:30 CST, simultaneous SO_2 and GOM spikes were observed (Figure 6). After 10:30 CST, the SO_2 level decreased sharply and remained low for the remainder of the day. GOM dropped from a peak after 10:30 CST, but increased again after 11:30 CST and remained elevated for the rest of the day before decreasing in the late afternoon and early evening. Similar variations of SO_2, GOM, BrO, and ozone levels were also observed on 6 May 2011 (Figure 6). Even though we cannot completely rule out transport from regional emissions as a source of GOM during the afternoon periods on these two days, these emissions would have to be large and widespread to produce GOM perturbations of the observed duration and amplitude, and by inference over a broadly diluted plume if regional sources are significantly involved. This suggests that the two different GOM production processes (*i.e.*, direct emissions and photochemical oxidation) can happen on the same day under certain conditions at this site.

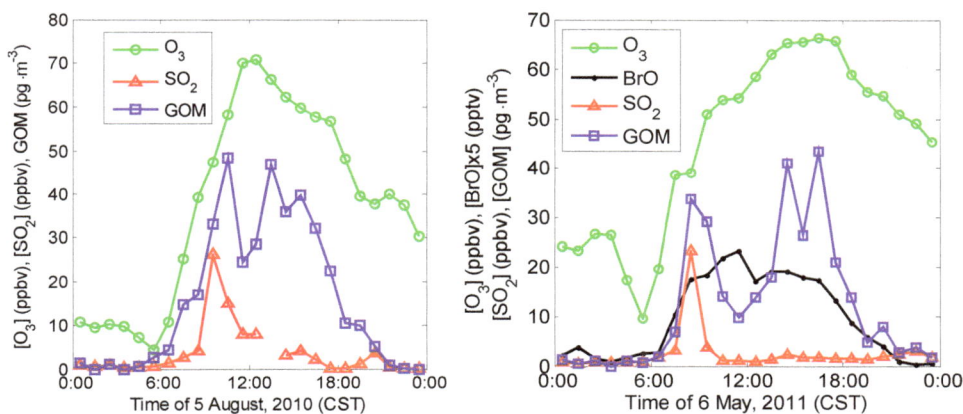

Figure 6. Diurnal variations of ozone, SO_2, and GOM observed on 5 August 2010 (**Left**) and 5 May 2011 (**Right**).

Enhanced levels of GOM (~60–80 pg·m^{-3}) and SO_2 (~20–30 ppbv) were also observed in winds from the west and southwest, as shown by the wind rose plots in Figure 7, further confirming intermittent sources of SO_2 and GOM shown in Figure 6 from the nearby petroleum refinery located in Pascagoula, Mississippi, ~10 km to the southwest of the monitoring site. The slightly elevated SO_2 and GOM levels are most likely due to source emissions, e.g., from the coal-fired power plant (Daniel) (Figure 1) located to the northwest of the site.

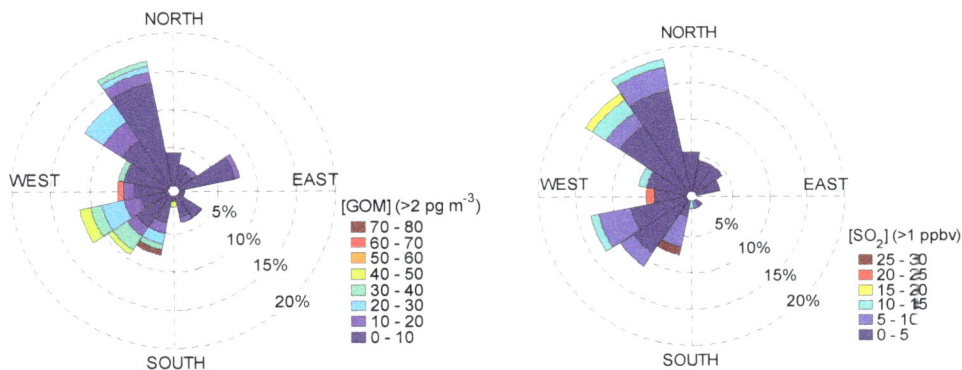

Figure 7. Wind rose plots for GOM with [GOM] > 2 pg·m^{-3} (**Left**) and SO$_2$ with [SO$_2$] > 1 ppbv (**Right**) during the summer 2010 campaign.

2.3.2. Correlation between GOM/GEM and CO

Correlation between GOM and CO in both studies reveals that the highest GOM levels were observed in air masses with CO concentrations centered at ~150 ppbv (Figure 8). This indicates that the high GOM levels observed during the studies likely existed in air masses associated with continental emissions, as opposed to marine-associated air masses, where CO concentrations are usually close to or below ~100 ppbv. It is possible that any GOM produced in marine air would be quickly deposited to the ocean surface or adsorbed onto sea salt aerosols. This is consistent with elevated PBM concentrations observed in the air masses transported from south during the spring 2011 intensive. It is also interesting to note that GOM concentrations were low in the strongest CO plumes, typically encountered in the early morning or at night when solar radiation was close to zero and photochemical processes were not active.

No significant correlation between CO and GEM was observed in either of the intensive studies (Figure 8), indicating that vehicle emissions are not a significant source of GEM. In addition, lower GEM concentrations were observed in summer 2010 than in spring 2011 (Figure 8). This is consistent with the seasonal GEM variation at most sites of northern mid-latitudes (e.g., [14]).

142

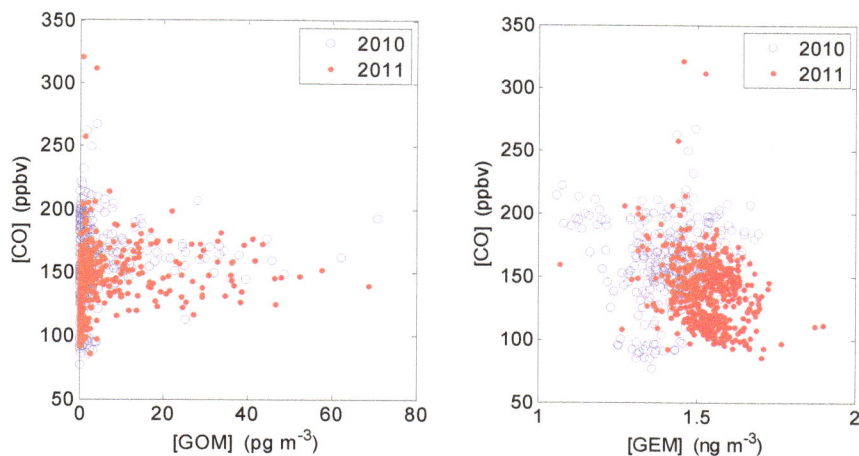

Figure 8. Correlation between CO and GOM (**Left**) and between CO and GEM (**Right**) during the summer 2010 intensive (blue circles) and the spring 2011 intensive (red dots).

2.3.3. GOM/PBM and Humidity

Strong negative correlation between absolute humidity (water volume mixing ratio) and GOM or PBM was observed in both the 2010 and 2011 intensive studies (Figure 9), with better correlation for PBM than for GOM, especially in spring 2011 (e.g., water volume mixing ratio *versus* PBM: $r^2 = 0.63$, (n = 332) for the spring 2011 intensive). We use water volume mixing ratio rather than relative humidity (RH) as the latter is influenced by temp: even at similar water volume mixing ratio RH sometimes approached 100% simply because of lower temperatures in the early morning. This indicates high moisture in the air could scavenge both GOM and PBM into particles or the ground surfaces in this environment. During the 2010 intensive the water volume mixing ratio stayed greater than 2%, the correlation between GOM and humidity was poor. Significant negative correlation was also found at two different sites in the same region [14].

2.3.4. Correlation between GOM/Ozone and Beryllium-7

Beryllium-7 (^{7}Be) is a cosmogenic radionuclide produced by spallation reactions of cosmic rays with N_2 and O_2. Because ^{7}Be is predominantly produced in the stratosphere and upper troposphere and high levels of GOM can be produced via oxidation of GEM in the upper troposphere and observed at some high elevation sites (e.g., [21]), it is reasonable to assume that high concentrations GOM and ^{7}Be observed at the surface may be due in part to transport from the upper troposphere and lower stratosphere.

143

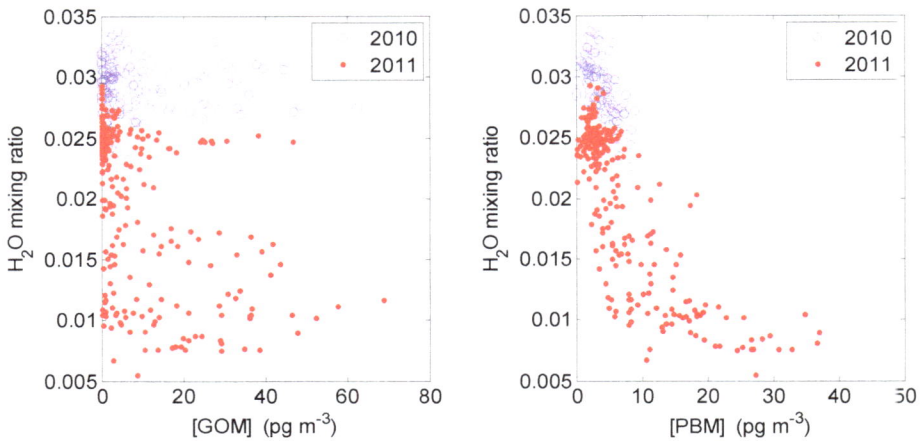

Figure 9. (**Left**) hourly averaged water volume mixing ratio *versus* GOM concentration during the 2010 intensive (blue circles) and the 2011 intensive (red dots). (**Right**) hourly averaged water volume mixing ratio *versus* PBM concentration during the 2010 intensive (blue circles) and the 2011 intensive (red dots). Data collected during the day with solar radiation greater than 10 W·m^{-2} are plotted.

It was surprising to find that there was little correlation between GOM and Beryllium-7 (^{7}Be) when the data plotted for the entire spring 2011 intensive (Figure 10) or considered only during post-frontal passage periods. HYSPLIT back trajectories suggest that downward mixing occurred during these post-frontal passage periods, with air masses transported from 2 km to 3 km to surface within a course of 2–3 days. This indicates that the downward mixing of air masses from aloft possibly with high GOM levels had little influence on surface GOM concentrations, although we did observe a slight positive correlation ($r^2 = 0.38$, n = 26) between O_3 and ^{7}Be during the nighttime (Figure 9), which may indicate a common source (*i.e.*, transport from the upper troposphere) for ozone and ^{7}Be. While ^{7}Be can be considered to be a stratospheric tracer [22,23], many studies have found that surface ^{7}Be concentrations can be highly affected by local meteorological variables and solar activities given the special characteristics of different sample sites [24,25]. Rain/washout is one of the reasons to complicate the GOM levels in the atmosphere and partially responsible for the poor correlation between GOM and ^{7}Be.

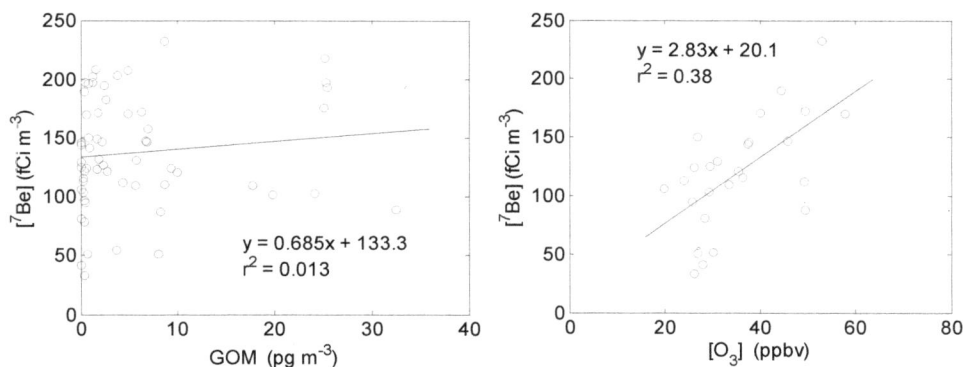

Figure 10. Correlation between [7]Be and GOM for the entire study (**Left**) and between [7]Be and ozone for nighttime periods when the solar radiation was less than 200 W·m^{-2} (**Right**) during the spring 2011 intensive. Hourly GOM and ozone measurements were averaged based on the time periods when [7]Be samples were collected.

2.4. Back Trajectory Frequency Analysis

In spring 2011, trajectory frequency analysis shows that periods with [GOM] < 2 pg·m^{-3} were largely associated with air masses coming from Gulf of Mexico (southeasterly), while events with [GOM] > 20 pg·m^{-3} were typically associated with frequent transport passing over the areas with high mercury emissions, mainly from power plants (Figure 11). A similar signature of back trajectory frequency was observed in Hg isotopes observed during the spring 2011 intensive [26]. Trajectory frequency analysis was also conducted for the summer 2010 intensive, but no clear trends arose, possibly due to stagnant conditions and variable light winds during this study. There were only a few short periods with wind speeds greater than 5 m·s^{-1} (Figure 2).

2.5. Principal Component Analysis

Principal component analysis (PCA) can be used to identify potential source-receptor correlationship at this site. PCA is a mathematical technique that reduces the dimensions of a data set based on covariance of variables, and has been applied to source identification in numerous air quality studies [27–30]. Our principal component analysis reveals that during the 2010 intensive, the measurements with highest loadings in the first three principal components include: GEM, O$_3$ and temperature in the first principal component, GOM and SO$_2$ in the second principal component, and PBM, NO and black carbon in the third principal component. During the 2011 intensive study, the first principal component includes GOM, O$_3$ and BrO; the second principal component includes GEM, CO and black carbon; and

145

the third principal component includes GEM and SO_2. This analysis reveals two possible factors that were clustered with GEM, GOM, and PBM: direct emissions and photochemical process. This study indicates that the receptor site, which is located in a coastal environment of the Gulf of Mexico, experienced impacts from mercury sources that are both local and regional in nature.

With these PCA results, we can qualitatively interpret the Hg source-receptor relationship at this site. High SO_2 events could be associated with plumes from nearby point sources such as power plants and refineries, resulting in the direct emission of GOM. High GOM and PBM events were typically observed in the passage of cold fronts, indicating the impact of northerly flow at the site, which brings in emissions from upwind sources with high GOM and PBM. Such episodes were also accompanied by high O_3 and ambient temperatures, and likely reflect a photochemical origin of GOM. High black carbon, NO and CO cases are indicators for biomass burning, industrial and mobile source activities.

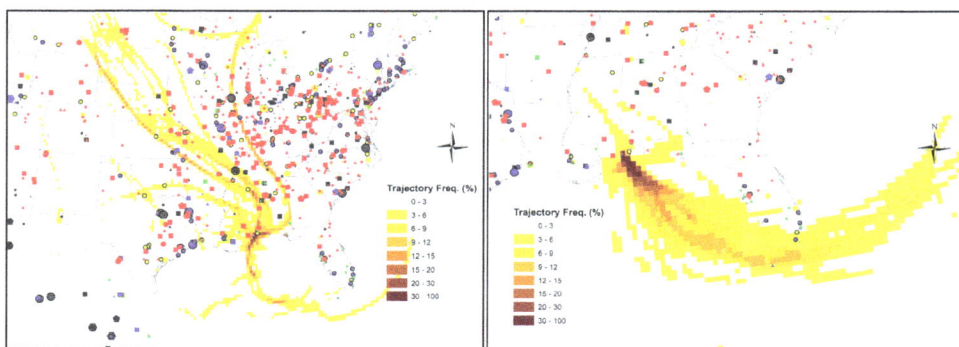

Figure 11. Back trajectory frequencies of high GOM (>20 pg·m^{-3}, (**Left**)) and low GOM (<2 pg·m^{-3}, (**Right**)) events during the spring 2011 intensive study. The color-coded trajectory frequency at each grid represents the percentage of trajectories passing through the gridded area. Symbols represent the major mercury emission sources with source types and strengths as the same as in Figure 1.

3. Experimental Section

3.1. Site Description

The monitoring station is located at the NOAA Grand Bay National Estuarine Research Reserve (NERR) in Moss Point, Mississippi (30.412°N, 88.404°W), one of the National Atmospheric Deposition Program's Atmospheric Mercury Network (AMNet) sites (site ID: MS12). The reserve is located in Southeastern Mississippi between Pascagoula, Mississippi, and the Alabama state line. The area contains a variety of wetland habitats, both tidal and non-tidal, as well as terrestrial habitats that are unique to the coastal zone, such as maritime forests of pine savannah (individual pine trees in mostly swampy grassland). The location of the monitoring site and major regional point sources of GOM are shown in Figure 1. The site is located about 5 km from the waters of Grand Bay, and approximately 30 km from the open waters of the Gulf of Mexico.

A 10-m walk-up tower was established at the edge of a marsh grass/bayou on the grounds of the Grand Bay NERR, while all chemical analyzers were housed in a climate-controlled shelter adjacent to the tower. Measurements of speciated Hg, ancillary chemical species, and meteorological parameters were made from the top of the tower. The two intensive studies were conducted from 29 July to 12 August 2010, and from 15 April to 9 May 2011.

3.2. Experimental Description

3.2.1. Surface Measurements

Two Tekran speciation systems (Tekran Instrument Corporation, Ontario, Canada) were used to measure GEM, GOM and PBM. Each system uses a Tekran 1130/1135 speciation unit coupled with a Tekran 2537 Cold Vapor Atomic Fluorescence Spectrometer (CVAFS). Details about the system are described by [31] and [32]. The standard protocols for AMNet were followed for data collection and reduction [21]. Briefly, as ambient air flows through the system, GOM is collected on a KCl-coated annular denuder followed by the collection of PBM (with particle diameter < 2.5 μm) on a regenerable quartz filter (RPF) and GEM on gold traps. The collected GOM on the denuder and PBM on the quartz filter are then thermally desorbed and analyzed by the Tekran 2537 as GEM. Every two hours each Hg speciation system provides 12 consecutive GEM measurements with a 5-min temporal resolution and one measurement of GOM and PBM in 1-hour integration time. Because air is only sampled during the first hour of the 2-hour period, the two speciation systems, operating out-of-phase by one hour, provided truly continuous measurements of atmospheric mercury speciation. The two collocated mercury speciation systems at the site also provided quality control and quality assurance information by comparing the concentrations measured with the two systems. The agreement between the two systems was good for GEM (with a slope of 1.05 ± 0.06 for 2010 and a slope of 0.99 ± 0.04 for 2011), GOM (with a slope of 0.98 ± 0.03 for 2010 and a slope of 0.81 ± 0.03 for 2011), and PBM (with a slope of 1.19 ± 0.05 for 2010 and a slope of 1.10 ± 0.03 for 2011).

The use of KCl-coated denuders, either as part of the automated Tekran speciation system or using manual analysis, is the approach that is currently accepted as the standard method for the measurement of GOM. We note, however,

147

that in recent work, Jaffe and co-workers have suggested that this approach does not quantitatively measure GOM and is biased low [33,34]. In a limited set of measurements during the RAMIX intercomparison, we used a laser induced fluorescence (LIF) technique [35] to simultaneously measure both ambient GEM and total mercury (TGM). The total mercury concentrations were obtained by converting GOM to GEM by pyrolysis. Our GOM concentrations, obtained from the difference between the GEM and TGM measurements, were in much better agreement with the measurements reported by the two Tekran speciation systems deployed at RAMIX compared with the DOHGS instrument deployed by [33] that also pyrolyzes the sample to calculate TGM. In this work we are assuming that the KCl denuder approach gives a quantitative measurement of GOM. It is clear that this issue of the discrepancy in the GOM measurements requires further investigation.

Measurements of ancillary chemical species, including ozone, NO, NO_Y, CO, SO_2, and black carbon were made with modified commercial analyzers (Thermo Fisher Scientific, Waltham, MA). Details of the trace gas measurements may be found in [36]. BrO was measured by chemical ionization mass spectrometry (CIMS) [37]. Meteorological sensors provided continuous measurements of temperature, pressure, relative humidity, wind speed and direction, precipitation, and solar radiation. Precipitation collectors gathered weekly rainfall for subsequent analysis of total and methyl mercury, major ions, and trace metals.

Activities of Beryllium-7 (^{7}Be) and lead-210 (^{210}Pb) were determined during the 2011 intensive by nondestructive gamma spectroscopy as described previously [38,39]. To collect ^{7}Be sample, aerosol particles were collected onto Whatman GF/A glass fiber filters using a high-volume (Hi-Vol) sampler in an open field next to the Grand Bay NERR building, which is located about 3 km to the northwest of the monitoring site. Either 24-h or 12-h Hi-Vol samples were collected during the spring 2011 intensive. Samples were usually analyzed for ^{7}Be radioactivity within a week of collection.

3.2.2. Aircraft Measurements

Measurements of GEM, ozone, SO_2, and condensation nuclei (CN) were made aboard the University of Tennessee Space Institute (UTSI) Piper Navajo aircraft during the research intensives. The airplane was based at Trent Lott Regional Airport, Moss Point, MS, about 14 km to the northwest of the monitoring site. The trace-gas and meteorological instrumentation was fully automated, and ran largely unattended on each flight. Meteorological parameters (temperature, pressure, relative humidity, solar radiation) were measured as part of the aircraft's standard instrumentation package. Water vapor was measured with a chilled mirror hygrometer. Concentrations of ozone (O_3) and sulfur dioxide (SO_2) were measured at 1s time resolution with modified commercial sensors (Thermo Fisher Scientific) equipped with custom electronics. Particles with diameter >0.014 μm were measured with a TSI Incorporated (Shoreview, MN) Model 3760 Condensation Nucleus Counter. Further descriptions of the instrumentation may be found in [40].

GOM was collected on KCl-coated denuders and uncoated quartz tubes, followed by thermal desorption and analysis with a Tekran Model 2537 ambient mercury vapor analyzer for the KCl-coated denuders and with a laser induced fluorescence (LIF) techniques for uncoated quartz tubes [41],. Sampling was conducted at flight altitudes ranging from the surface to 4.5 km above mean sea level (MSL). By characterizing the burden of primary and secondary trace gas and aerosol pollutants in the lower and middle troposphere, GEM and GOM measurements made from the Navajo can be interpreted.

3.2.3. Ozonesonde Launches

Ozonesondes were launched at the Grand Bay NERR monitoring site on several days. The GPS-enabled radiosonde, with FM-band data telemetry and an optional electrochemical cell ozonesonde (Ensci Corporation (now Droplet Measurement Technology), Boulder, CO, USA), transmitted data to a ground receiving station in the site trailer. The following variables were measured from the surface to the burst altitude of the sonde, typically above 30,000 meters: temperature, pressure, relative humidity, wind speed, wind direction, and O_3 mixing ratio.

3.3. HYSPLIT Back Trajectory Model

Five-day back trajectory simulations were conducted for high (>20 pg·m^{-3}) and low (<2 pg·m^{-3}) GOM events observed at the Grand Bay NERR monitoring site to establish the transport history of the associated air masses and source-receptor co-relationships. The back trajectories were simulated using the NOAA Hybrid Single-Particle Lagrangian Integrated Trajectory model (HYSPLIT, v4.9) [42] and high resolution meteorological data simulated using the WRF-ARW model (Version 3.2, [43]), with a horizontal resolution of 4 km and a time resolution of 3 h. Trajectories were initialized from the Grand Bay surface at the middle point of the mixing layer for the hours when the high and low GOM were observed.

3.4. Principal Component Analysis

Principal component analysis (PCA) was applied to try to identify potential source-receptor co-relationship at this site. A data matrix was first constructed by using 11 surface observations, including GEM, GOM, PBM, NO, CO, SO_2, O_3, BrO, black carbon, air temperature, and relative humidity. A MATLAB function of principal component analysis was then used to calculate the principal coefficients and scores. The original data were standardized and filtered for data points with solar radiation greater than 10 W·m^{-2} to represent daytime data only.

4. Conclusions

The two mercury intensive studies at Grand Bay, Mississippi in summer 2010 and spring 2011 show that the monitoring site typically exhibits rural/remote characteristics with generally low concentrations of anthropogenic chemical species, but with occasional transport-related episodes with higher concentrations. Measured GEM concentrations exhibited little variation, little or no dependence on wind direction, and no discernible diurnal pattern. PBM had more transport related episodes and a modest diurnal profile. GOM exhibits a more pronounced diurnal profile. Diurnal profiles of GOM show increases in daytime, coincident with O_3, BrO, and SO_2 peaks, illustrating the importance of photochemical production of oxidized mercury and direct emissions from local sources. Elevated GOM levels are associated with dryer air, characteristic of continental emissions ([CO] *ca* 150 ppbv). These results suggest GOM is transported from northerly continental sources following cold-frontal penetration in spring. There was no evidence of strong, substantial GOM production or transport in marine air masses.

Back-trajectory analysis of enhanced GOM events suggests that GOM concentrations at the sites are influenced episodically by local and regional sources, while low GOM levels were largely associated the trajectories passing through relatively clean areas. Principal component analysis reveals two main factors: direct emissions and photochemical processes that were clustered with high GOM and PBM. This study indicates that the receptor site which is located in a coastal environment of the Gulf of Mexico experienced impacts from mercury sources that are both local (within ~50 km) and regional (within ~50 km) in nature. Further modeling studies on atmospheric mercury in this region would be required to provide source-attribution information and estimated impacts of alternative future climate scenarios, while measurements from intensive studies as these can then be used to evaluate model performance.

Acknowledgments: The authors thank the NOAA Grand Bay NERR for cooperation in facilitating the field studies and the UTSI flight crew for their dedicated work to make the airborne measurements successful. This study was funded by NOAA (NA09OAR4600193 and NA10OAR4600209). Support for this research was also partially provided by the Cooperative Institute for Climate and Satellites agreement funded by NOAA's Office of Oceanic and Atmospheric Research under a NOAA Cooperative Agreement.

Author Contributions: Xinrong Ren wrote the majority of the manuscript and performed much of the data analysis. Winston Luke and Paul Kelley worked on data processing and quality control. Xinrong Ren, Winston Luke, Paul Kelley, Jake Walker, Steve Brooks, Christopher Moore, Phil Swartzendruber, Dieter Bauer, James Remeika, Anthony Hynes, Jack Dibb, John Rolison, Nishanth Krishnamurthy, William M. Landing, Arsineh Hecobian, Jeffreery Shook and L. Greg Huey collected data in the field intensives. Jack Dibb performed analysis of [7]Be samples and contributed valuable scientific insight and editing. Mark Cohen performed HYSPLIT trajectory simulations and provided scientific insight and editing.

Fong Ngan provided high-resolution WRF-ARM meteorological data used for HYSPLIT trajectory simulations.

Conflicts of Interest: The authors declare no conflict of interest.

References and Notes

1. Mason, R.; Fitzgerald, W.F.; Morel, F.M. The biogeochemical cycling of elemental mercury: Anthropogenic influences. *Geochim. Cosmochim. Act.* **1994**, *58*, 3191–3198.
2. Schroeder, W.H.; Munthe, J. Atmospheric mercury—An overview. *Atmos. Environ.* **1998**, *32*, 809–822.
3. Fitzgerald, W.F.; Engstrom, D.R.; Mason, R.P.; Nater, E.A. The case for atmospheric mercury contamination in remote areas. *Environ. Sci. Technol.* **1998**, *3*, 1–7.
4. Lin, C.-J.; Pehkonen, S.O. The chemistry of atmospheric mercury: A review. *Atmos. Environ* **1999**, *33*, 2067–2079.
5. Selin, N.E. Global Biogeochemical Cycling of Mercury: A Review. *Ann. Rev. Environ. Res.* **2009**, *34*, 43–63.
6. AMAP/UNEP. *Technical Background Report for the Global Mercury Assessment 2013*; Arctic Monitoring and Assessment Programme, Oslo, Norway/UNEP Chemicals Branch: Geneva, Switzerland, 2013.
7. Morel, F.M.M.; Kraepiel, A.M.L.; Amyot, M. The chemical cycle and bioaccumulation of mercury. *Annu. Rev. Ecol. Syst.* **1998**, *29*, 543–566.
8. Choi, A.L.; Grandjean, P. Methylmercury exposure and health effects in humans. *Environ. Chem.* **2008**, *5*, 112–120.
9. Sunderland, E. Mercury exposure from domestic and imported estuarine and marine fish in the U.S. Seafood Market. *Environ. Health Perspect.* **2007**, *115*, 235–242.
10. Gustin, M.S.; Lindberg, S.E.; Weisberg, P.J. An update on the natural sources and sinks of atmospheric mercury. *Appl. Geochem.* **2008**, *23*, 482–493.
11. UNEP. Global Mercury Assessment 2013: Sources, Emissions, Releases and Environmental Transport. UNEP Chemicals Branch: Geneva, Switzerland, 2013. Available online: http://www.unep.org/PDF/PressReleases/GlobalMercury Assessment2013.pdf (accessed on 25 February 2014).
12. Engle, M.A.; Tate, M.T.; Krabbenhoft, D.P.; Kolker, A.; Olson, M.L.; Edgerton, E.S.; DeWild, J.F.; McPherson, A.K. Characterization and cycling of atmospheric mercury along the central U.S. Gulf Coast. *Appl. Geochem.* **2008**, *23*, 419–437.
13. National Atmospheric Deposition Program's Mercury Deposition Network (MDN), Monitoring Mercury Deposition: A Key Tool to Understanding the Link between Emissions and Effects. Available online: http://nadp.sws.uiuc.edu/lib/brochures/ mdn.pdf (accessed on 15 February 2014).
14. Nair, U.S.; Wu, Y.; Walters, J.; Jansen, J.; Edgerton, E.S. Diurnal and seasonal variation of mercury species at coastal-suburban, urban, and rural sites in the southeastern United States. *Atmos. Environ.* **2012**, *47*, 499–508.

151

15. Weiss-Penzias, P.; Jaffe, D.E.; McClintick, A.; Prestbo, E.M.; Landis, M.S. Gaseous elemental mercury in the marine boundary layer: evidence for rapid removal in anthropogenic pollution. *Environ. Sci. Technol.* **2003**, *37*, 3755–3763.

16. Brunke, E.-G.; Labuschagne, C.; Ebinghaus, R.; Kock, H.H.; Slemr, F. Gaseous elemental mercury depletion events observed at Cape Point during 2007–2008. *Atmos. Chem. Phys.* **2010**, *10*, 1121–1131.

17. Schroeder, W.H.; Anlauf, K.G.; Barrie, L.A.; Lu, J.Y.; Steffen, A.; Schneeberger, D.R.; Berg, T. Arctic springtime depletion of mercury. *Nature* **1998**, *394*, 331–332.

18. Ebinghaus, R.; Kock, H.H.; Temme, C.; Einax, J.W.; Löwe, A.G.; Richter, A.; Burrows, J.P.; Schroeder, W.H. Antarctic springtime depletion of atmospheric mercury. *Environ. Sci. Technol.* **2002**, *36*, 1238–1244.

19. Tas, E.; Obrist, D.; Peleg, M.; Matveev, V.; Faïn, X.; Asaf, D.; Luria, M. Measurement-based modelling of bromine-induced oxidation of mercury above the Dead Sea. *Atmos. Chem. Phys.* **2012**, *12*, 2429–2440.

20. Ren, X.; Luke, W.T.; Kelley, P.; Cohen, M.; Tong, D.; Artz, R.; Olsen, M.L.; Schmeltz, D. Mercury speciation at a suburban site in the Mid-Atlantic United States: Seasonal and diurnal variations and source-receptor correlationship. *Atmos. Chem. Phys.* **2014**, in preparation.

21. Gay, D.A.; Schmeltz, D.; Prestbo, E.; Olson, M.; Sharac, T.; Tordon, R. The Atmospheric Mercury Network: measurement and initial examination of an ongoing atmospheric mercury record across North America. *Atmos. Chem. Phys.* **2013**, *13*, 11339–11349.

22. Tremblay, J.; Servranckx, R. Beryllium-7 as a tracer of stratospheric ozone: A case study. *J. Radioanal. Nucl. Chem.* **1993**, *172*, 49–56.

23. Dibb, J.E.; Talbot, R.W.; Lefer, B.L.; Scheuer, E.; Gregory, G.L.; Browell, E.V.; Bradshaw, J.D.; Sandholm, S.T.; Singh, H.B. Distributions of beryllium 7 and lead 2109, and soluble aerosol-associated ionic species over the western Pacific: PEM West B, February–March 1994. *J. Geophys. Res.: Atmos.* **1997**, *102*, 28287–28302.

24. Kikuchi, S.; Sakurai, H.; Gunji, S.; Tokanai, F. Temporal variation of [7]Be concentrations in atmosphere for 8 y from 2000 at Yamagata, Japan: Solar influence on the [7]Be time series. *J. Environ. Radioact.* **2009**, *100*, 515–521.

25. Piñero Garcíaa, F.; Ferro Garcíaa, M.A.; Azahrab, M. [7]Be behaviour in the atmosphere of the city of Granada January 2005 to December 2009. *Atmos. Environ.* **2012**, *47*, 84–91.

26. Rolison, J.M.; Landing, W.M.; Luke, W.; Cohen, M.; Salters, V.J.M. Isotopic composition of species-specific atmospheric Hg in a coastal environment. *Chem. Geol.* **2013**, *336*, 37–49.

27. Thurston, G.D.; Spengler, J.D. A quantitative assessment of source contributions to inhalable particulate matter pollution in metropolitan Boston. *Atmos. Environ.* **1985**, *19*, 9–26.

28. Buhr, M.; Parrish, D.; Elliot, J.; Holloway, J.; Carpenter, J.; Goldan, P.; Kuster, W.; Trainer, M.; Montzka, S.; McKeen, S.; Fehsenfeld, F. Evaluation of ozone precursor source types using principal component analysis of ambient air measurements in rural Alabama. *J. Geophys. Res.* **1995**, *100*, 22853–22860.

29. Statheropoulos, M.; Vassiliadis, N.; Pappa, A. Principal component and canonical correlation analysis for examining air pollution and meteorological data. *Atmos. Environ.* **1998**, *32*, 1087–1095.

30. Guo, H.; Wang, Tao; Louie, P.K.K. Source apportionment of ambient non-methane hydrocarbons in Hong Kong: Application of a principal component analysis/absolute principal component scores (PCA/APCS) receptor model. *Environ. Pollut.* **2004**, *129*, 489–498.

31. Landis, M.S.; Stevens, R.K.; Schaedlich, F.; Prestbo, E.M. Development and characterization of an annular denuder methodology for the measurement of divalent inorganic reactive gaseous mercury in ambient air. *Environ. Sci. Technol.* **2002**, *36*, 3000–3009.

32. Lindberg, S.; Brooks, S.; Lin, C.-J.; Scott, K.; Landis, M.; Stevens, R.; Goodsite, M.; Richter, A. Dynamic oxidation of gaseous mercury in the Arctic troposphere at polar sunrise. *Environ. Sci. Technol.* **2002**, *36*, 1245–1256.

33. Ambrose, J.L.; Lyman, S.N.; Huang, J.; Gustin, M.S.; Jaffe, D.A. Fast time resolution oxidized mercury measurements during the Reno Atmospheric Mercury Intercomparison Experiment (RAMIX). *Environ. Sci. Technol.* **2013**, *47*, 7285–7294.

34. Gustin, M.S.; Huang, J.; Miller, M.B.; Peterson, C.; Jaffe, D.A.; Ambrose, J.; Finley, B.D.; Lyman, S.N.; Call, K.; Talbot, R.; *et al.* Do we understand what the mercury speciation instruments are actually measuring? Results of RAMIX. *Environ. Sci. Technol.* **2013**, *47*, 7295–7306.

35. Bauer, D.; Campuzano-Jost, P.; Hynes, A.J. Rapid, ultra-sensitive detection of gas phase elemental mercury under atmospheric conditions using sequential two-photon laser induced fluorescence. *J. Environ. Monit.* **2002**, *4*, 339–343.

36. Luke, W.T.; Kelley, P.; Lefer, B.L.; Flynn, J.; Rappenglück, B.; Leuchner, M.; Dibb, J.E.; Ziemba, L.D.; Anderson, C.H.; Buhr, M. Measurements of primary trace gases and NOy composition in Houston, Texas. *Atmos. Environ.* **2010**, *44*, 4068–4080.

37. Liao, J.; Huey, L.G.; Tanner, D.J.; Brough, N.; Brooks, S.; Dibb, J.E.; Stutz, J.; Thomas, J.L.; Lefer, B.; Haman, C.; *et al.* Observations of hydroxyl and peroxy radicals and the impact of BrO at Summit, Greenland in 2007 and 2008. *Atmos. Chem. Phys.* **2011**, *11*, 8577–8591.

38. Dibb, J.E.; Talbot, R.W.; Klemm, K.I.; Gregory, G.L.; Singh, H.B.; Bradshaw, J.D.; Sandholm, S.T. Asian influence over the western North Pacific during the fall season: Inferences from lead 210, soluble ionic species, and ozone. *J. Geophys. Res.* **1996**, *101*, 1779–1792.

39. Dibb, J.E.; Talbot, R.W.; Scheuer, E.; Seid, G.; DeBell, L.; Lefer, B.; Ridley, B. Stratospheric influence on the northern North American free troposphere during TOPSE: [7]Be as a stratospheric tracer. *J. Geophys. Res.: Atmos.* **2003**.

40. Luke, W.T.; Arnold, J.R.; Gunter, R.L.; Watson, T.B.; Wellman, D.L.; Dasgupta, P.K.; Li, J.; Riemer, D.; Tate, P. The NOAA Twin Otter and its role in BRACE: Platform description. *Atmos. Environ.* **2007**, *41*, 4177–4189.

41. Ernest, C.T.; Donohoue, D.; Bauer, D.; Ter Schure, A.; Hynes, A.J. Programmable thermal dissociation of reactive gaseous mercury—A potential approach to chemical speciation: results from a field study. *Atmos. Chem. Phys. Discuss.* **2012**, *12*, 33291–33322.

42. Draxler, R.R.; Rolph, G.D. *HYSPLIT (HYbrid Single-Particle Lagrangian Integrated Trajectory) Model*; NOAA Air Resources Laboratory: Maryland, MD, USA, 2014. Available online: http://ready.arl.noaa.gov/HYSPLIT.php (accessed on 25 April 2013).

43. Ngan, F.; Cohen, M.; Luke, W.; Ren, X.; Draxler, R. Meteorological modeling using WRF-ARW model for Grand Bay Intensive studies of atmospheric mercury. *Atmosphere* **2014**, in preparation.

Programmable Thermal Dissociation of Reactive Gaseous Mercury, a Potential Approach to Chemical Speciation: Results from a Field Study

Cheryl Tatum Ernest, Deanna Donohoue, Dieter Bauer, Arnout Ter Schure and Anthony J. Hynes

Abstract: Programmable Thermal Dissociation (PTD) has been used to investigate the chemical speciation of Reactive Gaseous Mercury (RGM, Hg^{2+}). RGM was collected on denuders and analyzed using PTD. The technique was tested in a field campaign at a coal-fired power plant in Pensacola, Florida. Stack gas samples were collected from ducts located after the electrostatic precipitator and prior to entering the stack. An airship was used to sample from the stack plume, downwind of the stack exit. The PTD profiles from these samples were compared with PTD profiles of $HgCl_2$. Comparison of stack and in-plume samples suggest that the chemical speciation are the same and that it is possible to track a specific chemical form of RGM from the stack and follow its evolution in the stack plume. Comparison of the measured plume RGM with the amount calculated from in-stack measurements and the measured plume dilution suggest that the stack and plume RGM concentrations are consistent with dilution. The PTD profiles of the stack and plume samples are consistent with $HgCl_2$ being the chemical form of the sampled RGM. Comparison with literature PTD profiles of reference mercury compounds suggests no other likely candidates for the speciation of RGM.

Reprinted from *Atmosphere*. Cite as: Ernest, C.T.; Donohoue, D.; Bauer, D.; Ter Schure, A.; Hynes, A.J. Programmable Thermal Dissociation of Reactive Gaseous Mercury, a Potential Approach to Chemical Speciation: Results from a Field Study. *Atmosphere* **2014**, *5*, 575–596.

1. Introduction

The specific chemical speciation of mercury in flue gas emitted from coal-fired power plants (CFPPs) has important implications for its impact on the environment. The rate of wet and dry deposition and any potential atmospheric reactivity will depend both on the oxidation state of mercury and also on the specific chemical form of any oxidized mercury. The importance of understanding chemical speciation has been highlighted by work that suggests that oxidized mercury may be reduced to elemental mercury in power plant plumes.

155

The first observations of in-plume reduction were reported by Edgerton *et al.* [1] who measured Gaseous Elemental Mercury (GEM, Hg°), Reactive Gaseous Mercury (RGM, Hg^{2+}) and fine particulate mercury (Hg-P) at three sites in the southeastern United States (U.S.), using simultaneous measurements of SO_2 and NO_y to identify plumes from CFPPs. Their measurements suggested that total-Hg (*i.e.*, GEM + RGM) was essentially conserved from the point of emission to the sampling site; however, GEM was the dominant component with less than 20% present as RGM. This contrasted strongly with speciation estimates obtained from EPRI-ICR model equations for CFPPs that burn bituminous coal. The EPRI-ICR model [2] is an empirical correlation model developed with ICR (EPA Information Collection Rule) coal analysis data and stack test data and it predicted that RGM should be the dominant component of the total mercury in the observed plumes.

Lohman *et al.* [3] simulated nine power plant plume events with a reactive plume model that included a comprehensive treatment of plume dispersion, transformation, and deposition. Their study focused on observations at one of the sites, Yorkville, GA, USA, used in the Edgerton *et al.* [1] study. The EPRI-ICR model was used to predict the speciation of the mercury emitted from the CFPP's in the vicinity of the Yorkville site. The reactive plume model was then used to examine any change in the speciation of mercury as a result of in-plume chemistry. The model simulations failed to reproduce any depletion in RGM that could rationalize the observations of Edgerton *et al.* [1] and, as possible explanations they modeled RGM reduction to GEM in the plume by unknown chemistry, rapid reduction of RGM on ground surfaces, and/or an overestimation of the RGM fraction in the power plant emissions by the EPRI-ICR model. The incorporation of either a pseudo-first order decomposition of RGM or reaction with SO_2 as possible in-plume reduction processes produced better agreement with the observations.

Deeds *et al.* [4] reported on an aircraft study of mercury speciation in a CFPP plume at the Nanticoke generating station in Ontario, Canada. They found that the speciation in the plume was significantly different than the estimated stack speciation. Although they concluded that both elemental and particulate mercury levels were consistent with plume dilution, they did not observe a mass balance in total mercury and they saw significantly lower levels of RGM than would have been expected based simply on plume dilution alone. In contrast, a study carried out at the We Energies Pleasant Prairie Power Plant, Pleasant Prairie, Wisconsin, utilized aircraft and ground measurements and concluded that there was significant reduction in the fraction of Reactive Gaseous Mercury (RGM) (with a corresponding increase in the fraction of elemental mercury) as part of the Total Gaseous Mercury (TGM) emitted from the Pleasant Prairie stack [5]. Both studies demonstrate the difficulties of obtaining quantitative plume data using aircraft sampling.

In this context, the Electric Power Research Institute (EPRI) and the United States Environmental Protection Agency (USEPA) sponsored a major field campaign at CFPP Crist in Pensacola, Florida [6]. The objectives of the campaign were, by making simultaneous measurements of both GEM and RGM in the boiler stack and in the stack emission plume, to (A) determine if significant reduction of RGM was occurring in the plume; and (B) investigate the rate of reaction if significant reduction was observed. If GEM/RGM ratios in the plume were observed to be significantly higher than those ratios within the stack, then this would be evidence of such conversion occurring. The stack measurements were made using both a mercury continuous emission monitoring system (CEMS) and the Ontario Hydro method to inter-compare methods and provide confidence in the measured stack concentrations. Rather than using an aircraft that would traverse the plume, the Crist campaign utilized an airship as a sampling platform that could maintain the sampling instrumentation within the plume and measure at a variety of distances from the stack to follow the evolution of the plume chemistry. The plume measurements were designed to incorporate both a CEMS system that was identical to the stack system, avoiding instrumental bias, together with manual denuder measurements of RGM. In addition, a variety of ancillary measurements were made both in the stack and plume [6].

The chemistry of mercury in combustion emissions has been reviewed in detail by Schofield [7] who concluded that heterogeneous processes are responsible for mercury oxidation in flue gas, with homogeneous gas phase chemistry playing no role. This explains both the variability in the fraction of total mercury that is oxidized in CFPP and the inability of combustion models to predict this variability. Prior work on the speciation of mercury in CFPP flue gas has attempted to quantitatively measure the concentrations of GEM and total RGM. It is currently assumed that mercuric chloride, $HgCl_2$, is the major component of oxidized mercury in the combustor [7,8], and hence the major component of the oxidized mercury in the exhaust plume. However, to date no direct measurements of $HgCl_2$ in either stack or plume gases have confirmed this. Identification of the specific chemical components of RGM would improve our understanding of the mercury oxidation chemistry within power plant combustion systems and allow us to better understand the fate of RGM after it exits the power plant stack and interacts with the ambient atmosphere. Preliminary studies in our laboratory [9] suggested that collection of RGM on denuders coupled with analysis using programmable thermal dissociation (PTD) could provide information on the specific chemical speciation of RGM. As a result, PTD measurements were included as an innovative component of the Plant Crist campaign, but precluding any additional characterization of the PTD approach prior to the field campaign. In this sense, the field measurements were an attempt to characterize the potential of PTD analysis applied to samples collected in a realistic combustion environment. It was in effect a "proof of concept" experiment. The results are, however, of value because of the unique nature of this field experiment, with simultaneous sampling in the stack and the stack plume.

2. Results and Discussion

2.1. Prior PTD Studies

PTD has been used for the analysis of solids containing mercury compounds. The PTD profiles obtained here can be compared with prior published work on solid $HgCl_2$ samples, and other representative mercury compounds. In comparing it is important to note that the prior studies used much higher concentrations of RGM. Since they used small amounts of a solid sample, the oven temperature distribution was, in contrast to this study, not a problem. In addition, some techniques were designed to be able to distinguish between thermal desorption, *i.e.*, the evaporation or sublimation of the mercury compound, and decomposition to produce GEM. Bister and Scholz [10] used PTD to examine mercury speciation in contaminated soils. They used atomic absorption for detection of GEM. Calibration samples were diluted with quartz powder. To distinguish between desorption and decomposition the vapor passed through an 800 °C quartz pyrolysis tube before analysis thus measuring the sum of desorption and decomposition. By bypassing the pyrolyzer they could monitor decomposition alone. Their PTD profile of $HgCl_2$ shows decomposition starting at ~80 °C and peaking at 200 °C. They saw no difference in PTD profiles obtained by bypassing the pyrolyzer and it was concluded that decomposition was occurring exclusively.

Feng *et al.* [11] used ICPMS detection of GEM, examining aerosol samples collected in Toronto, Canada. The detection technique presumably measured the sum of desorption and decomposition although this was not explicitly discussed. Their calibration samples were diluted using fly ash. For $HgCl_2$ they showed a very broad PTD profile with decomposition/desorption starting at 100 °C and peaking at 300 °C.

Lopez-Anton *et al.* [12] used a commercially available thermal dissociation module coupled with a CVAFS detector. They used a pyrolyzer prior to analysis and would have detected both dissociated and desorbed/sublimed RGM. They examined fly ash and solid mercuric compounds generating standards by mixing pure compounds with powdered quartz and obtained a very sharp $HgCl_2$ PTD profile with appearance at ~70 °C and the peak at 125 °C.

Wu *et al.* [13] used mass spectrometry and this allowed them to simultaneously monitor decomposition, producing GEM, and desorption, monitoring $HgCl_2$ directly. Their standard compounds were diluted with several solid powders. Commercial SiO_2, TiO_2, Al_2O_3, and coconut shell AC were used as diluents and interestingly,

158

both the shape of the PTD profile and the ratio of GEM:HgCl$_2$ varied dramatically as a function of the diluent powders. For example, using the powdered quartz diluent a broad PTD profile was obtained with a peak at 150 °C and with GEM and HgCl$_2$ being detected in equal amounts. In contrast, a PTD profile with Al$_2$O$_3$ diluent was much sharper with a peak at 150 °C but consisting exclusively of GEM.

These studies also examined the PTD profiles of other Hg(II) and Hg(I) compounds, HgS, HgSO$_4$, HgO, HgBr$_2$, Hg$_2$Cl$_2$ and Hg$_2$SO$_4$. In general, the non-halide Hg(II) compounds are found to dissociate at higher temperatures than HgCl$_2$ although there is considerable variability between the studies. Lopez-Anton *et al.* [12] show a PTD profile for HgBr$_2$ that has its decomposition peak shifted slightly to lower temperature relative to HgCl$_2$. It is also clear that there is significant variability in the results of these studies.

In addition to using considerably higher mercury concentrations than our work, all of the studies used a linear temperature ramp in contrast to the ramp and hold sequence that was used here. As we note below we have now switched to a linear ramping sequence. Unfortunately, we were not aware of the two prior literature studies [10,11] at the start of the Crist campaign in 2008, hence our decision to use the ramp and hold sequences described as "oven 1" and "oven 2". The "oven 2" sequence is probably closest to a straight ramping sequence and the "oven 2" profiles are similar to published PTD profiles and to our more recent, better defined work shown in Supplemental Figure S1 The "oven 1" profiles show more variability and if the work of Wu *et al.* [13] is correct this may be due at least in part to the competition between sublimation and decomposition. We plan to modify our implementation of PTD by incorporating a pyrolyzer and second LIF detection cell to be able to simultaneously monitor both dissociation and desorption/sublimation profiles.

2.2. Laboratory HgCl$_2$ PTD Profiles

Laboratory PTD profiles of HgCl$_2$ were obtained after the completion of the Crist field experiment using two thermal ramping schemes that we denote as "oven 1" and "oven 2". Figure 1 shows the individual and averaged calibration profiles obtained at a single deposition flow for the "oven 1" thermal cycle. The LIF signal has been normalized by the signal resulting from the injection of known amounts of GEM, as described in Section 3.6.

Figure 1 indicates the reproducibility of 6 PTD profiles obtained by depositing pure HgCl$_2$ under nominally identical conditions. The HgCl$_2$ loadings varied considerably. Possible reasons include fluctuations in the gas phase concentration of HgCl$_2$, variations in denuder collection efficiency, and variation in competition between desorption of molecular HgCl$_2$, which we would not detect, and decomposition to produce GEM which we do detect. We have found it difficult to develop stable sources of $HgCl_2$ for calibration purposes and this represents a major challenge in this type of experiment.

As a result of time constraints we were only able to obtain a single calibration profile obtained for the "oven 2" profile which is shown in Figure 2.

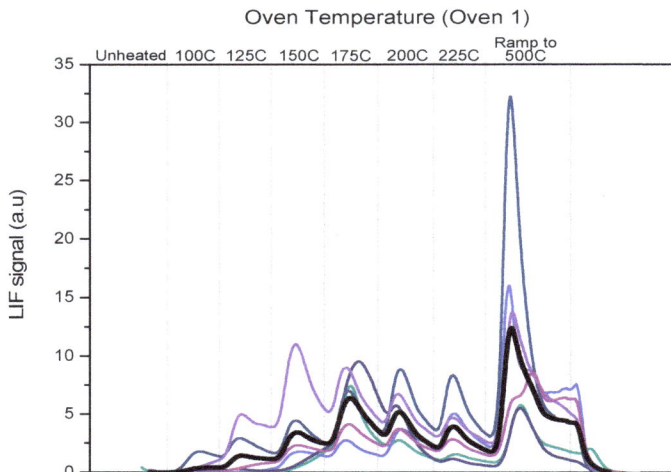

Figure 1. Six programmable thermal dissociation (PTD) profiles using Oven 1 program for an uncoated tubular denuder loaded with pure $HgCl_2$. Denuders were loaded by flowing N_2 over pure $HgCl_2$ for 1 min at 135 sccm. The average of the six profiles is also shown.

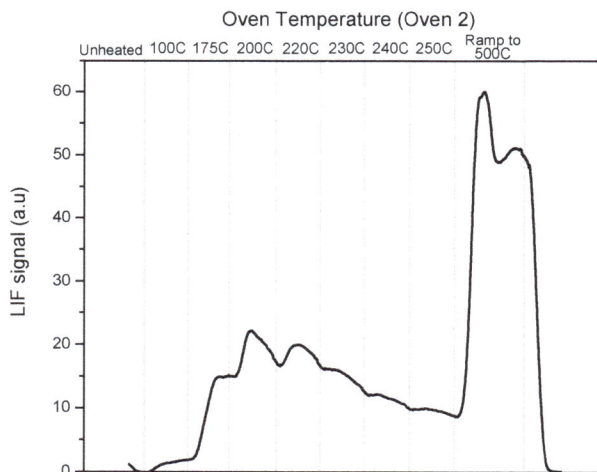

Figure 2. PTD profile using Oven 2 program for an uncoated tubular denuder loaded with pure $HgCl_2$.

Some of the variability in the shape and structure of the PTD profiles has been determined to be a function of the deposition pattern of RGM on the denuder, the very significant temperature gradient along the denuder during the heating cycle, and the direction of buffer gas flow during both sampling and analysis. During the Crist measurements RGM was sampled over the whole denuder surface including the ends that are located outside of the oven and the direction of gas flow was not noted during sampling. As discussed below, we have found that during the heating cycle the center portion of the denuder heats most rapidly and the initial production of GEM reflects dissociation of RGM in this section of the denuder. The temperature in the PTD profile figures is the temperature as read by the oven thermocouple and most closely reflects the temperature of the central section of the denuder; the ends of the denuder are significantly cooler. As the temperature increases RGM deposited further from the center reaches decomposition temperature and the final ramp produces decomposition over most of the length of the tube. The oscillations in the PTD profiles appear to be the result of this effect in conjunction with the "ramp and hold for 5 min" sequence that we used for temperature ramping at Crist.

In addition, RGM deposition appears to be a function of flow direction, presumably maximum deposition occurs as the flow enters the denuder and decreases as the RGM concentration is depleted as the gas flows down the denuder. The flow-effect becomes more significant as the RGM loading increases. One consequence of the flow-effect is that dissociation profiles are more reproducible if the direction of gas flow through the denuder is the same during sampling and analysis. Since the direction of the sampling flow was not noted during stack sampling, this variation, together with a lack of reproducibility in the oven cycling, contributes to variation in the PTD profiles. The "post-Crist" improvements that have eliminated these problems are discussed below. Nonetheless, these issues affect the work at Crist and the initial calibration profiles obtained at the end of the campaign. In spite of this, the unique nature of the sampling opportunity, with simultaneous measurements in the stack and plume do allow us to draw useful conclusions, as we discuss below.

2.3. Stack PTD Profiles

Figure 3b shows PTD profiles for several stack samples analyzed using the "oven 2" thermal cycle described below. The "oven 2" PTD profile for pure $HgCl_2$ (Figure 2) is shown again for comparison in Figure 3a. The stack profiles show a sharp onset of RGM dissociation at ~200 °C. GEM continues to evolve as the temperature increases, with a spike in RGM dissociation as the oven temperature is ramped to 500 °C. The profiles are similar to the calibration $HgCl_2$ profiles although the onset of dissociation occurs at a slightly lower temperature in the $HgCl_2$ profiles. Nevertheless, the profiles shown in Figure 3 are quite similar and consistent with the $HgCl_2$ calibration profile and the onset of RGM dissociation was quite consistent.

The observation that these profiles are consistent with the HgCl$_2$ profiles does not constitute a definitive identification of HgCl$_2$.

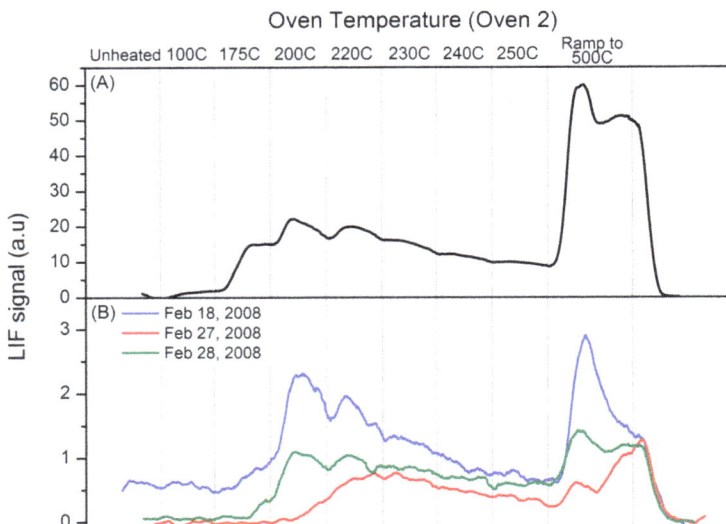

Figure 3. Panel (**A**). Calibration profile obtained using Oven 2 program for denuder loaded with pure HgCl$_2$; Panel (**B**). Dissociation profiles obtained using Oven 2 program for stack samples collected 18, 27 and 28 February 2008.

Figure 4b shows PTD profiles of stack samples analyzed with the "oven 1" cycle described below together with the averaged "oven 1" PTD profile of HgCl$_2$ shown in Figure 4a. The grey shading in this and subsequent "oven 1" PTD figures spans the minimum and maximum signals observed. The "oven 1" profiles have a slower temperature ramp between 100 and 200 °C; however, we saw no evidence for any RGM species with a lower peak decomposition temperature. The "oven 1" PTD profile for pure HgCl$_2$ shows measurable decomposition at 100 °C and typically peaks at ~200 °C. In the "oven 1" PTD profiles of stack samples, decomposition begins at a slightly higher temperature but the peak decomposition temperatures of the majority of the stack PTD profiles are consistent with HgCl$_2$ being the sampled component of the deposited RGM. We should emphasize that this does not identify or prove that the compound is HgCl$_2$; however, we do not see any evidence for a species with a peak decomposition temperature that is significantly lower and could clearly be identified as not being consistent with HgCl$_2$.

As we discuss above, the variation and reproducibility of the PTD profiles is a function of the deposition pattern, the temperature gradient, reproducibility of oven temperature cycling, and the direction of gas flow during both sampling and analysis. During stack sampling it is reasonable to assume that a variety of compounds that are present in the stack gas are being co-deposited on the denuders and this may have some impact both on the decomposition temperature and the competition between desorption and decomposition. It is also possible that chemical reactions can take place on the surface of the denuder between deposited RGM and other components of the stack gas. In-situ spiking of the stack gas with $HgCl_2$ could address the impact of co-deposition effects.

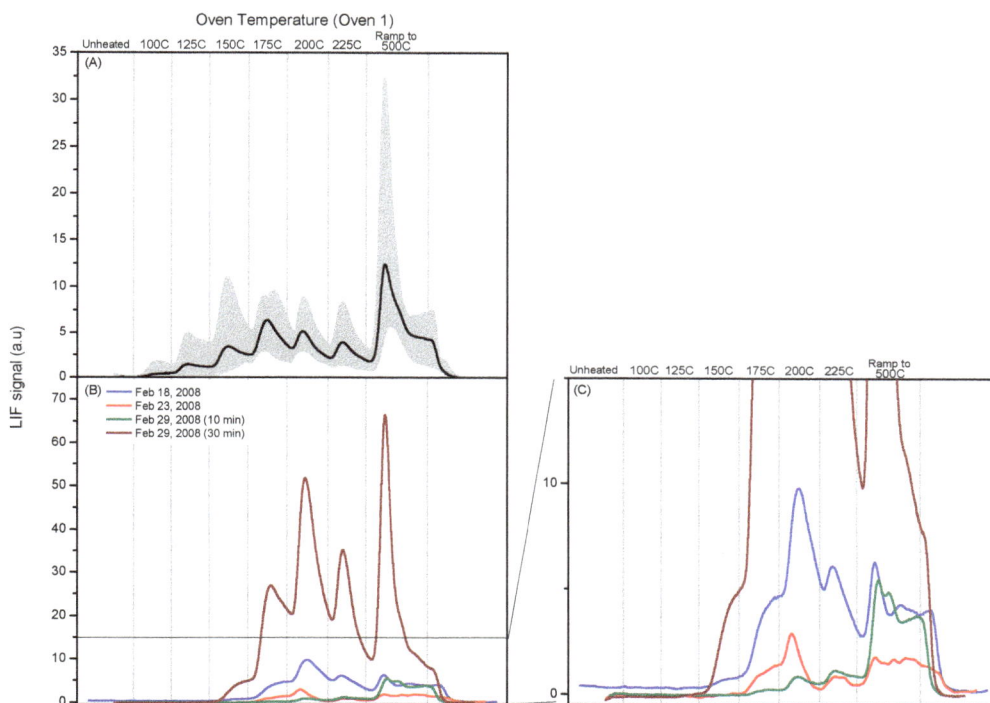

Figure 4. Panel (**A**): Calibration profile obtained from average of six PTD profiles using Oven 1 program for denuder loaded with pure $HgCl_2$. The grey shading spans the minimum and maximum signals observed; Panels (**B,C**): Dissociation profiles obtained using Oven 1 program for stack samples collected 18, 23 and 29 February 2008.

2.4. In-Plume Sampling

Figure 5b shows profiles from 28 February, showing the similarity in PTD profiles sampled in the stack and plume. The upwind sample was located approx. 4 km upwind of the Crist plant, Close-in was a sample collected approx. 0.6 km downwind of the Crist plant and further-out was a sample collected approx. 1.5 km downwind of the Crist plant.

163

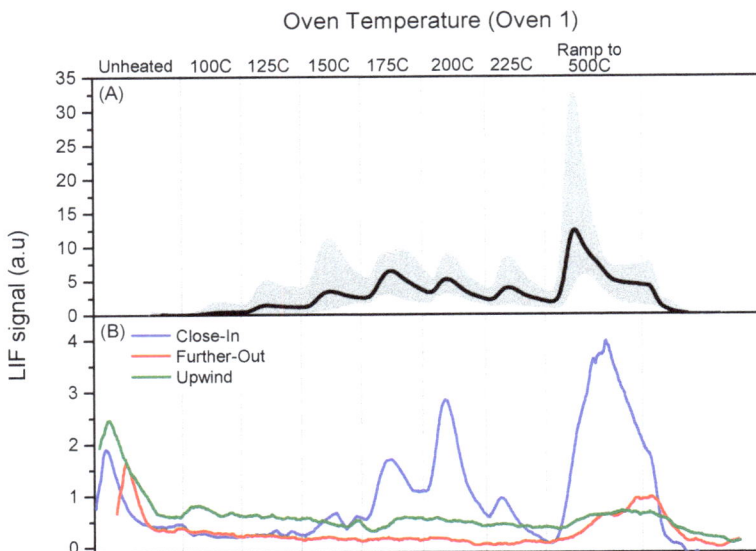

Figure 5. Panel (**A**): Calibration profile obtained from average of six PTD profiles using Oven 1 program for denuder loaded with pure $HgCl_2$. The grey shading spans the minimum and maximum signals observed; Panel (**B**): Dissociation profiles obtained using Oven 1 program for airship samples collected 28 February 2008.

The "close-in" sample shows a well-defined thermal dissociation profile that is similar to the stack profile of the morning of the 29 February 2008, as seen in Figure 4, burning an identical coal mixture. The airship profiles show a GEM signal that occurs prior to the heating cycle. This preheat mercury signal was observed in all of the airship samples including those that were not able to detect any plume RGM. No preheat mercury signal was observed in any of the stack samples. This could be a result of a contamination problem in the adapters used to plumb the screw-end denuders into the analysis gas flow system but its origin was not identified. Figure 6 shows the PTD profiles of $HgCl_2$, the "close-in" blimp sample and the stack sample from the morning of the 29 February 2008.

The similarity between the PTD stack and blimp profiles and their similarity to the $HgCl_2$ profile suggest that the species sampled from the stack and the plume are the same and are consistent with the chemical speciation of the sampled RGM being $HgCl_2$. The coal that was used during this sampling period contained the highest level of chlorine seen during the Crist campaign, constituting 97% of the total halogens with a bromine content of 0.16% and fluorine as the balance. Comparing PTD profiles of $HgCl_2$ with the PTD profiles of the stack and plume samples suggests they are consistent with $HgCl_2$ as the specific chemical speciation of sampled stack RGM but, as discussed above, this does not prove that the stack RGM is $HgCl_2$.

It could also consist of other oxidized mercury compounds that have a similar PTD profile to $HgCl_2$. One possible explanation of the observations of Edgerton *et al.* [1] is that a component of RGM undergoes slow thermal decomposition at ambient temperatures. No evidence for a significant component of the RGM sample that has a significantly lower decomposition temperature than $HgCl_2$ was observed.

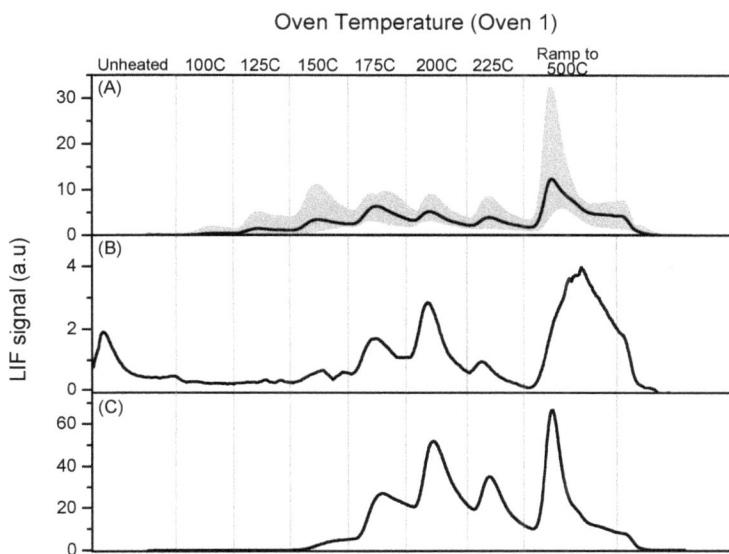

Figure 6. Panel (**A**): Calibration profile obtained from average of six PTD profiles using Oven 1 program for denuder loaded with pure $HgCl_2$. The grey shading spans the minimum and maximum signals observed; Panel (**B**): Dissociation profiles obtained using Oven 1 program for "close-in" airship sample collected 28 February 2008; Panel (**C**): Dissociation profiles obtained using Oven 1 program for stack sample collected 29 February 2008.

2.5. Denuder Sampling Efficiencies

The objective of this component of the Crist study was to investigate the potential of PTD for chemical speciation of RGM. The collection efficiency of the denuder is not known and PTD will not give a quantitative measurement of RGM concentration under these conditions. However, the PTD profiles were also calibrated for $Hg°$ as discussed in Section 2.6. This provides an approximate measure of the total amount of RGM collected on the denuder. Since the RGM stack concentrations were measured independently, we can obtain an approximate measure of the sampling efficiency of the denuders as a check on the integrity of the measurements. For example, a sampling efficiency in excess of 100% would imply major problems with denuder contamination.

All denuder sampling was performed under laminar flow conditions and a theoretical sampling efficiency can be estimated using the Gormley-Kennedy equation [14]. This assumes laminar flow and a sticking coefficient of 1 and requires a tube diameter and a flow rate, which are known, and a diffusion coefficient for the species of interest, which is not known. If a diffusion coefficient of $0.1 \text{ cm}^2 \cdot \text{s}^{-1}$ is assumed, a typical value for molecules in air, we calculate a sampling efficiency of ~60% for the stack sampling and ~70% for plume sampling. If we assume that the chemical identity of the RGM is $HgCl_2$, we can estimate a diffusion coefficient of $0.04 \text{ cm}^2 \cdot \text{s}^{-1}$ using the mass relationship from Schwarzenbach *et al.* [15] and this would give calculated sampling efficiencies of 35% and 43% at the stack and plume sampling flow rates.

To estimate the actual denuder sampling efficiencies, the total amount of RGM that was sampled was calculated from the stack RGM concentrations as measured by the CEMS instrumentation and sampling flow rate through the denuder. The RGM that was actually deposited on the denuder was taken from the total calibrated PTD profiles and included the final peak obtained during the temperature increase to 500 °C. The RGM actually deposited on the denuder was then divided by the calculated total sampled amount based on the flow and stack concentrations. This then gives an estimate of the denuder sampling efficiency, *i.e.*, the amount of RGM collected on the denuder compared with the total amount of RGM that was sampled from the stack and then flowed through the denuder, but did not necessarily deposit on the walls. These sampling efficiencies are estimates since some RGM was deposited on the ends of the denuders and was not measured. As noted above, it is possible that some of the deposited RGM evaporated before decomposition to Hg° and this would not have been measured.

The efficiencies for stack profiles shown in the figures are given in Table 1. A total of 28 stack profiles were analyzed with sampling efficiencies that ranged from ~1% to 55%. Sampling efficiencies from samples obtained with the particle filter in place were typically much lower than estimated from the Gormley–Kennedy equation, ranging from ~1% to 30%. Three PTD stack profiles were obtained without the particle filter in the sampling stream and the calculated sampling efficiencies were 50%, 55% and 37%, much closer to the estimated values so it appears that the presence of the filter significantly reduced the sampling efficiency. This does not appear to be due to collection of particulate mercury when the filter was removed. These samples were obtained downstream of the electrostatic precipitator and Ontario Hydro measurements found that the concentration of particulate mercury was below detection limits. We have looked at the variability in greater detail. Based on a total of 24 samples taken with the filter the mean and standard deviation of the RGM/m³ is 0.46 ± 0.66 ug/m³, taking only samples collected using etched denuders we obtain 0.25 ± 0.19 ug/m³ and those analyzed using the oven 2 cycle 0.35 ± 0.24 ug/m³. For the samples taken without the filter the average is 2.7 ± 1.0 ug/m^3. It is clear that the presence of the quartz filter significantly reduced the collection efficiency and we have no explanation for this. The significant variability is also clear and it does not appear to be explained by differences in analysis, *i.e.*, oven program, or tube etching.

In Figure 7 we show PTD profiles of two samples taken within a 24-h period burning identical coals. One, a 30 min stack sample from 29 February was taken with the filter in place, while the other, a 10 min stack sample from March 1st was taken without the filter. The similarity between the profiles suggests that the filter did not affect the speciation of the deposited RGM, while the difference in sampling time shows the reduced collection efficiency when the filter was in place. It is important to also recognize that the measured RGM in the stack gas was variable and there was disagreement of up to 40% in the stack RGM measurements using different approaches.

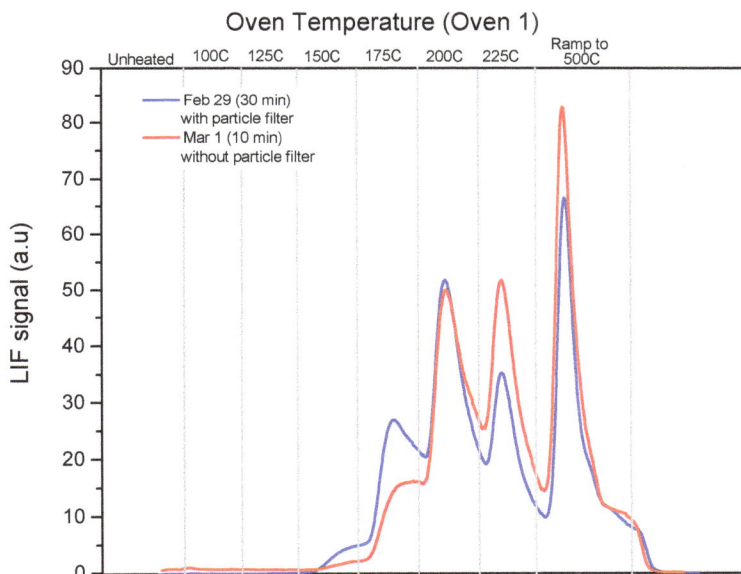

Figure 7. Comparison of PTD profiles (using Oven 1 program) of a 30 min stack sample obtained on 29 February 2008 with the particle filter in place and a 10 min stack sample obtained on 1 March 2008 without the particle filter.

An estimate of the collection efficiency of the in-plume samples requires a comparison between sampled RGM and that predicted using the measured stack concentrations and calculated plume dilution ratios. Dilution ratios were based on the measured concentrations of three tracer gases, SO_2, NO_y and CO_2, which were measured in stack and also in the airship. Figure 8 shows the "background corrected" concentrations of SO_2, NO_y and CO_2 as measured by the airship's sampling system during the 75-min "close-in" sampling run on 28 February.

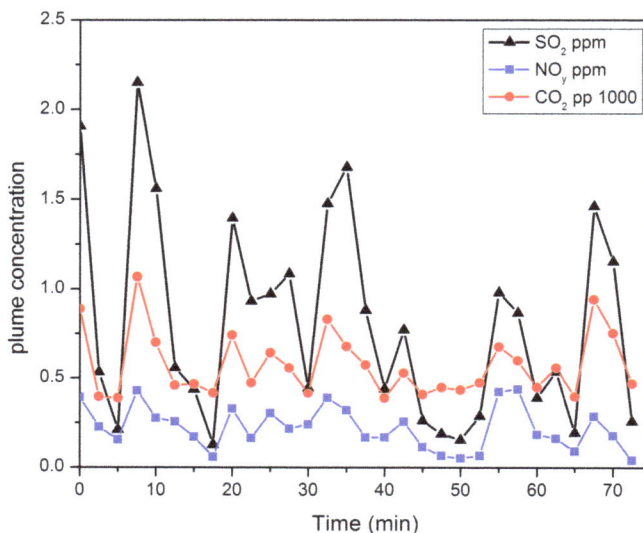

Figure 8. "Background corrected" concentrations of SO_2, NO_y and CO_2 as measured by the airship's sampling system during the "close-in" sampling run on 28 February.

The background correction was obtained by subtracting the concentrations measured upstream of the plume. The concentrations shown in Figure 8 reflect the increase in the concentration of these gases over ambient as the plume gases mix with ambient air. There was little variation in the stack concentrations during the sampling period and the observed variations reflect the difficulties of keeping the airship's sampling system in the plume. The measured airship concentrations of SO_2, NO_y and CO_2 can be used to calculate the dilution of the plume gases as the plume evolved and mixed with ambient air. Because of the variation in concentration, the dilution ratios were calculated for 2.5 min coincident time bins for each tracer gas. The calculated dilution ratio for each tracer gas was then multiplied by the stack RGM concentration measured by the CEMS instrumentation to calculate the mass of RGM sampled by the denuder during each 2.5 min bin. This was summed to calculate the total mass of RGM sampled by the denuder during the 75-min sampling period based on each tracer dilution ratio. Table 1 shows the calculated total amount of RGM sampled by the denuders at the "close-in" and "further-out" positions. These amounts are calculated from the RGM concentration measured in the stack by the CEMS instrument and the calculated dilution ratio from each tracer. The difference between the values calculated using the three tracers reflects the fact that the three

168

tracers give different integrated dilution ratios, an indication of the difficulty of the measurement.

Table 1. Estimated denuder collection efficiencies for in-plume samples obtained 28 February 2008.

Airship Sample	Denuder Type	Actual RGM * Deposited on Denuder (ng)	Total RGM ** Sampled by Denuder (ng) as Calculated from Dilution Ratios			Denuder Collection Efficiency (%) as Calculated from Dilution Ratios		
			SO_2	NO_y	CO_2	SO_2	NO_y	CO_2
"Close-in"	unetched quartz	1.19	1.78	2.16	1.19	67	55	100
"Further-out"	unetched pyrex	0.13	0.41	0.73	0.41	32	18	32

* As determined using PTD technique; ** Based on CEMS stack concentration and calculated dilution ratio for each tracer.

The observation of the early $Hg°$ signal and the drifting baseline complicate the analysis of the dissociation profiles in Figure 3. The "close-in" profile shows a well-defined dissociation profile between 150–225 °C and a larger area in the final ramp to 500 °C. Taking the total integrated sample gives ~1.3 ng of $Hg°$, 0.61 ng in the structured profile and 0.66 ng in the final ramp to 500 °C. If we take the 0.08 ng from the upwind sample as our field blank we calculate 0.13 ng of RGM deposited on the "further-out" denuder and 1.19 ng deposited on the "close-in" denuder. If we then compare this with the total RGM sampled by the denuders, as calculated from dilution ratios, we find sampling efficiencies of 67% and 55% using the SO_2 and NO_y dilution ratios and 100% from the CO_2 dilution ratio. For the "Further-out" denuder we calculate sampling efficiencies of 32%, 18% and 32% respectively. Given the large uncertainty in the dilution ratios these efficiencies suggest that the RGM sampled in-plume is consistent with the levels measured by the CEMS instrument in the stack. This, together with the PTD profiles of the stack and plume samples shown in Figure 6 suggests that it is possible to use PTD to measure a component of RGM in the stack and follow its evolution in the plume. It also suggests that the speciation of the RGM is not changed during this period and that it is consistent with $HgCl_2$ as the speciation of RGM. This can be compared with the observations of Deeds *et al.* [4] who found greatly decreased levels of RGM and no mass balance.

2.6. RGM Speciation

As we have noted above, it is important to emphasize that the generally held opinion that $HgCl_2$ is the species of RGM emitted from coal-fired power plants is not based on any experimental evidence, or on any reasonable mechanism for homogeneous formation of $HgCl_2$ in coal combustion. The PTD profiles shown here are clearly consistent with identifying the speciation of stack and plume RGM as $HgCl_2$ and represent the first actual experimental evidence that is consistent with $HgCl_2$ speciation. It is reasonable to ask if any other mercury species is consistent with the observed PTD profiles. As we have noted previously, all of the mercuric halides are stable gas phase species; however, for most of the coal burned during the Crist campaign more than 95% was present as Cl and less than 1% was present as Br, hence $HgBr_2$ is not viable as an RGM candidate. Sulfur typically constituted 1% by weight of the coal burned; however, HgS is not a stable gas phase species [16] and the PTD profiles of HgS are not consistent with our observations.

The reported PTD profiles for solid HgO [11–13] are much more variable with peak decomposition temperatures ranging from 260 °C to 600 °C, but are not consistent with the decomposition seen in oven 1 profiles. Earlier work [8] suggested that HgO could be a significant component of RGM at stack gas temperatures in excess of 450 °C. However, recent high level ab-initio calculations [17] have shown that HgO cannot exist as a stable diatomic molecule in the gas phase. The implications of this for the atmospheric chemistry of mercury have been discussed in detail elsewhere [18] but it is clear that even in lean flames HgO can play no role in Hg oxidation and cannot be a candidate for stack RGM. It should be noted that Donohoue [9] reported PTD profiles for HgO that were thought to be from gas phase deposition. These profiles showed decomposition at low temperatures but it is now clear that these profiles were an artifact that resulted from contamination in sample lines. We have measured PTD profiles of solid HgO and find that it decomposes at a significantly higher temperature than $HgCl_2$.

It is clear, therefore, that $HgCl_2$ is the only stable Hg(II) species with an observed PTD profile that is consistent with our stack and plume PTD profiles and is the most likely candidate for the actual species.

In spite of the caveats associated with the fact that this was the first attempt to do this kind of sampling and analysis in a working coal-fired power plant, our experimental evidence, sampling from both stack and plume, suggests that $HgCl_2$ is indeed the emitted form of RGM. To the best of our knowledge this is the first and only set of experimental evidence that addresses this speciation issue. It suggests that attempts to understand and model in-plume reduction of RGM to GEM should use $HgCl_2$ as the speciation of RGM.

2.7. Further Development of the PTD Approach

In continuing development of the PTD approach, focus was on the use of uncoated quartz tubular denuders with an etched central area. The control of the oven ramping temperature is critical for reproducible PTD profiles and we have eliminated the use of a PID temperature controller and the type of "ramp and hold" cycles that were used during the Crist campaign. In addition, by using an etched central portion of the denuders it is possible to completely eliminate the large peak that is associated with the final temperature ramp. Typically we now use 5 min of preheating at 50 °C and then a constant ramp from 50 to 500 °C. Supplemental Figure S1 shows PTD profiles of $HgCl_2$ deposited on etched quartz denuders that are obtained using this 50–500 °C ramping sequence. The mass of $HgCl_2$ deposited varies between 20 and 50 pg, *i.e.*, more than an order of magnitude lower than deposited during the Crist campaign. The profiles are quite reproducible and eliminate the "final ramp" peak and oscillations that are particularly evident in the "oven 1" profiles. The PTD profiles of $HgCl_2$ and $HgBr_2$ were compared and are identical on uncoated tubular denuders. However, this is not relevant to the Crist work because chlorine was the dominant halogen component of all the coal that was burned during the Crist campaign. In contrast, we find that solid samples of HgO and a compound deposited from the heterogeneous reaction of mercury and ozone have PTD profiles that decompose at significantly higher temperature than $HgCl_2$. In other work from this laboratory uncoated quartz denuders with an etched central surface were used to sample from a small aircraft over Mississippi and the Gulf of Mexico. Several samples show evidence for an RGM species that decomposes at a significantly lower temperature than $HgCl_2$ or $HgBr_2$ together with samples that decompose at temperatures that are consistent with the PTD profiles of these mercuric halides. Examples of two such profiles are shown in Supplemental Figures S2 and S3. Figure S2 shows a PTD profile of a high altitude sample with a total mass of 58 pg. Figure S3 shows a PTD profile of a sample taken at 500 ft in the marine boundary layer over the Gulf of Mexico with a total mass of 56 pg. In both cases the PTD profiles of the flight blanks are shown. These observations are significant in that they may imply there is a form of atmospheric RGM with a PTD profile which is not consistent with speciation as either $HgCl_2$ or $HgBr_2$; however, we saw no evidence for the presence of this species in the Plant Crist plume.

3. Experimental Section

3.1. Conventional Denuder Sampling

Since PTD is not a commonly used analytical method, it is useful to contrast the PTD technique with "conventional denuder sampling" as a route to the measurement of total RGM concentrations. The use of KCl coated annular denuder sampling coupled with thermal dissociation has been described by Landis *et al.* [19] and forms the basis of a commercially available RGM measurement system, the Tekran Model 1130 Mercury Speciation Unit (Tekran, 2010). In this instrument air is pulled through a KCl coated annular denuder that captures RGM but transmits elemental and particulate mercury. After a period of sampling, the denuder is flushed with zero air and the denuder is heated to 500 °C. The RGM is thermally decomposed producing elemental mercury that desorbs from the denuder surface and is then captured by a Tekran 2537 Mercury vapor analyzer. In the Tekran 2537 the elemental mercury is collected by amalgamation on a gold cartridge during a sampling phase. The instrument is then flushed with argon and the mercury desorbed by heating the gold cartridge. The desorbed gas phase mercury is detected by Cold Vapor Atomic Fluorescence Spectrophotometry (CVAFS).

3.2. Programmable Thermal Dissociation

In PTD, the evolution of GEM is monitored as a function of temperature in real time during pyrolysis of a sample. PTD could be considered to be somewhat analogous to thermogravimetric analysis. However, in PTD the evolution of a gas phase product is monitored, rather than monitoring the decrease in the weight of the sample. It has been used to examine mercury speciation in solids [10–13] and a comparison of the PTD profiles reported in these works and measured here was discussed in Section 2.3.

In contrast to the studies of mercury containing solids, the configuration of PTD utilized in this work is designed to sample gas phase RGM. During the thermal analysis we use laser-induced fluorescence to monitor the extent of RGM decomposition as a function of temperature in real time by measuring the evolution of GEM produced during decomposition. The RGM sample is obtained by pulling the analysis gas through a quartz or Pyrex tube that acts as a denuder and captures RGM but transmits elemental and particulate mercury. It should be emphasized that these are uncoated tubular denuders in contrast to the KCl coated annular denuders that are commonly used to quantify total RGM [19]. The collection efficiency of the denuder is not known and, in this configuration, PTD is not designed to produce a quantitative measurement of RGM concentration. After a period of sampling the denuder is transported to a laboratory, flushed with He and then heated in a series of temperature ramps in an oven. As the denuder temperature increases, RGM dissociates and the GEM product is desorbed from the denuder wall. The GEM evolution is monitored in real time using laser-induced fluorescence (LIF). Comparison of the PTD profiles of unknown samples with profiles of known oxidized mercury compounds may provide information on the chemical identity of the sample. Since PTD is an indirect technique and different compounds can have identical or very similar decomposition properties it is unlikely to provide a definitive molecular identification. However, it should be possible to indicate whether the PTD profile of a sample is consistent with a particular molecular species or class of species.

3.3. Denuder Sampling

3.3.1. Sampling during the Crist Campaign

Field sampling at CFPP Crist took place between 18 February and 1 March 2008. The plant has four coal-fired units, two of which were operational during the campaign. The operational units, #6 (320 MW) and #7 (500 MW) shared a common stack and mercury continuous emissions monitoring systems (CEMS) were located at the output of each unit prior to discharge into the stack. In addition to the mercury CEMS, independent measurements of mercury concentration were made using the Ontario Hydro method and sorbent traps. NO_x, SO_2 and CO_2 were measured by the Crist CEMS instrumentation. Mercury emissions were dominated by the exhaust from unit #7, which typically contained 5–7 $\mu g \cdot m^{-3}$ of mercury, more than 90% of which was present as RGM. The unit #6 exhaust typically contained less than 1 $\mu g \cdot m^{-3}$ of mercury and contained equal amounts of GEM and RGM. As a consequence, mercury in the stack gas was largely in the form of RGM. The Ontario Hydro and CEMS RGM stack measurements showed good agreement although the agreement between the GEM measurements was more problematic, reflecting the much lower concentrations of elemental mercury in the stack gas [6].

A preliminary laboratory assessment of the utility of PTD on KCl coated annular denuders is described by Donohoue [9]. Based on the prospect of sharper PTD profiles, together with time and cost limitations, uncoated tubular denuders were selected for PTD analysis during the campaign. A variety of quartz and Pyrex tubular denuders approximately 2.5 cm diameter and 50 cm long were used during the sampling campaign. Several of the denuders had a 25 cm central section that was etched to enhance the surface area. Denuders were prepared by rinsing with concentrated KOH, followed by distilled water, 10% nitric acid solution, distilled water, and finally methanol. They were dried in air without heating.

It was necessary to analyze the denuders and then clean them for additional sampling, making on-site analysis desirable. Analysis was performed at the University of West Florida (UWF), located approximately 7 m from Plant Crist and 5 m from Pensacola Airport, the base of operations for the airship. Denuder sampling experience prior to the Crist campaign had been a limited set of laboratory studies and high levels of RGM in the stack and blimp samples that could be analyzed using single photon LIF were anticipated.

During the sampling campaign the dye laser that was used to excite fluorescence showed short-term drift, due to a damaged output coupler that had been degrading the Nd-Yag pump laser over the course of the two week sampling period, and possibly also due to temperature variation in the laboratory at UWF. This short-term drift produced background shifts in a number of the dissociation profiles.

3.3.2. Stack Sampling

All stack sampling was performed by personnel from the Energy and Engineering Research Center (EERC), University of North Dakota in conjunction with their measurements of GEM and RGM using the Ontario Hydro method. Stack gas samples were collected from ducts located after the electrostatic precipitator and prior to entering the stack. The gas was pumped from the ducts and diluted to prevent condensation. Most of the samples were obtained with a quartz particle filter in the sampling line. The filter was located in the stack and had a capture efficiency of 99.95% for 0.3 μm particles. Typical sampling times were 1–30 min at the stack at 5 SLPM total flow (4.5 SLPM dilution air +0.5 SLPM stack gas). Typical stack gas temperatures at the sampling point were ~150 °C and the gas had cooled to ~25 °C before reaching the denuder. After sampling, the denuders were capped, and once sampling was completed for the day they were transported to UWF for analysis. Denuders were stored in the analysis laboratory at room temperature and typically analyzed the day following sampling.

3.3.3. In-Plume Sampling

An airship was used to sample the exhaust plume with the objective of holding the sampling modules within the centerline of the plume at approximately the same position for an extended period of time. The sampling PVC modules were suspended 60 feet below the airship's gondola and connected by umbilical tubes that included sample lines connected to pumps in the gondola. In addition to denuders and filters located in the sampling modules, a dual TEKRAN system located in the gondola ran in parallel and measured total gaseous mercury (RGM + GEM) and GEM separately at 2.5 min intervals. In addition, SO_2, NO_y, and CO_2 were measured continuously and used as plume-detection tracers.

EPA personnel performed the in-plume sampling. The sampling module for the PTD denuders pulled air directly from the plume at ambient temperature and this was drawn though the denuder at 3.5 SLPM. After a sampling period of typically one hour the module was pulled up into the airship's gondola, the denuder was removed and capped, a fresh denuder loaded into the module, and the module lowered back into the plume for another period of sampling. After the airship landed the denuders were transported to UWF for analysis. A total of 20 uncoated tubular denuder samples were obtained on six days during the campaign. The airship's sampling protocol involved taking one sample upwind of the stack to establish background conditions followed by a sample downwind and close to the stack, and then a final sample further downwind in-plume; referred to as "close-in" and "further-out," respectively.

174

3.4. Analysis by Programmable Thermal Dissociation (PTD)

Analysis was performed using PTD, heating the denuders in a series of steps and monitoring the evolution of GEM as a function of time using single photon LIF. Analysis was performed in He buffer to enhance detection sensitivity for GEM. For analysis the denuders were placed in a clamshell furnace and flushed with He at 0.45 L/min to remove all ambient gas and this flow was maintained for the full heating cycle. Two thermal cycles were used and are subsequently referred to as "oven 1" or "oven 2". In the "oven 1" cycle the temperature was raised to 100 °C and then increased in five 25 °C ramps up to 225 °C with each ramping cycle taking 5 min. Finally, the oven temperature was rapidly increased to 500 °C and held at this temperature for 5 min. In the other cycle "oven 2" the oven temperature was increased to 100 °C, 175 °C, 200 °C, four 10 °C ramps to 250 °C and then rapidly increased to 500 °C. Again each ramping step took 5 min. The "oven 1" profile was designed to capture species that dissociated at a significantly lower temperature than the mercuric halides. The "oven 2" profile, which began with a much steeper initial ramp from 100 °C to 170 °C and then with smaller increments to 250 °C, was designed to try and distinguish between compounds with similar decomposition temperatures in the 170–250 °C range. Throughout the initial flushing and the heating cycle the gas that left the denuder passed through a fluorescence cell and the concentration of GEM was monitored by single photon LIF. After the heating cycle was complete the LIF signal was calibrated by injecting a known amount of GEM into the gas flow through a septum as described below.

3.5. GEM Detection

GEM was detected by single photon, resonance LIF using excitation of the 6^3P_1-6^1S_0 transition at 253.7 nm. A frequency doubled dye laser pumped by the third harmonic of a Nd-Yag laser was used to generate the excitation beam. Resonance fluorescence was observed using a Hamamatsu 1P28 photomultiplier tube with a 253 nm filter. In this approach the detection PMT detects both LIF and laser scatter and thus sensitivity is limited by the ratio of intensity of the LIF signal to the laser scatter. Since the 6^3P_1 level is efficiently quenched by both O_2 and N_2 the thermal analysis was performed in He buffer gas to achieve good detection sensitivity. The PTD profiles of all the samples collected during the Crist campaign used uncoated tubular denuders and the analysis used single photon LIF.

3.6. Calibration of Absolute Hg° Concentration

The LIF signal was converted into an absolute Hg° concentration by injecting known amounts of GEM into the detection cell through a septum. A saturated calibration gas was obtained by taking a sample of elemental mercury and allowing

175

it to equilibrate at 6 °C for over 24 h before initial use. The elemental mercury was kept in a Pyrex vial capped with a septum in Ar bath gas. The transfer syringe was at room temperature and the saturated gas was withdrawn from the vial and directly injected slightly upstream of the fluorescence cell. The GEM concentration was calculated assuming saturation, correcting for the temperature difference between the syringe and the vial, and using the temperature dependence of the vapor pressure given by the Dumarey equation [20].

3.7. HgCl₂ Calibration Profiles

After the completion of the sampling campaign at Plant Crist, thermal dissociation profiles of $HgCl_2$ were measured for comparison with stack and plume samples. Samples were obtained by flowing N_2 over powdered $HgCl_2$ and passing the gas mixture through denuder tubes that were then analyzed using the "oven 1" or "oven 2" cycles. Calibration profiles from samples deposited at N_2 flows of 520 sccm and 135 sccm were measured although there were large fluctuations in the deposited amounts of $HgCl_2$ presumably as a result of fluctuating gas phase concentrations during sampling and variations in denuder collection efficiency. Six PTD profiles were measured using samples deposited at 135 sccm and two PTD profiles at 520 ccm deposition flow using the "oven 1" thermal ramp sequence. A single "oven 2" profile was measured, using a sample deposition flow of 520 sccm.

3.8. Characterization of the Oven Temperature Distribution

After the completion of the Plant Crist campaign and the laboratory H_3Cl_2 calibration experiments the temperature distribution along a denuder tube was measured by attaching thermocouples at the ends and center of the inside of a tube and at the center on the outside of the tube. With the oven controller reading 190 °C the temperature at the center of the denuder was 212 °C on the outside and 204 °C on the inside. The active heating section of the oven was 30 cm long and the temperature differential over the central 18 cm section was 30 °C with a maximum at the center, a drop of 10 °C at 9 cm downstream, *i.e.*, in the direction of gas flow, and a drop of 30 °C at 9 cm upstream at the point where cool gas is entering. The differential then increased considerably and was 30 °C cooler 12.5 cm upstream and 50 °C cooler at the 12.5 cm downstream point. The ability of the oven controller to produce reproducible temperature ramps was limited with the temperature typically overshooting the set point.

4. Conclusions

Measurements of the chemical speciation of RGM represent a significant analytical challenge. Lack of information on the chemical speciation is perhaps the single biggest obstacle to developing a detailed understanding of the chemical cycling of mercury in the atmosphere. A "proof of concept" experiment as a component of a major field campaign was conducted to test the use of PTD to chemically speciate RGM in a working combustion environment. Thermal dissociation profiles for RGM were obtained in both stack and in-plume samples that suggest that it is possible to track a specific chemical form of RGM from the stack and follow its evolution in the stack plume. In contrast to the recently published work of Deeds *et al.* [4], comparison of the measured plume RGM with the amount calculated from in-stack measurements and the measured plume dilution suggest that the stack and plume RGM concentrations are consistent with dilution. The PTD profiles of the stack and plume samples are consistent with $HgCl_2$ being the chemical form of the sampled RGM. It is also possible that the RGM consists of some other oxidized mercury species with a similar decomposition temperature. An example would be $HgBr_2$, but the absence of a significant amount of bromine in the feed coal makes this unlikely. Comparison with literature PTD profiles of reference mercury compounds suggests no other likely candidates for the speciation of RGM.

Acknowledgments: We wish to acknowledge the assistance of many of the other participants in the Plant Crist campaign for performing measurements, providing logistical support and releasing results prior to publication. This work was supported by the Electric Power Research Institute.

Author Contributions: Anthony Hynes, Cheryl Tatum Ernest and Dieter Bauer wrote the manuscript and performed much of the data analysis. Cheryl Tatum Ernest and Dieter Bauer performed the field experiments. Arnout Ter Schure performed the analysis of trace gas dilutions. Anthony Hynes, Cheryl Tatum Ernest, Dieter Bauer and Arnout Ter Schure were involved in the planning and preparation for the field deployment. Deanna Donohoue performed preliminary experiments and provided scientific insight and editing.

Conflicts of Interest: The authors declare no conflict of interest.

References

1. Edgerton, E.S.; Hartsell, B.E.; Jansen, J.J. Mercury speciation in coal-fired power plant plumes observed at three surface sites in the Southeastern U.S. *Environ. Sci. Technol.* **2006**, *40*, 4563–4570.
2. EPRI. *An Assessment of Mercury Emissions from U.S. Coal-Fired Power Plants*; EPRI Report No. 1000608; Electric Power Research Institute: Palo Alto, CA, USA, 2000.
3. Lohman, K.; Seigneur, C.; Edgerton, E.S.; Jansen, J.J. Modeling mercury in power plant plumes. *Environ. Sci. Technol.* **2006**, *40*, 3848–3854.
4. Deeds, D.A.; Banic, C.M.; Lu, J.; Daggupaty, S. Mercury speciation in a coal-fired power plant plume: An aircraft-based study of emissions from the 3640 MW Nanticoke Generating Station, ON, Canada. *J. Geophys. Res.: Atmos.* **2013**, *118*, 4919–4935.
5. Laudal, D.; Seigneur, C. *Mercury Reactions in Power Plant Plumes: Pleasant Prairie Experiment and Compliance Scenario Assessment*; Technical Report, Report No. 1010142; Electric Power Research Institute: Palo Alto, CA, USA, 2006.

6. Landis, M.; Ryan, J.; Oswald, E.; Jansen, J.; Monroe, L.; Walters, J.; Levin, L.; Ter Schure, A.; Laudal, D.; Edgerton, E. Plant crist mercury plume study. In Proceedings of Air Quality VII Conference, Arlington, VA, USA, 26–29 October 2009.

7. Schofield, K. Fuel-mercury combustion emissions: An important heterogeneous mechanism and an overall review of its implications. *Environ. Sci. Technol.* **2008**, *42*, 9014–9030.

8. Galbreath, K.C.; Zygarlicke, C.J. Mercury speciation in coal combustion and gasification flue gases. *Environ. Sci. Technol.* **1996**, *30*, 2421–2426.

9. Donohoue, D.A. Kinetic Studies of the Oxidation Pathways of Gaseous Elemental Mercury. Ph.D. Thesis, University of Miami, Miami, FL, USA, 2008.

10. Biester, H.; Scholz, C. Determination of mercury binding forms in contaminated soils: Mercury pyrolysis *versus* sequential extractions. *Environ. Sci. Technol.* **1997**, *31*, 233–239.

11. Feng, X.; Lu, J.; Gregoire, D.C.; Hao, Y.; Banic, C.M.; Schroeder, W.H. Analysis of inorganic mercury species associated with airborne particulate matter/aerosols: Method development. *Anal. Bioanal. Chem.* **2004**, *380*, 683–689.

12. Lopez-Anton, M.A.; Yang, Y.; Ron, P.; Maroto-Valer, M.M. Analysis of mercury species present during coal combustion by thermal desorption. *Fuel* **2010**, *89*, 629–634.

13. Wu, S.; Uddin, M.A.; Nagano, S.; Ozaki, M.; Sasaoka, E. Fundamental study on decomposition characteristics of mercury compounds over solid powder by temperature-programmed decomposition desorption mass spectrometry. *Energy Fuels* **2011**, *25*, 144–153.

14. Gormley, P.G.; Kennedy, M. Diffusion from a stream flowing through a cylindrical tube. *Proc. R. Irish Acad. Sect. A* **1949**, *52*, 163–169.

15. Schwarzenbach, R.P.; Gschwend, P.M.; Imboden, D.M. *Environmental Organic Chemistry*; John Wiley & Sons, Inc.: New York, NY, USA, 1993.

16. Goldfinger, P.; Jeunehomme, M. Mass spectrometric and Knudsen cell vaporization studies of 2B-6B compounds. *Trans. Faraday Soc.* **1963**, *59*, 2851–2867.

17. Shepler, B.C.; Peterson, K.A. Mercury monoxide: A systematic investigation of its ground electronic state. *J. Phys. Chem. A* **2003**, *107*, 1783–1787.

18. Hynes, A.J.; Donohoue, D.L.; Goodsite, M.E.; Hedgecock, I.M. Our current understanding of major chemical and physical processes affecting mercury dynamics in the atmosphere and at the air-water/terrestrial interfaces. In *Mercury Fate and Transport in the Global Atmosphere: Emissions, Measurements and Models*; Pirrone, N., Mason, R.P., Eds.; Springer: Berlin, Germany; pp. 427–457.

19. Landis, M.S.; Stevens, R.K.; Schaedlich, F.; Prestbo, E. Development and characterization of an annular denuder methodology for the measurement of divalent inorganic reactive gaseous mercury in ambient air. *Environ. Sci. Technol.* **2002**, *36*, 3000–3009.

20. Dumarey, R.; Brown, R.J.C.; Corns, W.T.; Brown, A.S.; Stockwell, P.B. Elemental mercury vapour in air: The origins and validation of the "Dumarey equation" describing the mass concentration at saturation. *Accredit. Qual. Assur.* **2010**, *15*, 409–414.

Regional Air Quality Model Application of the Aqueous-Phase Photo Reduction of Atmospheric Oxidized Mercury by Dicarboxylic Acids

Jesse O. Bash, Annmarie G. Carlton, William T. Hutzell and
O. Russell Bullock Jr.

Abstract: In most ecosystems, atmospheric deposition is the primary input of mercury. The total wet deposition of mercury in atmospheric chemistry models is sensitive to parameterization of the aqueous-phase reduction of divalent oxidized mercury (Hg^{2+}). However, most atmospheric chemistry models use a parameterization of the aqueous-phase reduction of Hg^{2+} that has been shown to be unlikely under normal ambient conditions or use a non mechanistic value derived to optimize wet deposition results. Recent laboratory experiments have shown that Hg^{2+} can be photochemically reduced to elemental mercury (Hg) in the aqueous-phase by dissolved organic matter and a mechanism and the rate for Hg^{2+} photochemical reduction by dicarboxylic acids (DCA) has been proposed. For the first time in a regional scale model, the DCA mechanism has been applied. The HO_2-Hg^{2+} reduction mechanism, the proposed DCA reduction mechanism, and no aqueous-phase reduction (NAR) of Hg^{2+} are evaluated against weekly wet deposition totals, concentrations and precipitation observations from the Mercury Deposition Network (MDN) using the Community Multiscale Air Quality (CMAQ) model version 4.7.1. Regional scale simulations of mercury wet deposition using a DCA reduction mechanism evaluated well against observations, and reduced the bias in model evaluation by at least 13% over the other schemes evaluated, although summertime deposition estimates were still biased by -31.4% against observations. The use of the DCA reduction mechanism physically links Hg^{2+} reduction to plausible atmospheric processes relevant under typical ambient conditions.

Reprinted from *Atmosphere*. Cite as: Bash, J.O.; Carlton, A.G.; Hutzell, W.T.; Bullock, O.R., Jr. Regional Air Quality Model Application of the Aqueous-Phase Photo Reduction of Atmospheric Oxidized Mercury by Dicarboxylic Acids. *Atmosphere* **2014**, *5*, 1–15.

1. Introduction

Current atmospheric mercury ambient concentrations are enriched by approximately a factor of three, due to centuries of mining operations using the element and emissions from fossil fuel combustion and industrial processes [1].

When mercury is deposited from the atmosphere, it can be converted into methylated mercury compounds in aquatic and terrestrial systems [2,3]. The primary vector of human mercury exposure is consuming fish with elevated levels of methylmercury, which can results in neurological damage and other health problems [2]. Bioaccumulation of methylmercury by piscivorous and insectivorous wildlife has been shown to adversely damage their nervous, excretory and reproductive systems [4].

Deposition of atmospheric mercury largely depends on its oxidation state. The atmospheric lifetime of gaseous elemental mercury (Hg°_g) is on the order of one year making it a global pollutant [5,6]. Oxidized mercury is more water soluble and reactive than Hg°_g and particulate bound mercury (PHg) is readily scrubbed by precipitation [6,7]. Oxidized gaseous (Hg^{2+}_g) and PHg mercury have an atmospheric lifetime on the order of a few days to weeks [8]. Hg^{2+}_g and PHg are emitted to the atmosphere by a number of industrial activities [9]. Emitted Hg°_g can be oxidized to produce Hg^{2+}_g and PHg compounds [5]. Gas-phase oxidation mechanisms effectively determine the rate at which Hg°_g in the global background pool deposits, while the Hg^{2+} reduction mechanisms, primarily in the aqueous-phase (e.g., cloud droplets), determine how much of the emitted Hg^{2+}_g and PHg are not regionally deposited and added to global background pool of Hg°_g.

Air quality models are the primary tools used to estimate how sources of atmospheric mercury deposit to sensitive aquatic and terrestrial ecosystems on regional to global scales. For these models to accurately estimate wet deposition and atmospheric concentrations, precipitation, mercury oxidation, reduction and partitioning to the particle and aqueous-phases must be accurately simulated. Model gas and aqueous chemistry are then critical to determine what fraction of deposited mercury comes from local regions and from the global background, while simulated precipitation determines the rate at which Hg^{2+}_g and PHg are scavenged from the atmosphere. Because atmospheric models over-predict mercury wet deposition without aqueous-phase reduction [10], aqueous-phase atmospheric reduction mechanisms for Hg^{2+} have been proposed [11–14]. A viable mechanism in regional and global models is needed to inform air-quality managers and policy makers with the best estimate of regional and global contributions to mercury deposition, the subsequent human and ecosystem exposure to MeHg and how the in changes chemical composition of the atmosphere may alter mercury deposition.

Many air quality models (AQMs) that simulate mercury fate and transport use the aqueous-phase reduction of divalent oxidized mercury (Hg^{2+}_{aq}). Some use reactions with the hyroperoxyl radical (HO_2) as the major reduction pathway of oxidized mercury [13,15]. Others use an aqueous-phase Hg^{2+} reduction rate scaled to match the estimated lifetime and seasonality of total gaseous mercury [16]. The aqueous-phase HO_2 reduction mechanism has been shown to be unlikely under normal environmental conditions [17]. Empirical aqueous-phase reduction schemes balance the gas-phase oxidation and emission sources of Hg^{2+}_g and may mask errors in the model chemistry, simulated precipitation, and/or emissions and may not capture changes in the Hg atmospheric chemistry due to emissions regulations or changes in atmospheric composition. Given that the aqueous reduction significantly controls the fate and transport of atmospheric mercury and that the current mechanisms have problems regarding model performance, a more physically plausible mechanisms is needed [10,15]. Photoreduction of $HgCl_2$ by dissolved organic acids has been reported in laboratory studies [11,13]. An atmospherically relevant mechanism has been demonstrated in laboratory experiments where C2-C4 dicarboxylic acids (DCAs) (oxalic, malonic and succinic acids) can readily complex with Hg^2_{+aq} to form $Hg°$ in low O_2 and Cl conditions [11]. Dicarboxylic acids demonstrated to reduce Hg are ubiquitous in the Earth's environment and are predicted to dominate DOC in cloud water [18,19]. In this study, the HO_2-Hg^{2+} reduction mechanism, a new DCA reduction mechanism proposed by Si and Ariya [11] using a rate of 1.2×10^4 $M^{-1} \cdot s^{-1}$, and no aqueous-phase reduction (NAR) of Hg^{2+} are evaluated against weekly wet deposition totals, concentrations and precipitation observations from the Mercury Deposition Network (MDN) [20] using the Community Multiscale Air Quality (CMAQ) model version 4.7.1 [5,21].

2. Results and Discussion

During the winter (January and February) and summer (July and August) simulations, the DCA reduction mechanism reduced the median model total mercury wet deposition bias by 13.2% and 14.3% when compared to the HO_2 case and by 39.4% and 25.9% for the NAR case for winter and summer simulations respectively (Figure 1). The NAR case led to normalized median model over-predictions of 53.4% and 43.7% in the summer and winter simulations respectively similar to Pongprueksa *et al.* [10] (Table 1). DCA and NAR case model simulations overestimated the wintertime wet deposition along the Gulf Coast (Figure 2). However, modeled wet deposition estimates in summer around the Gulf Coast were improved in the DCA case and were similar to the HO_2 case at other MDN sites (Figure 3). The NAR case over-predicted mercury wet deposition in the winter simulation by 43.7% while the DCA and HO_2 mechanism underestimated the wet deposition by −4.3% and −17.5% respectively (Table 1, Figure 1). The NAR case over-predicted the observed deposition in the winter by 43.7% with the largest over-predictions at coastal sites and in the Southeast and in the Ohio River Valley (Figure 2). Similarly, the DCA case overestimated the wet deposition along the Gulf Coast in the winter which also corresponded to a more frequent distribution of higher Hg concentrations in precipitation at those sites (Figure 2). This may be explained by elevated Hg^{2+}_g concentrations above the marine boundary layer in the southern

181

boundary conditions of this simulation as documented by Myers et al [22]. The normalized median error (NMdnE) in the winter DCA (69.4%) and HO$_2$ (61.5%) simulation cases were similar in magnitude but the model bias and Spearman rank correlation coefficient were improved (Table 1). Both summer and winter simulations using DCA reduction correlated better with the observations than using the HO$_2$ reduction (Table 1). The improved correlation was largely due to better capturing the observed deposition in the Southeast and Gulf Coast.

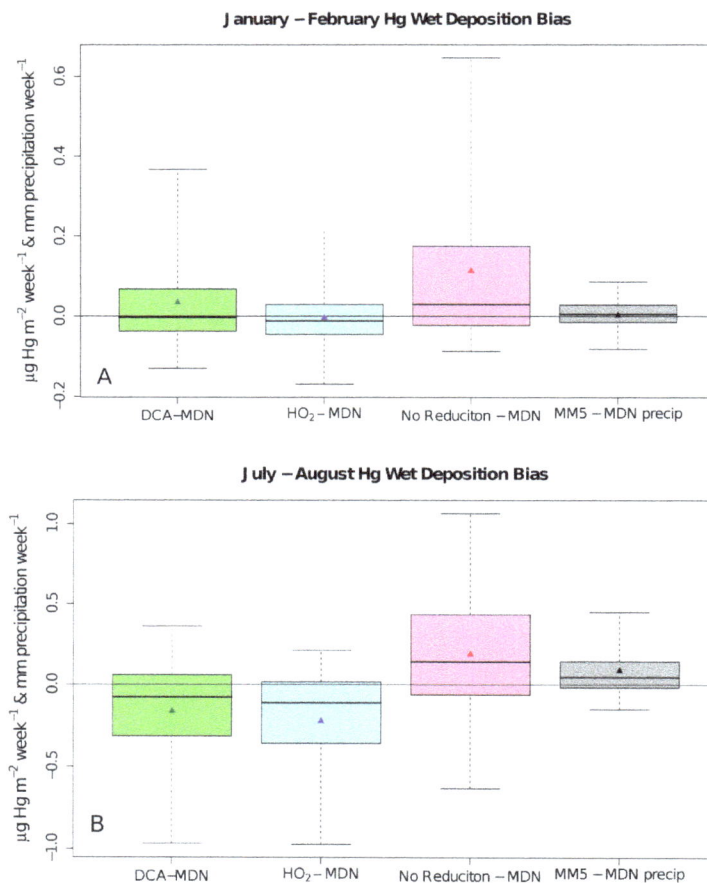

Figure 1. Boxplots of the Community Multiscale Air Quality (CMAQ) modeled bias in the Hg wet deposition and precipitation compared to mercury deposition network (MDN) observations using the DCA (green), HO$_2$ (blue), and NAR (red) cases and the MM5 precipitation (grey) for January and February 2002 (**A**) and July and August 2002 (**B**). The horizontal line through each box is the median bias, the box contains the 25% to 75% and the whiskers extend to the 5th and 95th percentile of the biases, and the triangle is the mean bias.

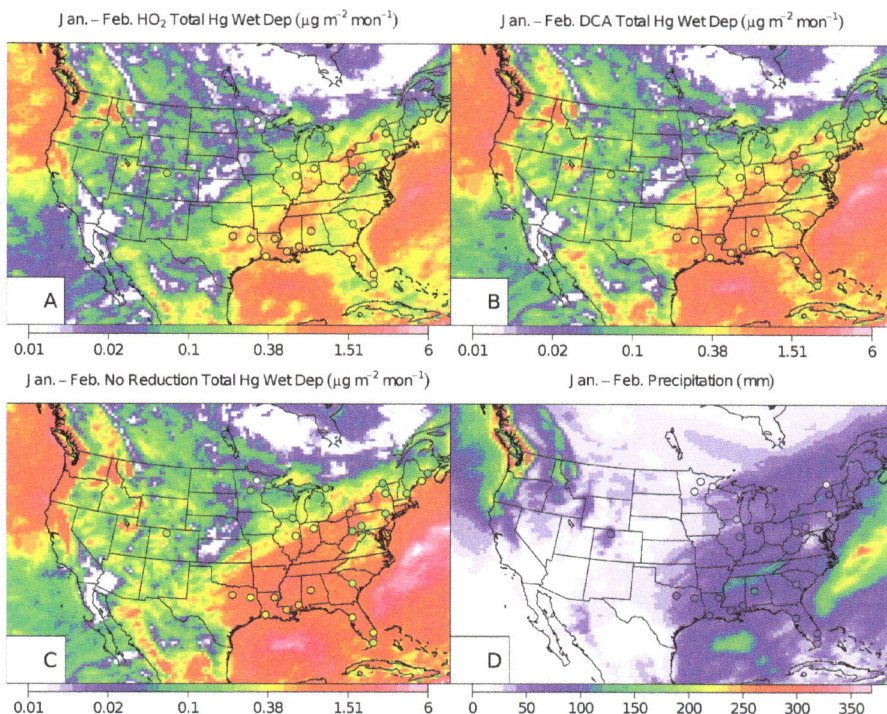

Jan. – Feb. HO$_2$ Total Hg Wet Dep (μg m^{-2} mon^{-1})

Jan. – Feb. DCA Total Hg Wet Dep (μg m^{-2} mon^{-1})

Jan. – Feb. No Reduction Total Hg Wet Dep (μg m^{-2} mon^{-1})

Jan. – Feb. Precipitation (mm)

Figure 2. CMAQ January and February 2002 wet deposition with MDN observations (points) for the HO$_2$ aqueous-phase reduction case (**A**), the DCA reduction case (**B**), NAR (**C**), and a map of the precipitation (**D**). Note, only the MDN measurement sites that had a complete set of observations over the two-month period were plotted.

Simulated intensity, duration and quantity of precipitation influences estimated wet deposition and the concentration of Hg in precipitation. The impact of the weekly precipitation quantity and weekly mean MDN and modeled Hg concentration (μg·m^{-3}) in the precipitation can be explored using MDN observations. MM5 precipitation estimates were relatively unbiased in the wintertime simulation (Figure 4). In the summer simulations, the bias was reversed by over-predicting the frequency of weeks with high precipitation and under-predicting the frequency of weeks with lower precipitation. The distribution of modeled Hg concentrations did not follow precipitation distributions in the summertime simulations except for the NAR case indicating that the HO$_2$ and DCA reduction cases may be too fast. Summer distributions of modeled concentrations were similar in the NAR case but did not correlate well (Pearman's r = 0.113; Spearman's rank correlation coefficient ρ = 0.060; N = 420) with the observations indicating that the distribution of concentrations over the domain was captured well but did not exhibit spatial or temporal patterns similar to the observations (Figure 4). Caution should be taken with the analysis of the probability distributions presented in Figure 4 because any given weekly sample may be an average of several events.

Table 1. Spearman's rank correlation coefficient (ρ), median bias (MdnB), median error (MdnE), normalized median bias (NMdnB), and normalized median error (NMdnE) of the MM5 modeled precipitation and CMAQ Hg wet deposition for total wet deposition, weekly (n = 372 in January-February, n = 420 in July-August) compared to MDN observations.

		ρ	MdnB	MndE	NMndB	NMndE
MM5	January–February	0.807	1.0 mm·week^{-1}	4.5 mm·week^{-1}	7.2%	33.2%
	July–August	0.424	8.5 mm·week^{-1}	16.2 mm·week^{-1}	53.4%	98.9%
No	January–February	0.586	29 ng·m^{-2}·week^{-1}	61 ng·m^{-2}·week^{-1}	43.7%	91.8%
	July–August	0.338	138 ng·m^{-2}·week^{-1}	269 ng·m^{-2}·week^{-1}	57.3%	111.3%
HO$_2$	January–February	0.570	−12 ng·m^{-2}·week^{-1}	41 ng·m^{-2}·week^{-1}	−17.5%	61.5%
	July–August	0.241	−110 ng·m^{-2}·week^{-1}	163 ng·m^{-2}·week^{-1}	−45.7%	67.5%
DCA	January–February	0.586	−3 ng·m^{-2}·week^{-1}	46 ng·m^{-2}·week^{-1}	−4.3%	69.4%
	July–August	0.252	−76 ng·m^{-2}·week^{-1}	175 ng·m^{-2}·week^{-1}	−31.4%	72.3%

CMAQ captures the seasonal and spatial variability of SOA well but has a known negative bias of approximately 40% during the summer months and DCA is cloud-formed SOA [23]. Cloud formed SOA precursors are likely under-predicted because recently identified water-soluble VOC precursors (e.g., glycoaldehyde [24] and methylacrolein [25]) are not included in SOA mechanism. Assuming that the bias in total SOA is proportional to cloud SOA precursors, a model sensitivity test was run where cloud SOA precursor concentrations were doubled for the July-August and January–February cases. A 100% increase in the DCA concentrations resulted in a 20% reduction and an 8% reduction in total wet deposition in the July–August and January–February cases respectively. The existing model bias was −45.7% and −4.3% and this sensitivity increased the model bias by 6 and 40 ng·m^{-2}·week^{-1} for the July–August and January–February cases respectively. Thus, "fixing" the ~40% underpredicted SOA (assuming a proportional ~40% increase in DCA) during the summer months resulted in wet deposition estimates similar to the HO$_2$ case in the summer and winter when using the DCA aqueous-phase reduction scheme. However, this sensitivity did not include the other changes in the atmospheric composition that would accompany a 40% increase in SOA. The DCA and HO$_2$ reduction mechanisms underpredicted the summer wet deposition observations and this seasonal model bias may be a result of another component of CMAQ's mercury chemical mechanism, the model boundary conditions, emissions, aqueous-phase oxidation if intermediate reduction products by O$_2$ occur, or gas-phase chemistry. The

HO$_2$ reduction case (the most rapid aqueous-phase reduction parameterization in the modeled sensitivities) shifted the probability distribution of average weekly aqueous-phase concentrations towards smaller concentrations than the slower mechanisms in all cases (Figure 4). This indicates that both reduction mechanisms may reduce too much Hg$^{2+}_{aq}$ in the warm months if the emissions of mercury species, modeled oxidant concentrations and the gas-phase chemistry are correct.

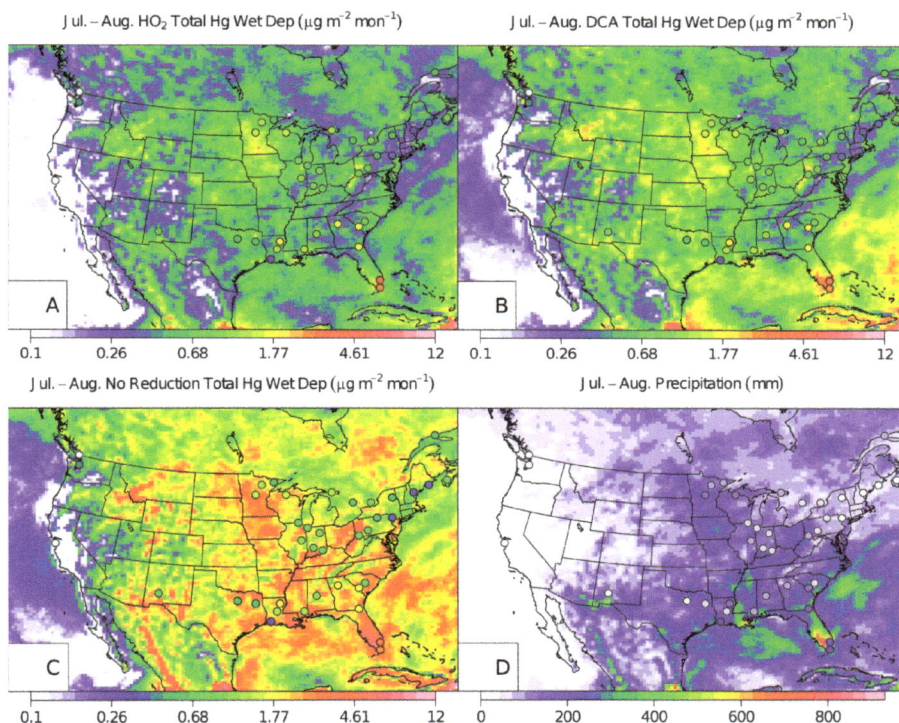

Figure 3. CMAQ July and August 2002 wet deposition with MDN observations (points) for the HO$_2$ aqueous-phase reduction case (**A**), the DCA reduction case (**B**), NAR case (**C**), and a map of the precipitation (**D**). Note, only the MDN measurement sites that had a complete set of observations over the two-month period were plotted.

Recent modeling studies using CMAQ and CAMx have documented biases in modeled Hg$^{2+}_g$ and PHg concentrations [26,27]. Some of these biases may originate with Hg$^{2+}_g$ underestimation of the measurements due to its affinity for KCl coated quartz denuder surfaces [28–30] and release in the presence of ozone [31]. However, it is important to quantify the impact that these aqueous-phase reduction mechanisms have on ambient Hg species. During the wintertime simulations, the

DCA reduction mechanism increased domain wide median $Hg°$, Hg^{2+}_g and PHg ambient concentrations by 0%, 6% and 7% respectively while the NAR case increased domain wide median Hg^{2+}_g and PHg ambient concentrations by −1%, 12% and 17% respectively over the HO_2 mechanism. During the summertime simulations where both Hg oxidant and DCA concentrations are higher, increased domain wide median $Hg°$, Hg^{2+}_g and PHg ambient concentrations by −1%, 8% and 5% respectively while the NAR case increased domain wide median Hg^{2+}_g and PHg ambient concentrations by −5%, 42% and 54% respectively over the HO_2 mechanism. The replacement of the HO_2 with the DCA reduction mechanism did not result in large increases in ambient Hg^{2+}_g or PHg concentrations while the removal of the aqueous-phase reduction mechanism resulted in elevated Hg^{2+}_g and PHg concentrations.

The improvements in model performance presented here may be influenced by areas of uncertainty in gas-phase or heterogeneous Hg chemistry. The aqueous reduction of Hg^{2+} by HO_2 in air-quality models is likely overestimated [32] and this could mask potential errors in $Hg°$ oxidation rates, Hg^{2+} emissions, and/or missing in-plume chemistry. A detailed review of Hg gas-phase chemistry is beyond the scope of this manuscript; however, in-plume reduction of Hg^{2+} is briefly discussed here. Currently, there are no mechanistic chemical mechanisms for in plume Hg^{2+} reduction and only empirical parameterizations have been applied in models [33]. In addition, there are recent observations that have indicated that $Hg°$ was being oxidized in plume [34,35]. Due to the lack of supporting experimental data, contrasting *in situ* observations and broad uncertainties in the complex heterogeneous chemistry governing the potential Hg^{2+} reduction by SO_2, the mechanisms of in-plume chemistry are unclear and the mechanistic parameterizations are not used in air-quality models [33,36]. As such the parameterization of in-plume chemistry is beyond the scope of this manuscript and was not considered in these sensitivity simulations.

The NAR case establishes that the model is sensitive to the aqueous reduction pathway as documented by Seignuer *et al.* [15] and Pongprueksa *et al* [10]. Similarities between the results from the HO_2 and DCA mechanisms indicate that the HO_2 mechanism may be a reasonable surrogate for reducing divalent mercury by DCA and may produce reasonable model wet deposition estimates. The DCA mechanism improved the predicted wet deposition for July and August by reducing the model bias and improving predicted spatial patterns shown by the Spearman's rank correlation coefficient for all four months simulated (Table 1). The improvement in the July and August deposition estimates removed much of the HO_2 deposition underestimation while being nearly unbiased in the January and February simulations. This will likely reduce the mercury wet deposition bias in annual simulations of mercury deposition with the DCA aqueous-phase reduction mechanism proposed by Si and Ariya [11].

January and February THg Wet Deposition Concentration

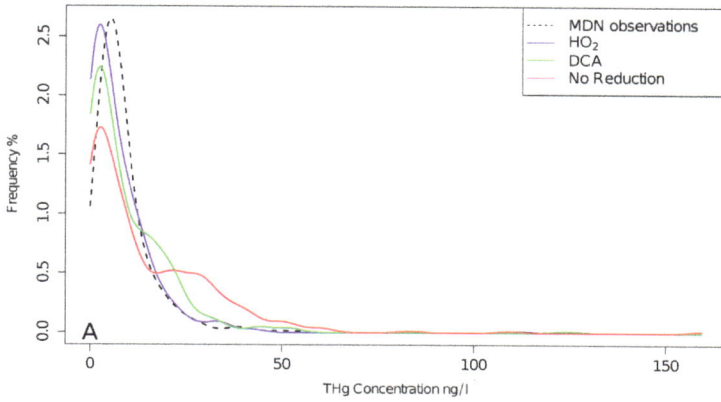

July and August THg Wet Deposition Concentration

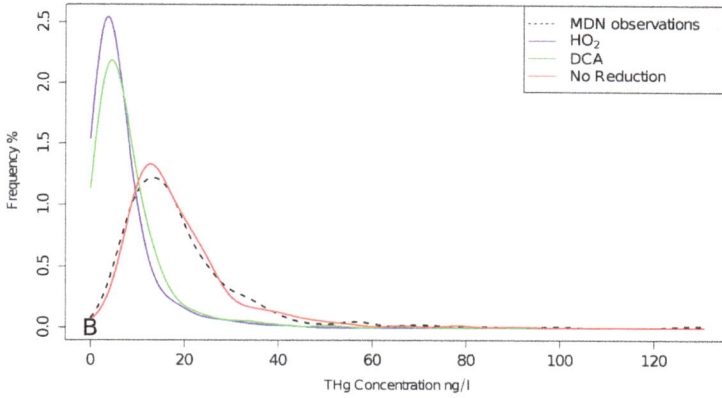

January and February Precipitation

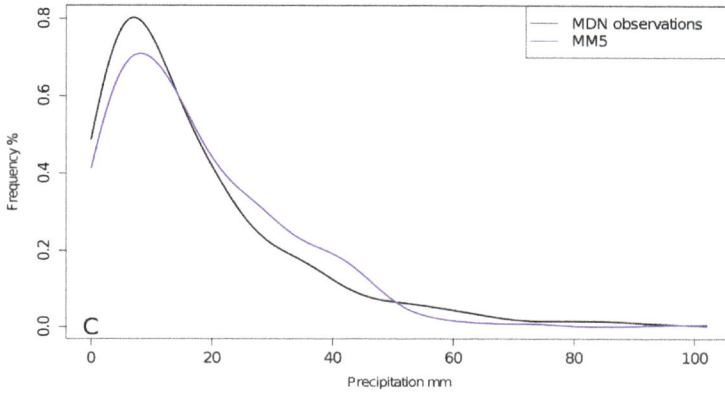

Figure 4. *Cont.*

187

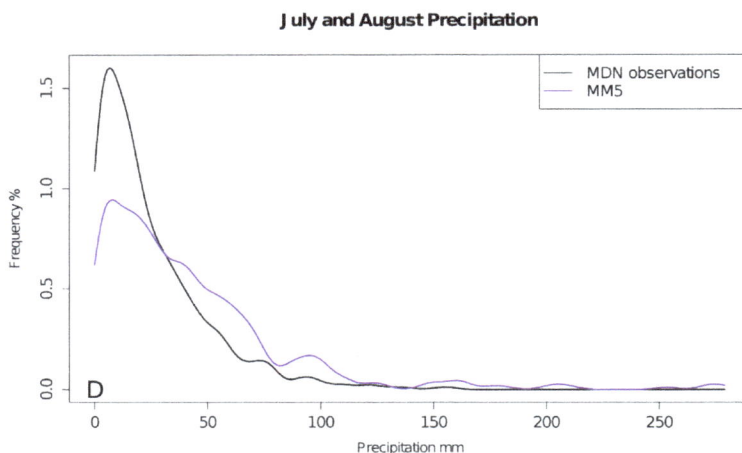

July and August Precipitation

Figure 4. Distribution of the weekly observed (black dashed line), HO$_2$ (blue), DCA (green) and NAR (red) cases for January and February simulations (**A**), July and August simulations (**B**), and for weekly observed (black) and modeled (blue) precipitation for January and February (**C**), and July and August (**D**).

3. Experimental Section

3.1. Application of an Aqueous-Phase Hg$^{2+}_{aq}$ DCA Reduction Scheme in CMAQ

CMAQ's aqueous chemistry mechanism was expanded to include organic reactions of glyoxal and methylglyoxal with ·OH to form cloud-produced SOA (SOA$_{cld}$) [37]. Species represented by SOA$_{cld}$ include oxalic, malonic, and succinic acids, as well as other carboxylic acids and larger humic-like substances (*i.e.*, organic compounds with oligomeric structure) demonstrated to form in aqueous-phase laboratory photooxidation experiments with glyoxal and methylglyoxal [38–41].

A rate constant for the reduction of Hg$^{2+}_{aq}$ by oxalic acid was implemented in the aqueous chemistry module in CMAQ. The oxalic acid rate is highest among DCAs and is used as an upper bound for mercury reduction *via* cloud processing. Additionally, oxalic acid is the dominant product for glyoxal [39] and methylglyoxal [42] oxidation in clouds. In other atmospheric aqueous environments, oxalic typically dominates over other DCAs [43,44]. At a typical cloud contact time (~10 min [45]), the product yield of Hg° does not appear to be significantly different among the different rates presented by Si and Ariya [11].

3.2. CMAQ Model Simulations

Simulations were run for January and February 2002 with a spin-up period from December 22nd to the 31st 2001, and July and August 2002 with a spin up period from

June 21st to the 30th. CMAQ version 4.7.1 configured with the carbon bond version 5 (CB05) chemical mechanism [46], version 5 of the aerosol model (AERO5) [23], in-line dry deposition, and 14 non-hydrostatic terrain-following vertical layers extending vertically to the 100 mb (~14 km) level was chosen as the base model. Initial and dynamic boundary conditions were provided from GEOS-Chem Hg [47]. Meteorological data were provided by the Penn Sate/NCAR fifth-generation mesoscale model (MM5) [48] with the P-X land surface scheme [49]. Anthropogenic Hg emissions estimates were used from version 3 of the 2002 EPA National Emissions Inventory (NEI, http://www.epa.goc/ttn/chief/net/critsummary.html) and mercury emissions from the recycling of deposited anthropogenic emissions and direct emissions from volcanoes and geologically enriched areas were estimated following Bullock *et al.* [50].

Three aqueous-phase Hg^{2+} reduction cases were tested: (1) the base case modeled the reduction of Hg^{2+}_{aq} by the $HO_2\cdot$ radical as described in Pehkonen and Lin [13] and implemented in CMAQ version 4.6 [50], (2) the DCA case modeled the reduction of Hg^{2+}_{aq} by dicarboxylic acids ($Hg^{2+}_{aq} + DCA_{aq} + h\nu \rightarrow Hg^{\circ}_{aq}$) proposed by Si and Ariya [11] using a rate of $1.2 \times 10^4 \ M^{-1}\cdot s^{-1}$, and (3) NAR case had no aqueous-phase reduction mechanism for Hg^{2+}_{aq}. Results of the three cases were evaluated against total mercury wet deposition measurements collected by the mercury deposition network (MDN) [20]. In 2002, there were few MDN monitors in operation in the Western US thus the evaluation presented here is primarily focused on the Eastern US. There were a total of 64 sites in operation in January and February with 372 valid weekly observations and 65 sites in July and August with 420 valid weekly observations. Model evaluation metrics used were the median bias (MdnB), normalized median bias (NMdnB), median error (MdnE), normalized median error (NMdnE), and Spearman's rank correlation coefficient (ρ). The MdnB, MdnE, NMdnB, NMdnE, and ρ are defined as follows:

$$MdnB = median(C_M - C_O)_N \tag{1}$$

$$MdnE = median \mid C_M - C_O \mid_N \tag{2}$$

$$NMdnB = \frac{median(C_M - C_O)_N}{median(C_O)_N} \tag{3}$$

$$NMdnE = \frac{median|C_M - C_O|_N}{median(C_O)_N} \tag{4}$$

$$\rho = 1 - \frac{6\sum_{i=1}^N D_i^2}{N(N^2 - 1)} \tag{5}$$

where, C_M and C_O are the model and observed deposition values respectively, N is the total number of model and observation pairs, D_i is the difference in ranks between the ith pair of deposition values. Note that in Equations (2) and (4) that estimate error, the vertical bars denote the absolute value between the model and observations. The median was chosen as the measure of central tendency and the Spearman's rank correlation coefficient was used to assess the fit between the model simulations and observations. Wet deposition data are not normally distributed and metrics based off the median offer a more robust representation of these comparison metrics and are less sensitive to outliers [51]. All model evaluations were done with observations and model results paired in space and time for each weekly MND observation, the highest temporal resolution of the observations, as recommended by Simon *et al.* [52].

4. Conclusions

The HO_2 aqueous-phase reduction mechanism is widely believed to be unlikely under normal environmental conditions [17,32,53]. These results from a more likely reduction mechanism, the DCA case, will be more useful and physically meaningful in future applications than applying tuned empirical reduction rates or a mechanism that is unlikely under typical ambient environmental conditions. Both aqueous-phase reduction mechanisms shift the distribution of weekly deposition towards smaller events. However, the aqueous-phase reduction mechanisms under-predict the large weekly deposition events in July and August due to an over-prediction of precipitation due to the dilution of the aqueous concentrations while over-predicting the frequency of small deposition events in January and February (Figure 4). Overall, the DCA case resulted in a reduction in the modeled Hg deposition biases and an increase in the model correlation over HO_2 case without large increases in the surface level ambient concentrations. Considerable uncertainty exists in the gas-phase mercury chemistry, emissions and boundary conditions preventing a conclusive identification of the cause of the seasonal Hg wet deposition biases. Biases in our simulation may originate from missing the temporal dynamics in the emissions of $Hg^{2+}{}_g$ and PHg. Alternatively, the reaction rate constants are missing a temperature dependence and so the $Hg^\circ{}_g$ oxidation rates may be too fast in the cool seasons and aloft or the $Hg^\circ{}_g$ oxidation mechanism may be incomplete. Recent annual simulations of CMAQ and GEOS-Chem compared to ambient speciated mercury observations indicate that the models are overestimating the fraction of oxidized gaseous mercury [26,54]. The sources of the seasonal biases should be further investigated.

Acknowledgments: Although this work was reviewed by EPA and approved for publication, it may not necessarily reflect official Agency policy. Mention of commercial products does not constitute endorsement by the Agency. This publication was supported in part by US EPA grant (83504101).

Conflicts of Interest: The authors declare no conflict of interest.

References and Notes

1.   Streets, D.G.; Devane, M.K.; Lu, Z.F.; Bond, T.C.; Sunderland, E.M.; Jacob, D.J. All-time releases of mercury to the atmosphere from human activities. *Environ. Sci. Tech.* **2011**, *45*, 10485–10491.

2.   Sunderland, E.M.; Krabbenhoft, D.P.; Moreau, J.W.; Strode, S.A.; Landing, W.M. Mercury sources, distribution, and bioavailability in the North Pacific Ocean: Insights from data and models. *Glob. Biogeochem. Cy.* **2009**, *23*, GB2010.

3.   Goulet, R.R.; Holmes, J.; Page, B.; Poissant, L.; Siciliano, S.D.; Lean, D.R.S.; Wang, F.; Amyot, M.; Tessier, A. Mercury transformations and fluxes in sediments of a riverine wetland. *Geochim. Cosmochim. Acta* **2007**, *71*, 3393–3406.

4.   Wolfe, M.F.; Schwarzbach, S.; Sulaiman, R.A. Effects of mercury on wildlife: A comprehensive review. *Environ. Toxicol. Chem.* **1998**, *17*, 146–160.

5.   Lin, C.J.; Pongprueksa, P.; Lindberg, S.E.; Pehkonen, S.O.; Byun, D.; Jang, C. Scientific uncertainties in atmospheric mercury models I: Model science evaluation. *Atmos. Environ.* **2006**, *40*, 2911–2928.

6.   Schroeder, W.H.; Munthe, J. Atmospheric mercury—An overview. *Atmos. Environ.* **1998**, *32*, 809–822.

7.   Fain, X.; Obrist, D.; Hallar, A.G.; McCubbin, I.; Rahn, T. High levels of reactive gaseous mercury observed at a high elevation research laboratory in the Rocky Mountains. *Atmos. Chem. Phys.* **2009**, *9*, 8049–8060.

8.   Selin, N.E. Global Biogeochemical Cycling of Mercury: A Review. *Annu. Rev. Environ. Resour.* **2009**, *34*, 43–63.

9.   Pacyna, E.G.; Pacyna, J.M.; Steenhuisen, F.; Wilson, S. Global anthropogenic mercury emission inventory for 2000. *Atmos. Environ.* **2006**, *40*, 4048–4063.

10. Pongprueksa, P.; Lin, C.J.; Lindberg, S.E.; Jang, C.; Braverman, T.; Bullock, O.R.; Ho, T.C.; Chu, H.W. Scientific uncertainties in atmospheric mercury models III: Boundary and initial conditions, model grid resolution, and Hg(II) reduction mechanism. *Atmos. Environ.* **2008**, *42*, 1828–1845.

11. Si, L.; Ariya, P.A. Reduction of oxidized mercury species by dicarboxylic acids (C(2)–C(4)): Kinetic and product studies. *Environ. Sci. Tech.* **2008**, *42*, 5150–5155.

12. Van Loon, L.; Mader, E.; Scott, S.L. Reduction of the aqueous mercuric ion by sulfite: UV spectrum of HgSO3 and its intramolecular redox reaction. *J. Phys. Chem. A* **2000**, *104*, 1621–1626.

13. Pehkonen, S.O.; Lin, C.J. Aqueous photochemistry of mercury with organic acids. *J. Air Waste Manag. Assoc.* **1998**, *48*, 144–150.

14. Xiao, Z.F.; Munthe, J.; Strömberg, D.; Lindqvist, O. *Photochemical Behaviour of Inorganic Mercury Compounds in Aqueous Solution*; Lewis Publishers: London, UK, 1994.

15. Seigneur, C.; Vijayaraghavan, K.; Lohman, K. Atmospheric mercury chemistry: Sensitivity of global model simulations to chemical reactions. *J. Geophys. Res.: Atmos.* **2006**, *111*.

16. Selin, N.E.; Jacob, D.J.; Park, R.J.; Yantosca, R.M.; Strode, S.; Jaegle, L.; Jaffe, D. Chemical cycling and deposition of atmospheric mercury: Global constraints from observations. *J. Geophys. Res.: Atmos.* **2007**, *112*.

17. Gardfeldt, K.; Jonsson, M. Is bimolecular reduction of Hg(II) complexes possible in aqueous systems of environmental importance. *J. Phys. Chem. A* **2003**, *107*, 4478–4482.

18. Sorooshian, A.; Lu, M.L.; Brechtel, F.J.; Jonsson, H.; Feingold, G.; Flagan, R.C.; Seinfeld, J.H. On the source of organic acid aerosol layers above clouds. *Environ. Sci. Tech.* **2007**, *41*, 4647–4654.

19. Liu, J.; Horowitz, L.W.; Fan, S.; Carlton, A.G.; Levy, H. Global in-cloud production of secondary organic aerosols: Implementation of a detailed chemical mechanism in the GFDL atmospheric model AM3. *J. Geophys. Res.: Atmos.* **2012**, *117*, D15303.

20. Lindberg, S.; Vermette, S. Workshop on sampling mercury in precipitation for the national atmospheric deposition program. *Atmos. Environ.* **1995**, *29*, 1219–1220.

21. Foley, K.M.; Roselle, S.J.; Appel, K.W.; Bhave, P.V.; Pleim, J.E.; Otte, T.L.; Mathur, R.; Sarwar, G.; Young, J.O.; Gilliam, R.C.; *et al.* Incremental testing of the Community Multiscale Air Quality (CMAQ) modeling system version 4.7. *Geosci. Model. Dev.* **2010**, *3*, 205–226.

22. Myers, T.; Atkinson, R.D.; Bullock, O.R.; Bash, J.O. Investigation of effects of varying model inputs on mercury deposition estimates in the Southwest US. *Atmos. Chem. Phys.* **2013**, *13*, 997–1009.

23. Carlton, A.G.; Bhave, P.V.; Napelenok, S.L.; Edney, E.D.; Sarwar, G.; Pinder, R.W.; Pouliot, G.A.; Houyoux, M. Model representation of secondary organic aerosol in CMAQv4.7. *Environ. Sci. Tech.* **2010**, *44*, 8553–8560.

24. Perri, M.J.; Seitzinger, S.; Turpin, B.J. Secondary organic aerosol production from aqueous photooxidation of glycolaldehyde: Laboratory experiments. *Atmos. Environ.* **2009**, *43*, 1487–1497.

25. El Haddad, I.; Liu, Y.; Nieto-Gligorovski, L.; Michaud, V.; Temime-Roussel, B.; Quivet, E.; Marchand, N.; Sellegri, K.; Monod, A. In-cloud processes of methacrolein under simulated conditions-Part 2: Formation of secondary organic aerosol. *Atmos. Chem. Phys.* **2009**, *9*, 5107–5117.

26. Baker, K.R.; Bash, J.O. Regional scale photochemical model evaluation of total mercury wet deposition and speciated ambient mercury. *Atmos. Environ.* **2012**, *49*, 151–152.

27. Holloway, T.; Voigt, C.; Morton, J.; Spak, S.N.; Rutter, A.P.; Schauer, J.J. An assessment of atmospheric mercury in the Community Multiscale Air Quality (CMAQ) model at an urban site and a rural site in the Great Lakes Region of North America. *Atmos. Chem. Phys.* **2012**, *12*, 7117–7133.

28. Huang, J.Y.; Miller, M.B.; Weiss-Penzias, P.; Gustin, M.S. Comparison of gaseous oxidized Hg measured by KCl-coated denuders, and nylon and cation exchange membranes. *Environ. Sci. Tech.* **2013**, *47*, 7307–7316.

29. Ambrose, J.L.; Lyman, S.N.; Huang, J.Y.; Gustin, M.S.; Jaffe, D.A. Fast time resolution oxidized mercury measurements during the Reno Atmospheric Mercury Intercomparison Experiment (RAMIX). *Environ. Sci. Tech.* **2013**, *47*, 7285–7294.

30. Gustin, M.S.; Huang, J.Y.; Miller, M.B.; Peterson, C.; Jaffe, D.A.; Ambrose, J.; Finley, B.D.; Lyman, S.N.; Call, K.; Talbot, R.; *et al.* Do we understand what the mercury speciation instruments are actually measuring? results of RAMIX. *Environ. Sci. Tech.* **2013**, *47*, 7295–7306.

31. Lyman, S.N.; Jaffe, D.A.; Gustin, M.S. Release of mercury halides from KCl denuders in the presence of ozone. *Atmos. Chem. Phys.* **2010**, *10*, 8197–8204.

32. Hynes, A.J.; Donohoue, D.L.; Goodsite, M.E.; Hedgecock, I.M. Our Current Understanding of Major Chemical and Physical Processes Affecting Mercury Dynamics in the Atmosphere and at the Air-Water/Terrestrial Iterfaces. In *Mercury Fate and Transport in the Global Atmosphere*; Springer: Berlin, Germany, 2009; pp. 427–457.

33. Subir, M.; Ariya, P.A.; Dastoor, A.P. A review of the sources of uncertainties in atmospheric mercury modeling II. Mercury surface and heterogeneous chemistry—A missing link. *Atmos. Environ.* **2012**, *46*, 1–10.

34. Kolker, A.; Olson, M.L.; Krabbenhoft, D.P.; Tate, M.T.; Engle, M.A. Patterns of mercury dispersion from local and regional emission sources, rural central wisconsin, USA. *Atmos. Chem. Phys.* **2010**, *10*, 4467–4476.

35. Timonen, H.; Ambrose, J.L.; Jaffe, D.A. Oxidation of elemental Hg in anthropogenic and marine airmasses. *Atmos. Chem. Phys.* **2013**, *13*, 2827–2836.

36. Kos, G.; Ryzhkov, A.; Dastoor, A.; Narayan, J.; Steffen, A.; Ariya, P.A.; Zhang, L. Evaluation of discrepancy between measured and modelled oxidized mercury species. *Atmos. Chem. Phys.* **2013**, *13*, 4839–4863.

37. Carlton, A.G.; Turpin, B.J.; Altieri, K.E.; Seitzinger, S.P.; Mathur, R.; Roselle, S.J.; Weber, R.J. CMAQ model performance enhanced when in-cloud secondary organic aerosol is included: Comparisons of organic carbon predictions with measurements. *Environ. Sci. Tech.* **2008**, *42*, 8798–8802.

38. Carlton, A.G.; Turpin, B.J.; Lim, H.J.; Altieri, K.E.; Seitzinger, S. Link between isoprene and secondary organic aerosol (SOA): Pyruvic acid oxidation yields low volatility organic acids in clouds. *Geophys. Res. Lett.* **2006**, *33*, L06822.

39. Carlton, A.G.; Turpin, B.J.; Altieri, K.E.; Seitzinger, S.; Reff, A.; Lim, H.J.; Ervens, B. Atmospheric oxalic acid and SOA production from glyoxal: Results of aqueous photooxidation experiments. *Atmos. Environ.* **2007**, *41*, 7588–7602.

40. Altieri, K.E.; Carlton, A.G.; Lim, H.-J.; Turpin, B.J.; Seitzinger, S.P. Evidence for oligomer formation in clouds: Reactions of isoprene oxidation products. *Environ. Sci. Tech.* **2006**, *40*, 4956–4960.

41. Altieri, K.E.; Seitzinger, S.P.; Carlton, A.G.; Turpin, B.J.; Klein, G.C.; Marshall, A.G. Oligomers formed through in-cloud methylglyoxal reactions: Chemical composition, properties, and mechanisms investigated by ultra-high resolution FT-ICR mass spectrometry. *Atmos. Environ.* **2008**, *42*, 1476–1490.

42. Tan, Y.; Carlton, A.G.; Seitzinger, S.P.; Turpin, B.J. SOA from methylglyoxal in clouds and wet aerosols: Measurement and prediction of key products. *Atmos. Environ.* **2010**, *44*, 5218–5226.

43. Legrand, M.; Preunkert, S.; Oliveira, T.; Pio, C.A.; Hammer, S.; Gelencser, A.; Kasper-Giebl, A.; Laj, P. Origin of C-2-C-5 dicarboxylic acids in the European atmosphere inferred from year-round aerosol study conducted at a west-east transect. *J. Geophys. Res.: Atmos.* **2007**, *112*, D23S07.

44. Sorooshian, A.; Varutbangkul, V.; Brechtel, F.J.; Ervens, B.; Feingold, G.; Bahreini, R.; Murphy, S.M.; Holloway, J.S.; Atlas, E.L.; Buzorius, G.; *et al.* Oxalic acid in clear and cloudy atmospheres: Analysis of data from international consortium for atmospheric research on transport and transformation 2004. *J. Geophys. Res.: Atmos.* **2006**, *111*, D23S45.

45. Ervens, B.; Feingold, G.; Frost, G.J.; Kreidenweis, S.M. A modeling study of aqueous production of dicarboxylic acids: 1. Chemical pathways and speciated organic mass production. *J. Geophys. Res.: Atmos.* **2004**, *109*, D15205.

46. Sarwar, G.; Luecken, D.; Yarwood, G.; Whitten, G.Z.; Carter, W.P.L. Impact of an updated carbon bond mechanism on predictions from the CMAQ modeling system: Preliminary assessment. *J. Appl. Meteorol. Clim.* **2008**, *47*, 3–14.

47. Strode, S.A.; Jaegle, L.; Selin, N.E.; Jacob, D.J.; Park, R.J.; Yantosca, R.M.; Mason, R.P.; Slemr, F. Air-sea exchange in the global mercury cycle. *Glob. Biogeochem. Cy.* **2007**, *21*, GB1017.

48. Grell, G.; Dudhia, J.; Stouffer, D. *A Description of the Fifth-Generation PENN State/NCAR Mesoscale Model (MM5)*; National Center for Atmospheric Research: Boulder, CO, USA, 1994.

49. Pleim, J.E.; Xiu, A. Development and testing of a surface flux and planetary boundary-layer model for application in mesoscale models. *J. Appl. Meteorol.* **1995**, *34*, 16–32.

50. Bullock, O.R.; Atkinson, D.; Braverman, T.; Civerolo, K.; Dastoor, A.; Davignon, D.; Ku, J.Y.; Lohman, K.; Myers, T.C.; Park, R.J.; *et al.* The North American Mercury Model Intercomparison Study (NAMMIS): Study description and model-to-model comparisons. *J. Geophys. Res.: Atmos.* **2008**, *113*, D17310.

51. Appel, K.W.; Roselle, S.J.; Gilliam, R.C.; Pleim, J.E. Sensitivity of the Community Multiscale Air Quality (CMAQ) model v4.7 results for the eastern United States to MM5 and WRF meteorological drivers. *Geosci. Model. Dev.* **2010**, *3*, 169–188.

52. Simon, H.; Baker, K.R.; Phillips, S. Compilation and interpretation of photochemical model performance statistics published between 2006 and 2012. *Atmos. Environ.* **2012**, *61*, 124–139.

53. Zheng, W.; Hintelmann, H. Mercury isotope fractionation during photoreduction in natural water is controlled by its Hg/DOC ratio. *Geochim. Cosmochim. Acta* **2009**, *73*, 6704–6715.

54. Amos, H.M.; Jacob, D.J.; Holmes, C.D.; Fisher, J.A.; Wang, Q.; Yantosca, R.M.; Corbitt, E.S.; Galarneau, E.; Rutter, A.P.; Gustin, M.S.; *et al.* Gas-particle partitioning of atmospheric Hg(II) and its effect on global mercury deposition. *Atmos. Chem. Phys.* **2012**, *12*, 591–603.

Decreases in Mercury Wet Deposition over the United States during 2004–2010: Roles of Domestic and Global Background Emission Reductions

Yanxu Zhang and Lyatt Jaeglé

Abstract: Wet deposition of mercury (Hg) across the United States is influenced by changes in atmospheric conditions, domestic emissions and global background emissions. We examine trends in Hg precipitation concentrations at 47 Mercury Deposition Network (MDN) sites during 2004–2010 by using the GEOS-Chem nested-grid Hg simulation. We run the model with constant anthropogenic emissions and subtract the model results from the observations. This helps to remove the variability in observed Hg concentrations caused by meteorological factors, including precipitation. We find significant decreasing trends in Hg concentrations in precipitation at MDN sites in the Northeast ($-4.1 \pm 0.49\%$ yr^{-1}) and Midwest ($-2.7 \pm 0.68\%$ yr^{-1}). Over the Southeast ($-0.53 \pm 0.59\%$ yr^{-1}), trends are weaker and not significant, while over the West, trends are highly variable. We conduct model simulations assuming a 45% decrease in Hg emissions from domestic sources in the modeled period and a uniform 12% decrease in background atmospheric Hg concentrations. The combination of domestic emission reductions and decreasing background concentrations explains the observed trends over the Northeast and Midwest, with domestic emission reductions accounting for 58–46% of the decreasing trends. Over the Southeast, we overestimate the observed decreasing trend, indicating potential issues with our assumption of uniformly decreasing background Hg concentrations.

Reprinted from *Atmosphere*. Cite as: Zhang, Y.; Jaeglé, L. Decreases in Mercury Wet Deposition over the United States during 2004–2010: Roles of Domestic and Global Background Emission Reductions. *Atmosphere* **2013**, *4*, 113–131.

1. Introduction

Mercury (Hg) is listed as a Hazardous Air Pollutant (HAP) by the US Environmental Protection Agency (EPA) because of its neurotoxicity [1]. Atmospheric Hg is emitted in two forms: gaseous elemental Hg (GEM or Hg(0)) and divalent oxidized Hg (Hg(II)), which can partition into both gaseous oxidized Hg (GOM) and particulate-bound Hg (PBM). Hg originates from both natural sources, mainly in the form of Hg(0), and anthropogenic sources, as Hg(0) and Hg(II) [2,3]. It is generally assumed that 1/3 of global Hg emissions are natural, 1/3 are anthropogenic and

1/3 are derived from re-emission of previously deposited anthropogenic Hg [4,5]. Hg(0) has an atmospheric lifetime of several months to a year and can thus be transported on large scales before being deposited or oxidized to Hg(II), while Hg(II) is readily deposited and thus has shorter transport distances [6]. Deposition of Hg to water surfaces and the resulting possibility of fish contamination is the main human exposure pathway to Hg in North America [7,8].

In the United States, the main anthropogenic sources of Hg include coal-fired power plants (CFPP), waste incineration, cement manufacturing, chemical production, primary metal production and metal mining [9]. The relative contribution of these domestic anthropogenic emissions to Hg deposition over the United States has been evaluated in several studies [10–12]. For example, Seigneur *et al.* [10] used a multi-scale modeling system and found North American anthropogenic emissions contributions to deposition to be 24% (wet deposition) and 43% (dry deposition). Selin and Jacob [11] found that North American anthropogenic sources contribute 27% (wet) and 17% (dry) to deposition fluxes over the US in 2004–2005. To account for the observed small fraction of Hg(II) in plumes downwind of power plants [13,14], Zhang *et al.* [12] modified the source profiles of Hg emitted from power plants, as well as waste incinerators and estimated lower contributions of 10% (wet) and 13% (dry) in 2008–2009.

Figure 1 summarizes US Hg anthropogenic emissions for 1990–2010 based on several inventories: the Toxics Release Inventory (TRI) [15], the National Toxics Inventory (NTI) [16] and the National Emission Inventory (NEI) [17]. We separate emissions into three categories: CFPP, waste incinerators and "others" (the remaining US anthropogenic source categories). The implementation of the 1990 Solid Waste Combustion Rule as part of the Clean Air Act has led to the decrease of Hg emissions from medical, municipal, commercial and industrial waste incinerators from 122 $Mg \cdot yr^{-1}$ in 1990 to 10.9 $Mg \cdot yr^{-1}$ in 1999 (Figure 1) [18]. Anthropogenic Hg emissions remained nearly constant between 1999 and 2005, with CFPPs accounting for 50% of Hg emissions. In 2005, the EPA issued the Clean Air Mercury Rule to cap and reduce CFPP emissions [19]. Although this rule was overturned in 2008, Hg emissions from the CFPP sector are estimated by EPA to have decreased by 43%, from 52 $Mg \cdot yr^{-1}$ in 2005 (NEI2005) to 29.5 $Mg \cdot yr^{-1}$ in 2008 (NEI2008) and remained constant during 2008–2010 based on the recent release of 2010 Information Collection Request (ICR) estimates by the EPA [20]. The causes for these large emission reductions are not fully understood, but likely include compliance with state Hg-specific rules, voluntary reductions by utility operators and co-benefits of Hg reductions from control devices installed for the reduction of other pollutants, such as SO_2 and particulate matter [9]. In parallel, between 2005 and 2008 Hg emissions from waste incinerators decreased from 20 $Mg \cdot yr^{-1}$ (NEI2005) to 7 $Mg \cdot yr^{-1}$ (NEI2008). Emissions from other sectors also decreased

197

from 39 Mg·yr^{-1} (NEI2005) to 24 Mg·yr^{-1} (NEI2008). With the announcement of the Mercury and Air Toxics Standard (MATS) in December 2011 [21], Hg emissions from CFPPs are projected to decrease to less than 7 Mg·yr^{-1} in 2016 [9,22]. The significant decreases in Hg emissions over the 1990–1999 and 2005–2010 time periods provide an opportunity to quantitatively evaluate the contribution of anthropogenic emissions to Hg deposition over the US.

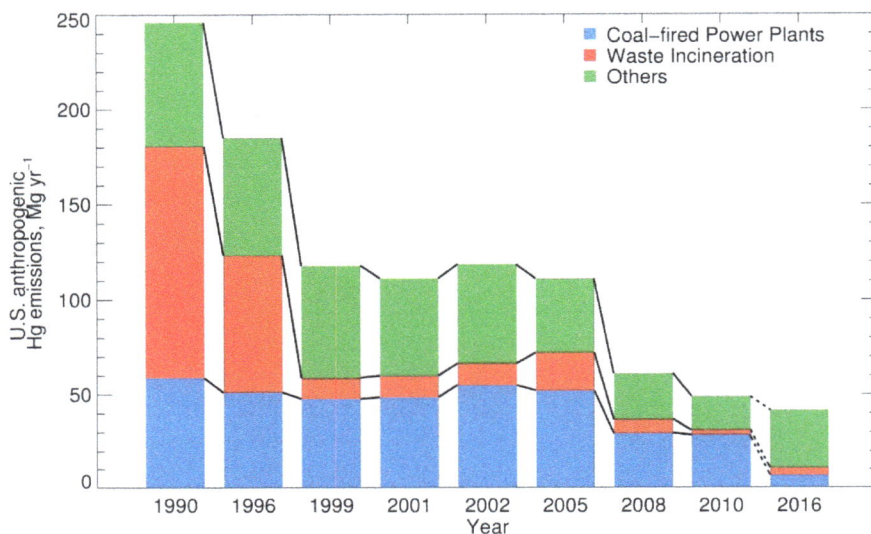

Figure 1. Anthropogenic Hg emissions in the US between 1990 and 2010 and projected emissions for 2016. Hg emissions are lumped into three source categories: coal-fired power plants (CFPPs), waste incineration (including hospital/medical/infectious, municipal, industrial/commercial and hazardous waste incineration and sewage sludge) and others (mainly stone/clay/glass/cement production, chemical production, primary metal production and metal mining). Emissions for 1990, 1999, 2002, 2005 and 2008 are from the National Emission Inventories (NEI) [17]. The 1996 emissions are from the National Toxics Inventories (NTI) [16]. The data for 2001 is from US EPA [23]. The CFPP emission for 2010 is based on Information Collection Request (ICR) by the EPA [20]. For 2010, non-CFPP emissions are estimated based on Toxic Release Inventories (TRI) for 2005 and 2010 [15]. Emissions for 2016 are projected by the Integrated Planning Model (IPM) [22].

Likely as a response to these anthropogenic emission reductions, the observed Hg wet deposition fluxes and Hg concentrations in precipitation at Mercury Deposition Network (MDN) sites over the US have decreased during these periods, although the reported trends vary depending on the study and time period, as summarized in Table 1. For example, Butler *et al.* [24] find trends ranging from -1.7% yr^{-1} to -3.5% yr^{-1} in Hg concentrations at most sites in the Northeast and Midwest for 1998–2005, but found no significant trend over the Southeast. Prestbo and Gay [25] also report decreasing Hg concentration trends (1–2% yr^{-1} during 1996–2005), but only at half of the MDN sites, particularly across Pennsylvania and extending up through the Northeast region. Risch *et al.* [26] find Hg wet deposition to be unchanged in the Great Lakes region during 2002–2008 by evaluating five monitoring networks in the US and Canada. Furthermore, they find that the small decreases in Hg concentrations were offset by increases in precipitation. Gratz *et al.* [27] describe how declines in precipitation Hg concentrations at remote sites in the Northeast are linked to local-scale meteorological variability rather than to reductions in emissions. In Florida, trends in Hg wet deposition have been minimal or non-existent, despite large reductions in regional emissions of Hg [24,28]. One issue associated with these different studies is the significant interannual variation of Hg concentrations in precipitation and Hg wet deposition fluxes due to variability in precipitation. Indeed, an increase in precipitation results in an increase in the Hg deposition flux. This is accompanied by a decrease in Hg concentrations in rain due to the dilution of the washout loading [25,29]. Thus, interannual changes in meteorological conditions, especially precipitation, complicate the interpretation of MDN observations and might mask any trends due to changes in Hg emissions.

Table 1. Summary of trends in Hg concentrations in precipitation and atmospheric concentrations from selected studies in the literature.

Period	Trends (% yr^{-1})				Reference
	Hg concentrations in precipitation over North America				
	Northeast	Midwest	Southeast	West	
1998–2005	-1.7 ± 0.51[a]	-3.5 ± 0.74	*0.01 ± 0.71*		24
1996–2005	-2.1 ± 0.88[b]	-1.8 ± 0.28	-1.3 ± 0.30	-1.4 ± 0.42	25
2002–2008	0.84 ± 2.9[b]	-2.0 ± 3.8			26
2004–2010	-4.1 ± 0.49[a]	-2.7 ± 0.68	*−0.53 ± 0.59*	*2.6 ± 1.5*	This study
	Atmospheric Hg concentrations				
1995–2005	Canada		-1.3 ± 1.9		33
1996–2009	Mace Head, Ireland		-1.4		34
	Cape Point		-2.7		
1990–2009	North Atlantic		-2.5 ± 0.54		35
	South Atlantic		*−1.9 ± 0.91*		
2000–2009	Temperate Canada		-1.9 ± 0.3		36

Significant trends are indicated in bold fonts, trends that are insignificant are indicated in italics, while trends that lack of the information for significance are indicated in normal fonts. [a] Mean ± standard error, calculated by random coefficient model with site as a random effect; [b] calculated by averaging trends at individual sites.

Over the last two decades, anthropogenic Hg emissions over East Asia, especially in China, have rapidly increased [3,30]. Streets *et al.* [31] estimate that Hg emissions from Asia and Oceania increased from 975 Mg·yr^{-1} in 1996 to 1,318 Mg·yr^{-1} in 2006, equivalent to an annual increase rate of +3.1% yr^{-1}. Wu *et al.* [30] found that anthropogenic Hg emissions from China have increased at an annual average growth rate of +2.9% yr^{-1} during 1995–2003, reaching 696 Mg·yr^{-1} in 2003. In particular, emissions from CFPPs in China increased from 63 to 100 Mg yr^{-1} with a +5.9% annual growth rate during 1995–2003 [30] and reached 132 Mg·yr^{-1} in 2007 [32].

Despite increases in anthropogenic emissions in Asia, there are indications that global background atmospheric Hg concentrations are decreasing (Table 1). Temme *et al.* [33] evaluated the trend of total gaseous mercury (TGM, defined as the sum of Hg(0) and gas phase Hg(II)) at 11 sites from the Canadian Atmospheric Mercury Network (CAMNet), finding a decreasing trend of −1.3±1.9% yr^{-1} with large site by site variation for the 1996–2005 period. Cole *et al.* [34] found that TGM concentrations measured at three temperate Canadian sites displayed trends ranging from −1.6 % yr^{-1} to −2.2% yr^{-1} (mean of −1.9% yr^{-1}) during 2000–2009. Slemr *et al.* [35] evaluated long-term observations at stations in the Southern and Northern Hemisphere combined with cruise measurements over the Atlantic Ocean and found that TGM concentrations decreased by 20–38% between 1996 and 2009. In particular, TGM concentrations at Mace Head, Ireland, have decreased from 1.75 ng m^{-3} in 1996–1999 to 1.4 ng m^{-3} in 2009, showing a trend of −0.024 ± 0.005 ng m^{-3} yr^{-1} (equivalent to −1.4% yr^{-1}). Soerensen *et al* [36] analyzed 1977–2010 trends in atmospheric Hg from 21 ship cruises over the North Atlantic Ocean (NA) and 15 over the South Atlantic Ocean (SA). During the 1990–2009 period, they found a decline of −0.046 ± 0.010 ng m^{-3} yr^{-1} (−2.5% yr^{-1}) over the NA and a smaller trend of −0.024 ± 0.005 ng m^{-3} yr^{-1} (−1.9% yr^{-1}) over the SA. The causes for these decreasing trends are not fully understood, but might be due to decreasing reemissions from the legacy of historical mercury emissions in soils and ocean [35,36].

In this study, we will evaluate trends in observations of Hg concentrations in precipitation from MDN sites across the United States for the 2004–2010 period. We will use a Hg simulation in the GEOS-Chem chemical transport model to take into account the influence of variability in precipitation on the detected trends. Our objectives are (1) to evaluate recent variations and trends in Hg wet deposition and precipitation concentration over the United States; (2) to attribute the observed trends to domestic and global background emission changes.

2. Observations and Model

Wet deposition of Hg has been monitored over the US at MDN sites as part of the National Atmospheric Deposition Program since 1995 [37]. The network initially consisted of 17 sites and currently includes more than 100 active sites. MDN sites collect weekly integrated precipitation samples and report Hg wet deposition flux and the Hg concentration in precipitated water/snow. In this study, we select MDN sites with at least 75% data coverage each year during the 2004–2010 period, resulting in a group of 47 sites (see Figure S1 for the locations of these sites). To account for the lower collection efficiency of snow by MDN samplers [24,25], we correct the weekly deposition fluxes based on the average snow capture efficiency and the snow fraction over total precipitation as described in Zhang *et al.* [12].

In this paper, we will calculate trends based on observed volume-weighted mean (VWM) Hg concentrations in precipitated rain/snow instead of trends in observed Hg wet deposition fluxes. While both variables are affected by changes in precipitation, we found that observed VWM Hg concentrations are significantly less sensitive to precipitation than the deposition fluxes. This is discussed in more detail in the Supporting Information (see Figure S2 and related text).

We use the GEOS-Chem multi-scale nested-grid Hg simulation over North America [12]. The GEOS-Chem chemical transport model itself is described in Bey *et al.* [38]. The model is driven by assimilated meteorological fields from the NASA Goddard Earth Observing System (GEOS). We use here GEOS-5 meteorological fields with a horizontal resolution of $1/2°$ latitude by $2/3°$ longitude and 72 hybrid eta levels from the surface to 0.01 hPa. The lowest 2,000 m are resolved with 13 layers.

The GEOS-Chem global atmospheric Hg simulation is described by Selin *et al.* [39], with recent updates in Hg oxidant chemistry by Holmes *et al.* [40] and gas-particulate partitioning of Hg(II) by Amos *et al.* [41]. The model has two atmospheric mercury species: Hg(0) and Hg(II). The Hg chemistry includes Hg(0) oxidation by Br atoms with kinetic parameters from Donohoue *et al.* [42], Goodsite *et al.* [43] and Balabanov *et al.* [44], as well as aqueous-phase photochemical reduction of Hg(II), scaled to NO_2 photolysis [40].

Over the US and Canada, we use Hg anthropogenic emissions from the 2005 EPA NEI inventory [17] and the 2005 Canadian National Pollutant Release Inventory [46]. Over the rest of the world, anthropogenic emissions are from the Global Emission Inventory Activity (GEIA) 2005 inventory of Pacyna *et al.* [45]. The NEI2005 and Canadian anthropogenic emission inventories assume a 55%/45% partitioning for Hg(0)/Hg(II), with 56.8%/43.2% for CFPP emissions. However, observations in power plant plumes suggest that most of the Hg(II) emitted is quickly reduced to Hg(0) within a few kilometers downwind [13, 14]. In Zhang *et al.* [12], we demonstrated that using a 83%/17% partitioning for N. American anthropogenic Hg emissions (89%/11% in CFPPs) significantly improves model performance in reproducing observations of Hg wet depositior and atmospheric Hg speciation, especially over the Ohio River Valley region, where numerous CFPPs are situated. In this study, we use the same anthropogenic Hg partitioning as in Zhang et al. [12]. For global anthropogenic emissions, we assume a partitioning of 87%/13% (89%/11% Hg(0)/Hg(II) from CFPP emissions). Re-emission of Hg(0) from soil and ocean are also considered based on legacy Hg concentrations in these media [39,47]. GEOS-Chem simulates wet scavenging of Hg(II) following the scheme of Liu et al. [48], with recent updates by Amos et al [41] and Wang et al. [49]. The model simulates the dry deposition of Hg(0) and Hg(II) based on the resistance-in-series scheme of Wesely [50]. Loss of Hg(II) via uptake onto sea-salt aerosol in the marine boundary layer is also considered [51].

The nested-grid model preserves the original resolution of the GEOS-5 meteorological fields($1/2°$ latitude $\times$ $2/3°$ longitude) over North America ($10°–70°$N and $40°–140°$W) and uses results from a global ($4° \times 5°$) simulation as initial and boundary conditions with self-consistent chemistry, deposition and meteorology between the nested and global domains. A full evaluation of the nested-grid Hg simulation is given in Zhang et al. [12]. The model reproduces the magnitude of MDN Hg wet deposition over the US with a small negative bias (-16%) and captures both the spatial distribution and seasonal cycle of wet deposition quite well. Furthermore, the model shows no bias in its simulation of ground-based and aircraft observations of GEM over the US The model reproduces the seasonal cycle of observed PBM. The model overestimates GOM observations by a factor of two, which is not unreasonable given the large variability and uncertainty of GOM measurements.

We run this nested-grid Hg simulation over North America for 2004–2010 assuming constant anthropogenic emissions based on the 2005 EPA NEI inventory ("BASE simulation"). We also conduct three additional sensitivity simulatiors. In the "US simulation", we use the 2008 NEI inventory, which shows a 45% decrease in domestic anthropogenic Hg emissions compared to the 2005 NEI inventory. If we assume that US domestic anthropogenic Hg emissions remained nearly unchanged during 2004–2005 and 2008–2010 (Figure 1), this corresponds to a -7.5% yr^{-1} trend for 2004–2010. In the "NH simulation", we apply a uniform 12% decrease in the nested-grid lateral boundary concentrations of Hg(0) and Hg(II). This reflects the decrease in background Northern Hemisphere TGM concentrations observed by Slemr et al. [35] at Mace Head, Ireland (from 1.60 ng·m^{-3} in 2003 to 1.40 ng·m^{-3} in 2009, corresponding to 12% decrease or a -2% yr^{-1} trend). Finally, the last sensitivity simulation "EA simulation" assumes a 32% increase in anthropogenic Hg emissions from East Asia ($70–150°$E, $8–45°$N) [30,32], corresponding 5.4% yr^{-1} annual increase. In the EA simulation, US anthropogenic emissions remain unchanged at their 2005 levels (NEI2005). Each of these sensitivity simulations (US, NH and EA) is run for

202

2008–2009, and we compare the VWM Hg concentrations in these two years against the results in the same year range for the BASE simulation. The modeled trend, in % yr^{-1}, is calculated by difference with the BASE simulation ($\frac{simulation-base}{base} \times \frac{100}{6}$). We divide the difference by six because we assume that the change occurs over the six years elapsed between 2004 and 2010.

We calculate the trend in observed monthly VWM Hg concentrations using two methods. In the "direct regression" method, we first remove the seasonal cycle in Hg concentrations by subtracting the mean 2004–2010 seasonal cycle at each of the 47 individual MDN sites. We then conduct a least-square linear regression of the resulting time series against month number ($n = 84$ for seven years), and the slope of the regression line is the trend in VWM Hg concentrations. We express the calculated slope into a percent trend (in % yr^{-1}) relative to the multi-year mean VWM Hg concentration of each site. The accompanying probability (p) for the slope is calculated based on the t-score of the regression. In a second method ("model subtraction"), we further subtract the deseasonalized modeled time series of VWM Hg concentrations from the observed deseasonalized time series before calculating the trend. We use our BASE simulation with constant anthropogenic emissions. In this manner, we take into account the influence of varying meteorological conditions, in particular precipitation, on observed VWM Hg concentrations. We thus expect the variability and trend of the residual after subtraction to be caused by changes in domestic anthropogenic emissions and/or in background Hg concentrations (including changes in natural sources and non-domestic anthropogenic sources). The model subtraction approach assumes that the model has no bias in precipitation, which as we will show below is generally the case. We do acknowledge that model errors in precipitation can introduce additional uncertainty in this model subtraction approach; however, we will demonstrate below that this approach leads to more consistent and statistically significant trends relative to the direct regression approach. To calculate the overall regional trend in Hg concentrations for multiple sites in a specific region (see next section for the four regions we use), we employed the random coefficient model with site as a random effect following Butler et al. [24]. We determine the significance of the trends by using $p < 0.1$ for the regressions at individual sites and $p < 0.05$ for the regional trends calculated with the random coefficient models.

3. Results and Discussion

3.1. Seasonal and Interannual Variability in Hg Wet Deposition

Figure 2 shows the observed monthly MDN VWM Hg concentrations in precipitation (left column) for 2004–2010. We aggregate observations at the 47 MDN sites into four regions: Northeast (16 sites), Midwest (13 sites), Southeast (13 sites) and West (five sites). Regional boundaries are defined in Figure 4. Hg concentrations in precipitation display a summer maximum and wintertime minimum. Summertime concentrations are larger than the wintertime concentrations by a factor of 2–3. This seasonality has been reported in many different studies [52–55] and reflects the stronger oxidation of Hg(0) to Hg(II) during summer, providing more Hg(II) for scavenging in rain [12,40]. The GEOS-Chem BASE simulation of VWM Hg concentration is within one standard deviation of MDN observations (Figure 2, left panels). The model shows a mean negative bias of −3.0% (Northeast), −15% (Midwest), −6.3% (Southeast) and −36% (West). The model negative bias in the West takes place in the summer, when there is little precipitation. The model captures the observed seasonal cycle, with temporal correlation coefficients (r) of 0.87 (Northeast), 0.75 (Midwest), 0.76 (West). In the Southeast region, we find a lower correlation ($r = 0.34$). This lower correlation in the Southeast is caused by a systematic model underestimate of the observed VWM Hg concentrations during summer, when convective scavenging from the free troposphere dominates wet deposition. We found previously [12] that this underestimate is likely the result of model errors in the height of convective precipitation. Compared to observations, the GEOS-5 precipitation fields used to drive the GEOS-Chem model have no bias over the Northeast and Southeast, but show a small underestimate of MDN precipitation over the Midwest (−17%) and an overestimate over West (+17%) (Figure 2, right panels). The r values for monthly precipitation depths are larger than 0.8 in each of our four regions.

Observed annual VWM Hg concentrations display some interannual variations, with the range ($\equiv \frac{maximum - minimum}{mean}$) varying from 20 to 40%. The coefficients of variance (ratios of standard deviation to the mean) are 12% (Northeast), 8% (Midwest), 5% (Southeast) and 12% (West). The MDN observed precipitation amounts also have significant interannual variability, with coefficients of variance of 5.0% (Northeast), 7% (Midwest), 12% (Southeast) and 20% (West). Although Hg concentrations in different years have similar seasonal patterns, the peak month and degree of seasonality can vary. For example, the West region displays much higher concentrations in August in 2008 and 2010 relative to other years. Over the Midwest, elevated Hg concentrations are measured during August 2006, 2008 and 2009, because of the relatively lower precipitation amount in these months, while the peak occurs earlier in other years. Lower Hg concentrations are observed during the summer of 2009 compared with multi-year mean in the Northeast and Midwest. The BASE simulation generally captures these features and has r values for observed and modeled annual VWM Hg concentrations of 0.54 (Northeast), 0.26 (Midwest), 0.73 (Southeast) and 0.29 (West). However, we find that the model tends to be on the high side of observations for the last three years and on the low side for the first three years, especially in the Northeast and Midwest. As we will show below, this indicates a trend in observed VWM Hg concentrations not captured by this BASE model simulation with constant anthropogenic emissions.

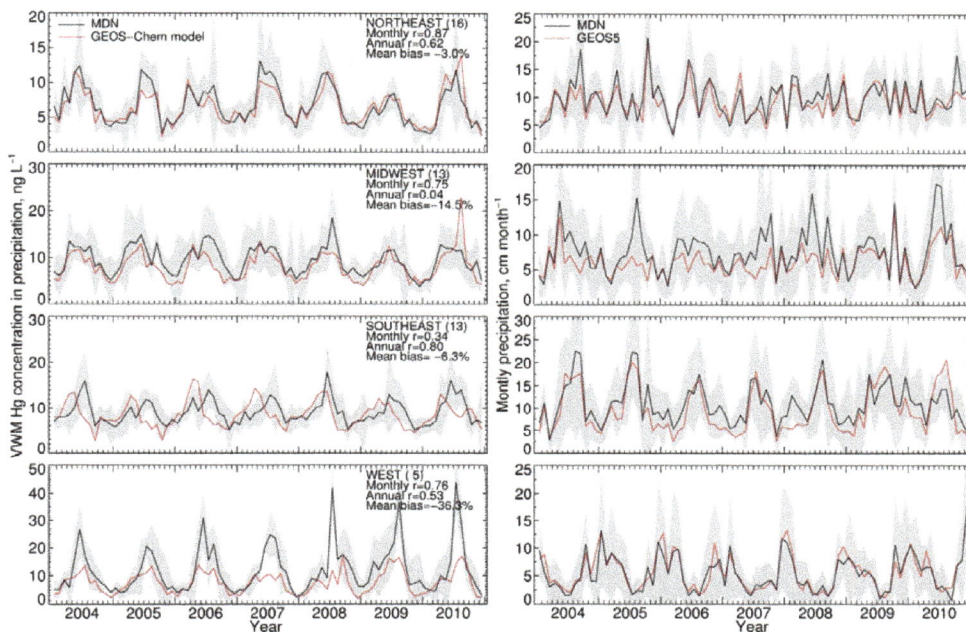

Figure 2. Monthly volume-weighted mean (VWM) Hg concentrations in precipitation (**left**) and precipitation depths (**right**) from Mercury Deposition Network (MDN) observations and the GEOS-Chem BASE simulation (constant anthropogenic emissions) for 2004–2010. The grey shaded area shows the ± standard deviation envelope for the MDN observations.

Table 2. Regional trends in VWM Hg concentration and precipitation depth.

Region	Number of Sites	VWM Hg Concentrations in Precipitation				Precipitation Depth (Observed)	
		Model Subtraction		Direct Regression			
		% yr^{-1}	ns [a]	% yr^{-1}	ns	% yr^{-1}	ns
NORTHEAST	16	**−4.1 ± 0.49**	13	**−3.8 ± 0.50**	9	+0.089 ± 0.26	3
MIDWEST	13	**−2.7 ± 0.68**	6	**−2.0 ± 0.64**	4	**+0.69 ± 0.38**	2
SOUTHEAST	13	−0.53 ± 0.59	4	−0.36 ± 0.54	3	−0.35 ± 0.32	6
WEST	5	+2.6 ± 1.5	3	**+4.0 ± 1.5**	3	**+1.8 ± 0.68**	1

Significant trends ($p < 0.05$) are indicated in bold fonts, trends that are insignificant are indicated in italics; [a] number of sites with significant trends ($p < 0.1$).

3.2. Trend in Hg Precipitation Concentrations for 2004–2010

Compared to the direct regression method, we find that the model subtraction approach reduces the noise of the time series of Hg concentrations, because it removes the variability associated with changes in precipitation. The number of sites with significant trends ($p < 0.1$) increases from 19 for the direct regression approach to 26 for the model subtraction approach. The values of these trends at individual MDN sites are listed in Table S1. Table S1 also includes a comparison to the trends calculated with the direct regression approach, as well as trends in modeled Hg concentrations (BASE simulation) and in precipitation depths.

Figure 3 shows the deseasonalized time series of VWM Hg concentrations calculated with our two approaches at four MDN sites. At PA30 and KY10, the direct regression method yields trends of -1.64% yr^{-1} and -0.91% yr^{-1}, respectively, which are not significant ($p = 0.13$ and 0.30). In contrast, the model subtraction approach results in stronger (-3.6% yr^{-1} and -2.9% yr^{-1}) and more significant trends ($p < 0.01$ and $p = 0.06$). At these sites, it appears that changes in local precipitation (-2.1% yr^{-1} and -3.4% yr^{-1}, Table S1) yield a positive trend in the VWM Hg concentrations calculated in the BASE model simulation ($+2.0\%$ yr^{-1} for both sites). At the NC42 site, the direct regression method yields no trend (0.01% yr^{-1}, $p = 0.50$). However, a strong decrease of precipitation occurred during 2004–2010 (-7.1% yr^{-1}, $p = 0.02$). The BASE model thus predicts a significant increasing trend of Hg concentration in precipitation ($+4.8\%$ yr^{-1}, $p < 0.01$). By subtracting this trend, the model subtraction approach yields a significant negative trend at NC42 (-4.8% yr^{-1}, $p = 0.02$). At SC05, although the direct regression approach yields a significant increasing trend in Hg concentration ($+3.5\%$ yr^{-1}, $p = 0.02$), this trend appears to be caused by decreasing precipitation (-2.5% yr^{-1}, $p = 0.25$). By taking this into account, the model subtraction approach results in no trend (-0.07% yr^{-1}, $p = 0.49$). More generally, we find that at 28 sites out of the 47 MDN sites, the model subtraction method increases the absolute value of the trend and for almost all these 28 sites, the p-value of the trend is reduced (more significant trends).

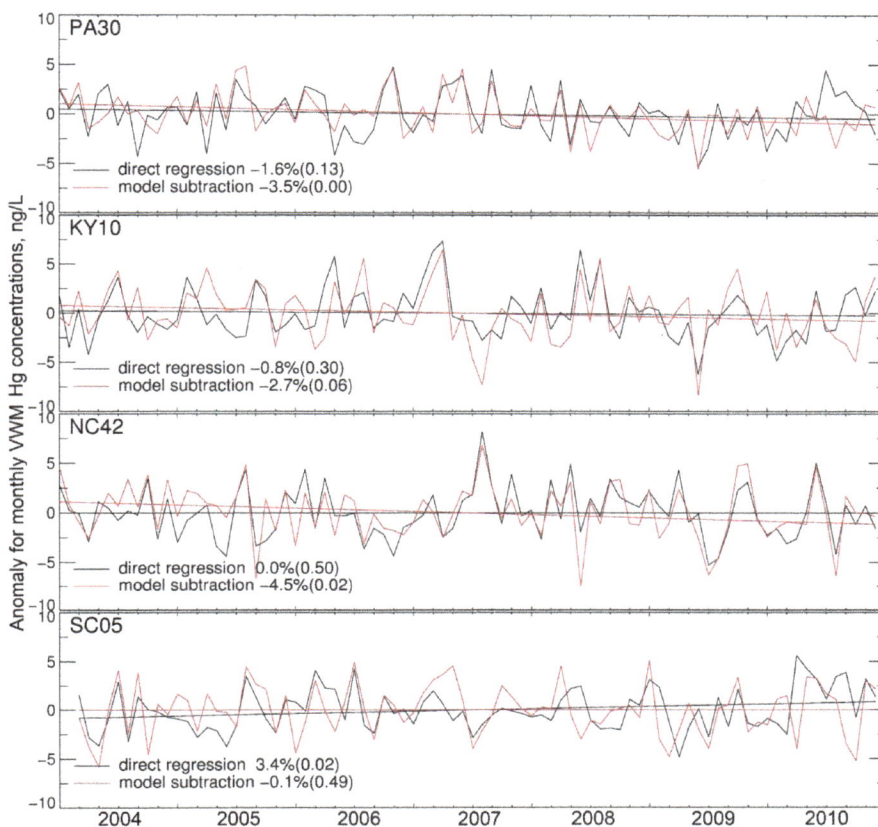

Figure 3. Monthly deseasonalized time series of VWM Hg concentrations in precipitation for the direct regression method (black) and the model subtraction method (red). Four MDN sites are shown, from top to bottom: Erie site, Pennsylvania (PA30); Mammoth Cave National Park-Houchin Meadow, Kentucky (KY10); Pettigrew State Park, North Carolina (NC42) and Cape Romain National Wildlife Refuge, South Carolina (SC05). See Figure S1 for the location of these sites. The resulting trends, in % yr−1 and p-scores (in parenthesis) are indicated in the legend.

Figure 4 shows the spatial distribution of the regression trends (model subtraction method) in monthly VWM Hg concentrations for all 47 MDN sites. Sites with a significant trend ($p < 0.1$) are shown as colored circles (26 sites), while sites with insignificant trend are in triangles (21 sites). The fractions of sites with significant trends vary by region: Northeast (13 significant sites out of a total of 16), Midwest (six out of 13), Southeast (four out of 13) and West (three out of five). The mean trends for the sites in each region are summarized in Table 2. The result of random coefficient model indicates that the calculated trends for Northeast and Midwest are significant ($p < 0.05$), while the trends are not significant over the Southeast and West. Over the Northeast, the 13 sites with significant trends display trends ranging from -8% yr^{-1} to -2.4% yr^{-1} (Table S1). As the trend in precipitation is weak and non-significant ($+0.089 \pm 0.26\%$ yr^{-1}, $p = 0.73$), there is not much difference in the trend calculated with the model subtraction and direct regression methods for this region (Table 2). In the Midwest, the six significant sites display trends in VWM Hg concentrations ranging from -2.9% yr^{-1} to -4.8% yr^{-1}. On the other hand, only three of 13 sites in the Southeast show significant negative trends, and those sites are located near the northern bound of this region. One site in Eastern Texas (Longview, TX 21) displays a significant increasing trend of nearly $+5\%$ yr^{-1}. Most of the other sites with non-significant trends show neutral or slightly decreasing trends. Two of five sites in the West (CO99 in Colorado and NV02 in Nevada) show significant and strong increasing trends, while the sites in Washington (WA18) and California (CA75) show decreasing trends. The overall trend of all the 47 sites is $-1.7 \pm 0.36\%$ yr^{-1} ($p < 0.01$), with large spatial variability.

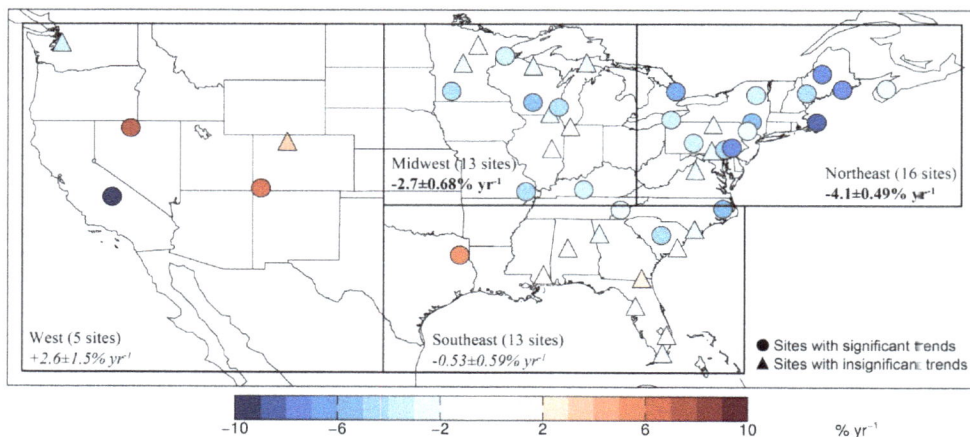

Figure 4. Trend in monthly VWM Hg concentrations at 47 MDN sites during 2004–2010. Sites with significant trend ($p < 0.1$) are shown as circles, while sites where the trend is not significant are shown as triangles. The symbols are color-coded based on the magnitude of the trend (in % yr^{-1}). We use the results of the model subtraction method to calculate the trends (see details in Section 2). The regional trends calculated by the random coefficient model (Northeast, Midwest Southeast and West) are also shown, with significant trends ($p < 0.05$) shown in bold font and insignificant ones shown in italics.

The precipitation shows a general increasing trend during 2004–2010 ranging from $+0.089 \pm 0.26\%$ yr^{-1} ($p = 0.73$) (Northeast) to $+1.8 \pm 0.68\%$ yr^{-1} ($p = 0.065$) (West), except in the Southeast, where it decreased ($-0.35 \pm 0.32\%$ yr^{-1}, $p = 0.27$). At most of the sites, these precipitation trends are not statistically significant, with only 12 sites (out of 47) displaying significant trends in precipitation depth. This overall positive trend in precipitation would cause a decrease in Hg concentration even without any change in Hg emissions. The Hg concentration trends calculated by direct regression approach are thus more difficult to interpret, because of the underlying trend in precipitation in these regions. Furthermore, the direct regression approach tends to miss the decreasing trend in Hg concentration if the precipitation trend is also decreasing, which would cause an increasing trend in Hg concentrations. For example, the direct regression approach predicts a weak decreasing trend of $-0.36 \pm 0.54\%$ yr^{-1} over the Southeast region. Removing the decreasing trend of precipitation with the model subtraction approach brings the calculated trend down by ~50% to $-0.53 \pm 0.59\%$ yr^{-1}.

Over the Northeast, Butler $et\ al.$ [24] calculated a trend of $-1.7 \pm 0.51\%$ yr^{-1} during 1998–2005 by using random coefficient models with site as a random effect (Table 1). Prestbo and Gay [25] obtained similar trends in the Northeast ($-2.1 \pm 0.88\%$ yr^{-1}) during 1996–2005 by using a non-parametric statistic method. Compared with these previous studies, our trends over the Northeast for 2004–2010 are stronger (model subtraction approach: $-4.1 \pm 0.49\%$ yr^{-1}; direct regression: $-3.8 \pm 0.50\%$ yr^{-1}). Over the Midwest, our results ($-2.7 \pm 0.68\%$ yr^{-1} and $-2.0 \pm 0.64\%$yr^{-1} for the two approaches) are comparable to previous studies, which ranged from -1.8% yr^{-1} to -3.5 %yr^{-1} (Table 1). Over the Southeast region, we found a weak and non-significant trend ($-0.53 \pm 0.59\%$ yr^{-1}) in Hg concentration during 2004–2010. Similar results are also obtained by previous studies focusing on the trend before 2005 and extending back for 8–10 years (Table 1). Prestbo and Gay [25] argued that the nearly steady Hg concentration in precipitation over the Southeast is because of the steady local emissions after an abrupt emission decrease by the closure or emission controlling of municipal and waste incinerators in the early 1990s. Butler $et\ al.$ [24] attributed the steady Hg concentrations over Southeast region to the influence of global background Hg, which increased or remained constant during that time period. Butler $et\ al.$ [24] also argued that higher halogen concentrations in the coastal environment in Southeast facilitate rapid transformation of Hg(0) from global pool to Hg(II), which can enhance Hg wet deposition.

3.3. Attribution of Observed Trends

We examine here whether these observed trends can be explained by changes in emissions from domestic and global sources. Figure 5 shows the expected trends in VWM Hg concentrations for the four model simulations (see Section 2 for a description of each model simulation). The US simulation (Figure 5(a)) suggests a 2–4% yr^{-1} decrease in VWM Hg concentrations near source regions over the Ohio River Valley region. The largest modeled decrease occurs near the borders of Ohio, West Virginia and Pennsylvania (-6% yr^{-1}). In the Southeast and Midwest regions the influence of decreasing domestic emissions is weaker and not as concentrated, with decreases ranging from -1% yr^{-1} to -2% yr^{-1} near point sources. The NH simulation (Figure 5(b)) displays uniform decreases in Hg concentrations across the US (1.7–2.0% yr^{-1}). The decreases are somewhat larger in the West ($-1.9 \pm 0.80\%$ yr^{-1}) and Southeast ($-1.9 \pm 0.18\%$ yr^{-1}) and slightly smaller over the Northeast ($-1.7 \pm 0.1\%$ yr^{-1}) and Midwest ($-1.7 \pm 0.47\%$ yr^{-1}), where domestic emissions are more important. The EA simulation (Figure 5(d)) leads to a broad increase in Hg concentrations ranging from 1% yr^{-1} in the Northwest to 0.5% yr^{-1} in the Ohio River Valley region, with a factor of two for the west-to-east gradient. A similar gradient was modeled by Strode et al. [56] when evaluating the trans-Pacific transport of Asian Hg emissions. They found that the contribution of Asia to the Hg(II) total (dry + wet) deposition over the US ranged from more than 30% in the West to less than 15% in the Northeast. Note that the EA simulation is contained in the envelope of the NH simulation, i.e., the observed atmospheric North Hemispheric background Hg concentration decreases despite of the increase of emission over East Asia [3].

Table 3 summarizes the results of the three simulations sampled at the locations of MDN sites in each region. The combined US and NH simulations (US + NH, see also Figure 5(c)) explain the magnitude of the observed decreasing trend over the Northeast (obs.: $-4.1 \pm 0.49\%$ yr^{-1}; model US + NH: $-4.1 \pm 0.52\%$ yr^{-1}) and Midwest (obs.: $-2.7 \pm 0.68\%$ yr^{-1}; model US + NH: $-3.1 \pm 0.60\%$ yr^{-1}) regions. We find that over the Northeast, decreases in domestic emissions account for 58% of the observed decrease, with the remaining 42% contributed by background Hg concentration decreases. Over the Midwest, the role of domestic emission reductions is slightly smaller, accounting for 46% of the decrease.

The US simulation predicts a decreasing trend of $-0.55 \pm 0.10\%$ yr^{-1} over the Southeast, which is similar to the observed trend ($-0.53 \pm 0.59\%$ yr^{-1}). However, adding in the contribution from the NH simulation leads to a US + NH decreasing trend of $-2.5 \pm 0.21\%$ yr^{-1}, which is a factor of 4–5 higher than the observed trend. This suggests that atmospheric Hg concentrations over the free troposphere in this region may have remained nearly constant or decreased much more slowly compared to the trend over Mace Head (Table 1), which is located at a higher latitude (53°N). The three sites with significant decreasing trends in the Southeast region (in North Carolina, South Carolina and Tennessee) have the highest latitudes among sites in this region (Figure 4) and for these sites the results from the NH + US scenarios ($-2.8 \pm 0.26\%$ yr^{-1}) are more in line with observed trends ($-3.5 \pm 1.5\%$ yr^{-1}). This could indicate a latitudinal dependence of the trend in background concentrations. Our results are consistent with a recent model study by Soerensen *et al.* [36]. They show that the decreasing trend of surface atmospheric Hg(0) concentrations due to the decrease in subsurface ocean Hg concentrations over the North Atlantic is 2–3 times higher over Mace Head than over the southeast US Thus, our assumption of uniformly decreasing background TGM concentrations for the NH scenario overestimates the trend over lower latitudes.

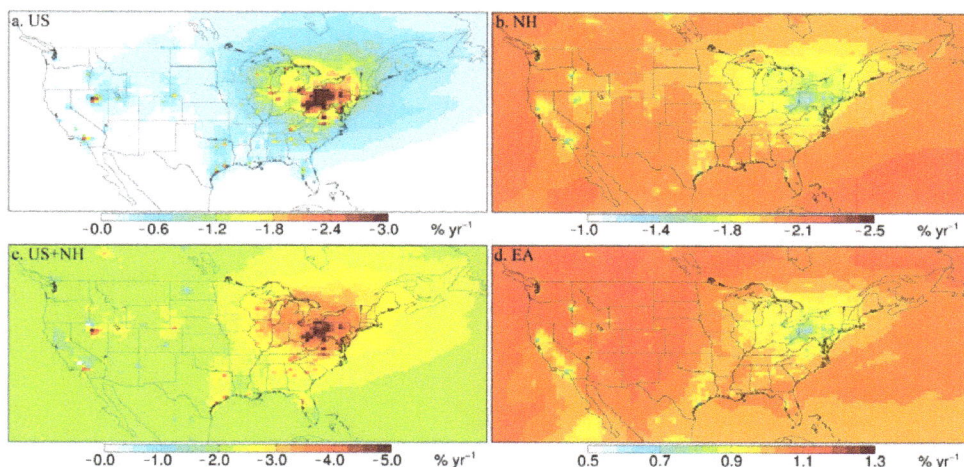

Figure 5. Calculated trends (% yr^{-1}) in VWM Hg concentrations over the US for scenarios US, NH and EA, compared with the BASE simulation. (**a**) Trend for US scenario (45% reduction of Hg emissions from domestic sources); (**b**) trend for NH scenario (12% decrease of Hg background atmospheric concentrations); (**c**) combined US and NH scenarios; (**d**) trend for EA scenario (32% increase in Hg emissions from East Asia).

For the West region, although the US + NH scenarios show decreasing trends ($-1.8 \pm 0.82\%$ yr^{-1}), the observed trend is positive ($+2.6 \pm 1.5\%$) with extremely large variations among sites. Our first hypothesis for this increasing trend is that it is caused by the recent increase of East Asian emissions. The EA scenario predicts an increasing trend over the whole US, as well as a west-to-east gradient ($+1.0 \pm 0.41\%$ yr^{-1} in West *vs.* $+0.84 \pm 0.048\%$ yr^{-1} in Northeast), but this gradient is not strong enough to explain the observed unique positive trend in the West region. Prestbo and Gay [25] also failed to detect the influence of Asian emissions after analyzing the Hg deposition flux and concentration data in precipitation observed at the MDN site WA18 located in Seattle, which is thought to be distant from large domestic Hg sources. We find similar results, with a non-significant trend of -0.23% yr^{-1} at WA18 for 2004–2010. We have also evaluated the trend of springtime

GEM concentrations observed at the Mount Bachelor Observatory (MBO) Oregon (44°N, 122°W, 2,700 m above sea level) during 2005–2009 (Jaffe, unpublished data) and found no significant trend. Considering the small number of sites in this region, as well as the large spatial variation, the trend at individual sites is likely influenced by local emissions and high variability in precipitation. For example, the site CO99 at Mesa Verde National Park, Colorado, near the Four Corners region shows an increasing trend of +6.4% yr^{-1}. A closer look at the weekly deposition flux at this site shows that the strong trend is caused by several strong episodic large deposition events occurring in the summers of 2009 and 2010. Wright and Nydick [57] evaluated these events by back trajectories and high-resolution dispersion modeling and found them to be due to transport of local CFPP emissions. Thus, more observations in the West, especially background sites without strong local source influence, are needed to evaluate the trend of Hg wet deposition over this region.

Table 3. Comparison between observed regional trends in VWM Hg concentrations (model subtraction approach) and three model simulations assuming changing emissions or background conditions.

		Observations	Model Scenarios			
Region	n	% yr^{-1}	US[a] % yr^{-1}	NH[b] % yr^{-1}	US+NH % yr^{-1}	EA[c] % yr^{-1}
NORTHEAST	16	−4.1 ± 0.49[d]	−2.4 ± 0.51	−1.7 ± 0.10	−4.1 ± 0.52	0.84 ± 0.048
MIDWEST	13	−2.7 ± 0.68	−1.4 ± 0.38	−1.7 ± 0.47	−3.1 ± 0.60	0.89 ± 0.22
SOUTHEAST	13	−0.53 ± 0.59	−0.55 ± 0.10	−1.9 ± 0.18	−2.5 ± 0.21	0.92 ± 0.086
WEST	5	+2.6 ± 1.5	+0.068 ± 0.19	−1.9 ± 0.80	−1.8 ± 0.82	1.0 ± 0.41

Significant trends ($p < 0.05$) are indicated in bold fonts, trends that are insignificant are indicated in italics. [a] Simulation assuming a 45% decrease in anthropogenic emissions over the US between 2004–2010; [b] simulation assuming a 12% decrease in Northern Hemisphere background TGM concentrations during 2004–2010; [c] simulation assuming a 32% increase in anthropogenic Hg emission from East Asia during 2004–2010; [d] mean ± standard error of trends at MDN sites.

4. Conclusions

We calculated 2004–2010 trends in VWM Hg concentrations in precipitation at 47 Mercury Deposition Network (MDN) sites across the United States. We used the GEOS-Chem nested-grid mercury simulation to take into account the influence of meteorology on observed trends and to assess the role of domestic emission reductions in explaining the trends. The GEOS-Chem model agrees relatively well with the observed VWM Hg concentrations in precipitation, with negative bias of −3.0% (Northeast), −15% (Midwest), −6.3% (Southeast) and −36% (West). The model captures the observed seasonal cycle and interannual variability.

We removed the effect of meteorological changes on VWM Hg concentrations by subtracting from the observed monthly observations results from the GEOS-Chem model simulation with constant anthropogenic emissions. The trend associated with the remaining part is independent from meteorological fluctuations and can be directly associated with local and/or global background emission changes. We demonstrated that this approach yields more robust and consistent results than directly calculating the trend from MDN observations. We found regional mean trends for VWM Hg concentrations at MDN sites during 2004–2010 of $-4.1\% \pm 0.49$ yr^{-1} (Northeast), $-2.7 \pm 0.68\%$ yr^{-1} (Midwest), $-0.53 \pm 0.59\%$ yr^{-1} (Southeast) and $+2.6 \pm 1.5\%$ yr^{-1} (West). Over the Northeast, 13 out of 16 sites have significant trends, while over the Midwest six out of 13 sites have significant trends. By applying a random coefficient model for the overall trend each region, we find that trends over both the Northeast and Midwest are significant ($p < 0.05$), with overall decreases of $-25 \pm 2.9\%$ (Northeast), $-16 \pm 4.1\%$ (Midwest) for 2004–2010. The trends over the Southeast and West regions are not significant because of the smaller numbers of sites with significant trends: four out of 13 in Southeast and three out of five for the West.

We conducted sensitivity simulations for three separate scenarios: a 45% decrease in domestic emissions (US), a 12% decrease in global boundary condition of atmospheric Hg concentrations (NH) and a 32% increase in East Asian anthropogenic Hg emissions (EA). The US simulation predicts a decrease in VWM Hg concentrations in the Northeast ($-2.4 \pm 0.51\%$ yr^{-1}), as well as in the Midwest ($-1.4 \pm 0.38\%$ yr^{-1}) and Southeast ($-0.55 \pm 0.10\%$ yr^{-1}). The NH scenario displays large decreases in the West ($-1.9 \pm 0.80\%$ yr^{-1}) and Southeast ($-1.9 \pm 0.18\%$ yr^{-1}) and somewhat smaller decreases over Northeast ($-1.7 \pm 0.10\%$ yr^{-1}) and Midwest ($-1.7 \pm 0.47\%$ yr^{-1}). The EA simulation generates a nationwide increase in Hg wet deposition ranging from ~1.0% yr^{-1} in the West to ~0.5% yr^{-1} over the Ohio River Valley region.

We found that the combined US and NH simulations explains most of the observed decreasing trend over the Northeast (model US + NH: $-4.1 \pm 0.52\%$ yr^{-1}; MDN: $-4.1 \pm 0.49\%$ yr^{-1}) and Midwest (model US + NH: $-3.1 \pm 0.60\%$ yr^{-1}; MDN: $-2.7 \pm 0.68\%$ yr^{-1}). The model predicts that domestic emission reductions account for 58% of the decreasing trend in the Northeast and 46% of the trend in the Midwest. The US simulation predicts a decreasing trend of $-0.55 \pm 0.10\%$ yr^{-1} over the Southeast, which is similar to the observed trend ($-0.53 \pm 0.59\%$ yr^{-1}). However, adding the effect of decreasing background concentrations (US + NH simulation) resulted in an overestimate of the observed trend by a factor of 4–5 over the Southeast region. For three sites, located at the northern edge of the Southeast region, the US + NH simulation agrees with the observed trends. This suggests that the trend in background concentration might have a latitudinal dependence, with weaker decreases at southern latitudes. Our NH simulation might thus be overestimating the changes due to background emissions. For the West region, although the observed trend is positive ($+2.6 \pm 1.5\%$), the US + NH scenarios show decreasing trends ($-1.8 \pm 0.82\%$ yr^{-1}). Considering the small number of sites in this region, more observations, especially background sites away from strong local source influence, are needed to evaluate the trend of Hg concentrations in precipitation over this region.

By combining our analysis of MDN observations and the GEOS-Chem simulations, we thus find that 58% of the observed 25% decrease in VWM Hg concentration over the Northeast for 2004–2010 can be explained by reductions in domestic Hg emissions, with the remaining 42% due to decreasing background concentrations. Over the Midwest, our model simulation suggests that 46% of the observed 16% decrease is explained by domestic emission reductions. In Zhang *et al.* [12], we found that domestic Hg emissions contributed to 16% of wet deposition over the Northeast in 2005. Between 2005 and 2010, domestic emissions have decreased by 60% over that region and thus now account for only 8.3% of wet deposition. Further domestic Hg emissions decreases are expected as part of the 2011 Mercury and Air Toxics Standards (MATS). While implementation of MATS will lead to significant decreases in local Hg wet deposition directly downwind of the sources, the regional influence of these decreases is likely to be small and more difficult to detect. We expect that future regional trends in wet deposition in the US will be strongly influenced by changes in background Hg concentrations.

Acknowledgments: This work was supported by funding from EPRI under contract EP-P43461/C18853. We thank EPRI program manager Leonard Levin for his support during this study. We are grateful to Dan Jaffe for his comments on this study. We would like to acknowledge and thank all the site operators for the NADP/MDN network.

Conflicts of Interest: The authors declare no conflict of interest.

References and Notes

1. The Original List of Hazardous Air Pollutants. Available online: http://www.epa.gov/ttn/atw/188polls.html (accessed on 13 March 2013).

2. Mason, R.A. Mercury Emissions from Natural Processes and their Importance in the Global Mercury Cycle. In *Mercury Fate and Transport in the Global Atmosphere: Emissions, Measurements and Models*; Pirrone, N., Mason, R.A., Eds.; Springer: Dordrecht, The Netherlands, 2009; pp. 173–191.

3. Streets, D.G.; Devane, M.K.; Lu, Z.; Bond, T.C.; Sunderland, E.M.; Jacob, D.J. All-time releases of mercury to the atmosphere from human activities. *Environ. Sci. Technol.* **2011**, *45*, 10485–10491.

4. Pirrone, N.; Cinnirella, S.; Feng, X.; Finkelman, R.B.; Friedli, H.R.; Learner, J.; Mason, R.; Mukherjee, A.B.; Stracher, G.; Streets, D.G.; Telmer, K. Global mercury emissions to the atmosphere from natural and anthropogenic sources. In *Mercury Fate and Transport in the Global Atmosphere: Emissions, Measurements and Models*; Pirrone, N., Mason, R.A., Eds.; Springer: Dordrecht, The Netherlands, 2009; pp. 3–50.

5. Selin, N.E. Global biogeochemical cycling of mercury: A review. *Annu. Rev. Environ. Resour.* **2009**, *34*, 43–63.

6. Holmes, C.D.; Jacob, D.J.; Yang, X. Global lifetime of elemental mercury against oxidation by atomic bromine in the free troposphere. *Geophys. Res. Lett.* **2009**, *33*, L20808.

7. Clarkson, T.W. The three modern faces of mercury. *Environ. Health Perspect.* **2002**, *110*, 11–23.

8. Mergler, D.; Anderson, H.A.; Chan, L.H.M.; Mahaffey, K.R.; Murray, M.; Sakamoto, M.; Stern, A.H. Methylmercury exposure and health effects in humans: A worldwide concern. *Ambio* **2007**, *36*, 3–11.

9. Houyoux, M.; Strum, M. *Memorandum: Emissions Overview: Hazardous Air Pollutants in Support of the Final Mercury and Air Toxics Standard*; EPA-454/R-11-014; Emission Inventory and Analysis Group Air Quality Assessment Division: Research Triangle Park, NC, USA, 2011.

10. Seigneur, C.; Vijayaraghavan, K.; Lohman, K.; Karamchandani, P.; Scott, C. Global source attribution for mercury deposition in the United States. *Environ. Sci. Technol.* **2004**, *38*, 555–569.

11. Selin, N.E.; Jacob, D.J. Seasonal and spatial patterns of mercury wet deposition in the United States: Constraints on the contribution from North American anthropogenic sources. *Atmos. Environ.* **2008**, *42*, 5193–5204.

12. Zhang, Y.; Jaeglé, L.; van Donkelaar, A.; Martin, R.V.; Holmes, C.D.; Amos, H.M.; Wang, Q.; Talbot, R.; Artz, R.; Brooks, S.; *et al.* Nested-grid simulation of mercury over North America. *Atmos. Chem. Phys.* **2012**, *12*, 6095–6111.

13. Edgerton, E.S.; Hartsell, B.E.; Jansen, J.J. Mercury speciation in coal-fired power plant plumes observed at three surface sites in the southeastern US. *Environ. Sci. Technol.* **2006**, *40*, 4563–4570.

14. Weiss-Penzias, P.S.; Gustin, M.S.; Lyman, S.N. Sources of gaseous oxidized mercury and mercury dry deposition at two southeastern U.S. sites. *Atmos. Environ.* **2011**, *45*, 4569–4579.

15. Toxics Release Inventory. Available online: http://www.epa.gov/tri/ (accessed on 13 March 2013).

16. National Toxics Inventory. Available online: http://epa.gov/air/data/netemis.html (accessed on 13 March 2013).

17. National Emission Inventory. Available online: http://www.epa.gov/ttn/chief/net/2008inventory.html (accessed on 13 March 2013).

18. Solid Waste Combustion Rule. Available online: http://www.epa.gov/airtoxics/129/gil2.pdf (accessed on 13 March 2013).

19. Clean Air Mercury Rule. Available online: http://www.epa.gov/camr/ (accessed on 13 March 2013).

20. Information Collection Request. Available online: http://www.epa.gov/ttn/atw/utility/utilitypg.html (accessed on 13 March 2013).

21. Mercury and Air Toxics Standard. Available online: http://www.epa.gov/mats (accessed on 13 March 2013).

22. Integrated Planning Model. Available online: http://www.epa.gov/airmarkets/progsregs/epa-ipm/index.html (accessed on 13 March 2013).

23. USEPA, *Emissions Inventory and Emissions Processing for the Clean Air Mercury Rule (CAMR)*; Office of Air Quality Planning and Standards Emissions, Monitoring and Analysis Division: Research Triangle Park, NC, USA, 2005.

24. Butler, T.J.; Cohen, M.D.; Vermeylen, F.M.; Likens, G.E.; Schmeltz, D.; Artz, R.S. Regional precipitation mercury trends in the eastern USA, 1998–2005: Declines in the Northeast and Midwest, no trend in the Southeast. *Atmos. Environ.* **2008**, *42*, 1582–1592.

25. Prestbo, E.M.; Gay, D.A. Wet deposition of mercury in the US and Canada, 1996–2005: Results and analysis of the NADP mercury deposition network (MDN). *Atmos. Environ.* **2009**, *43*, 4223–4233.

26. Risch, M.R.; Gay, D.A.; Fowler, K.K.; Keeler, G.J.; Backus, S.M.; Blanchard, P.; Barres, J.A.; Dvonch, J.T. Spatial patterns and temporal trends in mercury concentrations, precipitation depths, and mercury wet deposition in the North American Great Lakes region, 2002–2008. *Environ. Pollut.* **2012**, *161*, 261–271.

27. Gratz, L.E.; Keeler, G.J.; Miller, E.K. Long-term relationships between mercury wet deposition and meteorology. *Atmos. Environ.* **2009**, *43*, 6218–6229.

28. Vijayaraghavan, K.; Stoeckenius, T.; Ma, L.; Yarwood, G.; Morris, R.; Levin, L. Analysis of Temporal Trends in Mercury Emissions and Deposition in Florida. In Proceedings of the 10th International Conference on Mercury as Global Pollutant, Halifax, NS, Canada, August 2011.

29. Landis, M.S.; Vette, A.F.; Keeler, G.J. Atmospheric mercury in the Lake Michigan Basin: influence of the Chicago/Gary urban area. *Environ. Sci. Technol.* **2002**, *36*, 4508–4517.

30. Wu, Y.; Wang, S.; Streets, D.G.; Hao, J.; Chan, M.; Jiang, J. Trends in Anthropogenic Mercury Emissions in China from 1995 to 2003. *Environ. Sci. Technol.* **2006**, *40*, 5312–5318.

31. Streets, D.G.; Zhang, Q.; Wu, Y. Projections of Global Mercury Emissions in 2050. *Environ. Sci. Technol.* **2009**, *43*, 2983–2988.

32. Tian, H.; Wang, Y.; Xue, Z.; Qu, Y.; Chai, F.; Hao, J. Atmospheric emissions estimation of Hg, As, and Se from coal-fired power plants in China, 2007. *Sci. Total Environ.* **2011**, *409*, 3078–3081.

33. Temme, C.; Blanchard, P.; Steffen, A.; Banic, C.; Beauchamp, S.; Poissant, L.; Tordon, R.; Wiens, B. Trend, seasonal and multivariate analysis study of total gaseous mercury data from the Canadian atmospheric mercury measurement network (CAMNet). *Atmos. Environ.* **2007**, *41*, 5423–5441.

34. Cole, A.S.; Steffen, A.; Pfaffhuber, K.A.; Berg, T.; Pilote, M.; Poissant, L.; Tordon, R.; Hung, H. Ten-year trends of atmospheric mercury in the high Arctic compared to Canadian sub-Arctic and mid-latitude sites. *Atmos. Chem. Phys.* **2013**, *13*, 1535–1545.

35. Slemr, F.; Brunke, E.G.; Ebinghaus, R.; Kuss, J. Worldwide trend of atmospheric mercury since 1995. *Atmos. Chem. Phys.* **2011**, *11*, 4779–4787.

36. Soerensen, A.L.; Jacob, D.J.; Streets, D.G.; Witt, M.L.I.; Ebinghaus, R.; Mason, R.P.; Andersson, M.; Sunderland, E.M. Multi-decadal decline of mercury in the North Atlantic atmosphere explained by changing subsurface seawater concentrations. *Geophys. Res. Lett.* **2012**, *39*, L21810.

37. National Atmospheric Deposition Program, Mercury Deposition Network Information. Available online: http://nadp.sws.uiuc.edu/mdn/ (accessed on 13 March 2013).

38. Bey, I.; Jacob, D.J.; Yantosca, R.M.; Logan, J.A.; Field, B.D.; Fiore, A. M.; Li, Q.B.; Liu, H.G.Y.; Mickley, L.J.; Schultz, M.G. Global modeling of tropospheric chemistry with assimilated meteorology: Model description and evaluation. *J. Geophys. Res.-Atmos.* **2001**, *106*, 23073–23095.

39. Selin, N.E.; Jacob, D.J.; Park, R.J.; Yantosca, R.M.; Strode, S.; Jaegle, L.; Jaffe, D. Chemical cycling and deposition of atmospheric mercury: Global constraints from observations. *J. Geophys. Res.* **2007**, *112*, D02308.

40. Holmes, C.D.; Jacob, D.J.; Corbitt, E.S.; Mao, J.; Yang, X.; Talbot, R.; Slemr, F. Global atmospheric model for mercury including oxidation by bromine atoms. *Atmos. Chem. Phys.* **2010**, *10*, 12037–12057.

41. Amos, H.M.; Jacob, D.J.; Holmes, C.D.; Fisher, J.A.; Wang, Q.; Yantosca, R.M.; Corbitt, E.S.; Galarneau, E.; Rutter, A.P.; Gustin, M.S.; *et al.* Gas-particle partitioning of atmospheric Hg(II) and its effect on global mercury deposition. *Atmos. Chem. Phys.* **2012**, *12*, 591–603.

42. Donohoue, D.L.; Bauer, D.; Cossairt, B.; Hynes, A.J. Temperature and pressure dependent rate coefficients for the reaction of Hg with Br and the reaction of Br with Br: A pulsed laser photolysis-pulsed laser induced fluorescence study. *J. Phys. Chem. A* **2006**, *110*, 6623–6632.

43. Goodsite, M.E.; Plane, J.M.C.; Skov, H. A theoretical study of the oxidation of Hg^0 to $HgBr_2$ in the troposphere. *Environ. Sci. Technol.* **2004**, *38*, 1772–1776.

44. Balabanov, N.B.; Shepler, B.C.; Peterson, K.A. Accurate global potential energy surface and reaction dynamics for the ground state of $HgBr_2$. *J. Phys. Chem. A* **2005**, *109*, 8765–8773.

45. Pacyna, E.G.; Pacyna, J.M.; Sundseth, K.; Munthe, J.; KinNHom, K.; Wilson, S.; Steenhuisen, F.; Maxson, P. Global emission of mercury to the atmosphere from anthropogenic sources in 2005 and projections to 2020. *Atmos. Environ.* **2010**, *44*, 2487–2499.

46. National Pollutant Release Inventory. Available online: http://www.ec.gc.ca/inrp-npri/ (accessed on 13 March 2013).

47. Soerensen, A.L.; Sunderland, E.M.; Holmes, C.D.; Jacob, D.J.; Yantosca, R.M.; Skov, H.; Christensen, J.H.; Strode, S.A.; Mason, R.P. An Improved Global Model for Air-Sea Exchange of Mercury: High Concentrations over the North Atlantic. *Environ. Sci. Technol.* **2010**, *44*, 8574–8580.

48. Liu, H.; Jacob, D.; Bey, I.; Yantosca, R.M. Constraints from Pb210 and Be7 on wet deposition and transport in a global three-dimensional chemical tracer model driven by assimilated meteorological fields. *J. Geophys. Res.* **2001**, *106*, 12109–12128.

49. Wang, Q.; Jacob, D.J.; Fisher, J.A.; Mao, J.; Leibensperger, E.M.; Carouge, C.C.; Le Sager, P.; Kondo, Y.; Jimenez, J.L.; Cubison, M.J.; Doherty, S.J. Sources of carbonaceous aerosols and deposited black carbon in the Arctic in winter-spring: implications for radiative forcing. *Atmos. Chem. Phys.* **2011**, *11*, 12453–12473.

50. Wesely, M.L. Parameterization of surface resistances to gaseous dry deposition in regional-scale numerical-models. *Atmos. Environ.* **1989**, *23*, 1293–1304.

51. Holmes, C.D.; Jacob, D.J.; Mason, R.P.; Jaffe, D.A. Sources and deposition of reactive gaseous mercury in the marine atmosphere. *Atmos. Environ.* **2009**, *43*, 2278–2285.

52. Guentzel, J.L.; Landing, W.M.; Gill, G.A.; Pollman, C.D. Processes influencing rainfall deposition of mercury in Florida: The FAMS Project (1992–1996). *Environ. Sci. Technol.* **2001**, *35*, 863–873.

53. Hoyer, M.; Burke, J.; Keeler, G.J. Atmospheric sources, transport and deposition of mercury in Michigan: two years if event precipitation. *Water Air Soil Pollut.* **1995**, *80*, 199–208.

54. Landis, M.S.; Keeler, G.J. Atmospheric Mercury Deposition to Lake Michigan during the Lake Michigan Mass Balance Study. *Environ. Sci. Technol.* **2002**, *36*, 4518–4524.

55. Dvonch, J.T.; Keeler, G.J.; Marsik, F.J. The influence of meteorological conditions on the wet deposition of mercury in southern Florida. *J. Appl. Meteorol.* **2005**, *44*, 1421–1435.

56. Strode, S.A.; Jaeglé, L.; Jaffe, D.A.; Swartzendruber, P.C.; Selin, N.E.; Holmes, C.; Yantosca, R.M. Trans-Pacific transport of mercury. *J. Geophys. Res.* **2008**, *113*, D15305.

57. Wright, W.G.; Nydick, K. *Sources of Atmospheric Mercury Concentrations and Wet Deposition at Mesa Verde National Park, Southwestern Colorado, 2002–08*; Mountain Studies Institute Report: Silverton, CO, USA, 2010.

Airborne Vertical Profiling of Mercury Speciation near Tullahoma, TN, USA

Steve Brooks, Xinrong Ren, Mark Cohen, Winston T. Luke, Paul Kelley, Richard Artz, Anthony Hynes, William Landing and Borja Martos

Abstract: Atmospheric transport and *in situ* oxidation are important factors influencing mercury concentrations at the surface and wet and dry deposition rates. Contributions of both natural and anthropogenic processes can significantly impact burdens of mercury on local, regional and global scales. To address these key issues in atmospheric mercury research, airborne measurements of mercury speciation and ancillary parameters were conducted over a region near Tullahoma, Tennessee, USA, from August 2012 to June 2013. Here, for the first time, we present vertical profiles of Hg speciation from aircraft for an annual cycle over the same location. These airborne measurements included gaseous elemental mercury (GEM), gaseous oxidized mercury (GOM) and particulate bound mercury (PBM), as well as ozone (O_3), sulfur dioxide (SO_2), condensation nuclei (CN) and meteorological parameters. The flights, each lasting ~3 h, were conducted typically one week out of each month to characterize seasonality in mercury concentrations. Data obtained from 0 to 6 km altitudes show that GEM exhibited a relatively constant vertical profile for all seasons with an average concentration of 1.38 ± 0.17 ng·m^{-3}. A pronounced seasonality of GOM was observed, with the highest GOM concentrations up to 120 pg·m^{-3} in the summer flights and lowest (0–20 pg·m^{-3}) in the winter flights. Vertical profiles of GOM show the maximum levels at altitudes between 2 and 4 km. Limited PBM measurements exhibit similar levels to GOM at all altitudes. HYSPLIT back trajectories showed that the trajectories for elevated GOM (>70 pg·m^{-3}) or PBM concentrations (>30 pg·m^{-3}) were largely associated with air masses coming from west/northwest, while events with low GOM (<20 pg·m^{-3}) or PBM concentrations (<5 pg·m^{-3}) were generally associated with winds from a wider range of wind directions. This is the first set of speciated mercury vertical profiles collected in a single location over the course of a year. Even though there are current concerns that the KCl denuders used in this study may under-collect GOM, especially in the presence of elevated ozone, the collected data in this region shows the strong seasonality of oxidized mercury concentrations throughout the low to middle free troposphere.

Reprinted from *Atmosphere*. Cite as: Brooks, S.; Ren, X.; Cohen, M.; Luke, W.T.; Kelley, P.; Artz, R.; Hynes, A.; Landing, W.; Martos, B. Airborne Vertical Profiling of Mercury Speciation near Tullahoma, TN, USA. *Atmosphere* **2014**, *5*, 557–574.

1. Introduction

Airborne measurements of mercury speciation in the free troposphere are rare due to the expense of aircraft operation, the sensitive nature of the sampling and the need to pre-concentrate the mercury species. Swartzendruber *et al.* [1] conducted five summertime flights over Washington State and measured the vertical distributions of gaseous elemental mercury (GEM) and gaseous oxidized mercury (GOM). GOM concentrations were determined both by using the difference between total gaseous Hg (TGM) and GEM and by GOM collection onto KCl coated quartz denuders. For both techniques, the GOM concentration was lowest close to the surface (<1 km) and had a broad peak between 2 and 4 km altitude. The two techniques for measuring GOM concentrations were well correlated, but there was a nearly consistent factor of two difference in concentrations, with KCl denuders showing lower GOM concentrations. Over the five flights, the maximum GOM concentrations at 2–4 km altitude were ~150 and 75 $pg \cdot m^{-3}$ for the difference and denuder methods, respectively.

Murphy *et al.* [2] and, most recently, Lyman and Jaffe [3] conducted aircraft campaigns showing that Hg in the tropopause and lower stratosphere is predominately in an oxidized form. Lyman and Jaffe [3] reported GEM depletions near the tropopause and lower stratosphere, suggested *in situ* oxidation of GEM to GOM and particulate bound mercury (PBM) and reported that subsidence was likely a significant source of oxidized mercury in the free troposphere.

Long-term ground-based measurements near mountain peaks suggest that under certain conditions sampling of free tropospheric air is possible. Fain *et al.* [4] at Storm Peak Laboratory, Colorado, observed episodic elevated GOM concentrations to levels of 137 $pg \cdot m^{-3}$. These events were attributed to deep vertical mixing to the ground level of middle tropospheric air enriched in GOM [4]. Similarly, Swartzendruber *et al.* [5] at Mt. Bachelor Observatory, Oregon, attributed observed nighttime decreases in GEM concentration and enhancements in GOM concentration to air transported from the free troposphere during nocturnal down slope winds. Most recently, at the Mauna Loa Observatory, Hawaii, enhancements of GOM up to ~400 $pg \cdot m^{-3}$ have been observed and attributed to transport from the free troposphere [6].

Vertical modeling of GEM and GOM in the troposphere over North America was summarized by Bullock *et al.* [7]. The various models included were the Atmospheric and Environmental Research (AER), Inc., Global Chemical Transport Model for Mercury (CTM-Hg) [8], the Goddard Earth Observing System (GEOS-Chem) model [9] and the global/regional atmospheric heavy metals model (GRAHM) [10]. Over our study area in Tennessee, the three models predict low GOM concentrations (<20 $pg \cdot m^{-3}$) in the near-surface air with enhancements beginning above the boundary layer with concentrations in the range of ~30 to 120 $pg \cdot m^{-3}$ at an altitude of

220

3 km [7]. Other modeling by Holmes *et al.* [11] suggests that 47% of GEM tropospheric oxidation takes place above 5.5 km, 32% in the middle troposphere between 2.1 and 5.5 km and 21% below 2.1 km.

In this paper, we present, for the first time, monthly vertical profiles of Hg speciation and ancillary measurements from aircraft for an annual cycle over the same location. This study was conducted to examine potential transport from the upper troposphere to the lower troposphere and potential correlations of Hg species aloft to Hg concentrations at the surface. Additional goals were to provide data to evaluate and improve models, identify potential oxidation pathways and assess the relative contributions of natural and anthropogenic processes to local, regional and global Hg burdens.

2. Results and Discussion

2.1. A Typical Flight

Daily flights, with the exception of a single nighttime flight, departed at a local time of 10:00 (am) $\pm$ 45 min. During a typical 3-h flight, GEM, ozone, SO_2, condensation nuclei (CN) and meteorological parameters were measured continuously, while the GOM and PBM samples were collected for 30-min periods in altitude "blocks" during ascent and descent, as shown in Figure 1. During the flight, the average ground speed was ~69 $\pm$ 10 m·s^{-1}, so the distance the airplane traveled during the 30 min period was on the order of 120 km.

Typical sampling data, including GEM, GOM, PBM, O_3, SO_2, CN, temperature, pressure, dew point and solar radiation from a single flight on 23 November 2012, are shown in Figure 2. GOM and PBM were collected in 30-min samples on the ascent and descent. In an unpressurized aircraft, the normal rate of ascent and descent is about 150 m·min^{-1}; thus requiring ~40 min to climb or descend ~6 km. Given the high hourly cost of aircraft, we made use of the ascent and descent times in our sampling.

2.2. Overall Hg Speciation

An overall statistical summary of the airborne and ground-based mercury measurements during this study is shown in Table 1. Average concentrations were 1.38 $\pm$ 0.17 ng·m^{-3} for GEM, 34.3 $\pm$ 28.9 pg·m^{-3} for GOM and 29.6 $\pm$ 29.5 pg·m^{-3} for PBM. Maximum concentrations were 2.05 ng·m^{-3} for GEM, 125.6 pg·m^{-3} for GOM and 194.9 pg·m^{-3} for PBM.

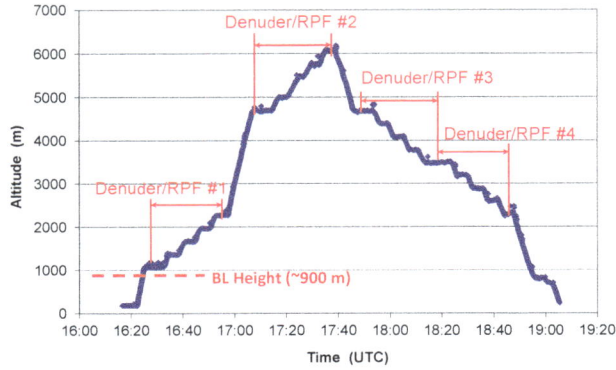

Figure 1. Time series of flight altitude during a typical flight on 13 November 2012 Denuder and regenerable particulate filter (RPF) samples were collected while the airplane was ascending or descending with 30-min sampling intervals for each denuder/RPF set. Samples were usually collected above the boundary layer.

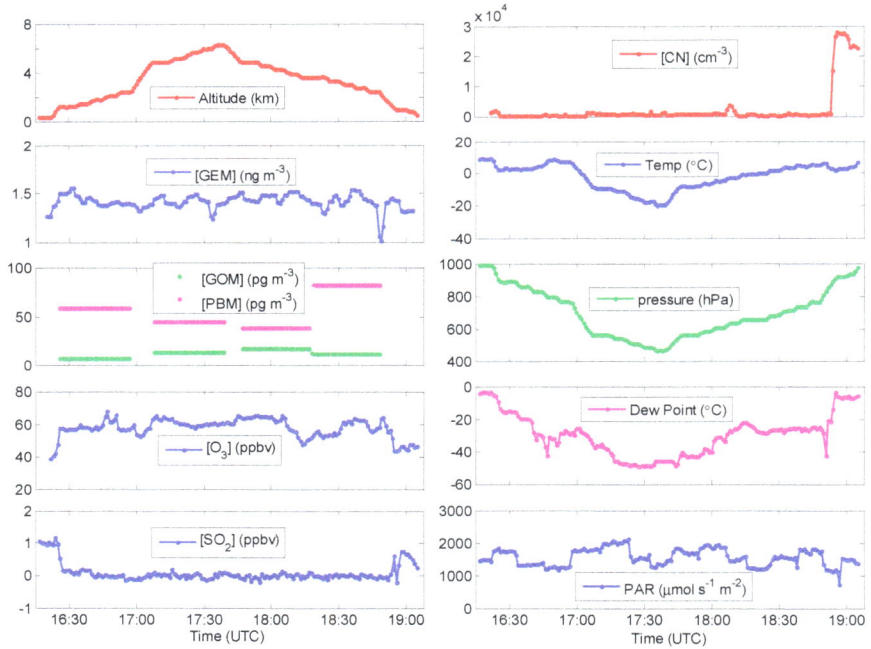

Figure 2. Time series of altitude, gaseous elemental mercury (GEM), gaseous oxidized mercury (GOM), particulate bound mercury (PBM), O_3, and SO_2 (**Left**), as well as condensation nuclei (CN), temperature, pressure, dew point and photosynthetically active radiation (PAR) (**Right**) during the flight on 13 November 2012.

222

Table 1. Statistical summary of airborne and ground measurements of mercury species.

Species	Sample #	Max (pg·m^{-3})	Min (pg·m^{-3})	Mean ± SD (pg·m^{-3})	Median (pg·m^{-3})
GEM_air *	1813	2050	750	1380 ± 174	1400
GOM_air *	106	125.6	3.1	34.3 ± 28.9	22.5
PBM_air *	53	194.9	4.4	29.6 ± 29.5	25.3
GEM_gnd **	26	1610	1170	1350 ± 109	1320
GOM_gnd **	27	12.3	0.6	2.3±2.4	1.80

* Airborne measurements; ** ground measurements.

2.2.1. GEM

The GEM vertical profile for each month showed relatively uniform vertical profiles from August 2012 to February 2013, from the surface to 6 km (Figure 3). Three individual GEM profiles for April, May and June 2013, represented just a single flight per month and showed decreasing GEM above the boundary layer (>1.5 km). However, the GEM range of these three flights was within the variation of some individual flights during other months, and therefore, the GEM difference of these three months from other months could be due to normal day-to-day variation (Figure 3). Also of note is the substantial GEM enhancement above 4 km for two flights in September 2012, which was coincident with enhanced condensation nuclei (CN). This might be due to regional or long-range transport of air pollution rich in both CN and GEM at this altitude. More measurements and modeling work would be needed in order to confirm this speculation.

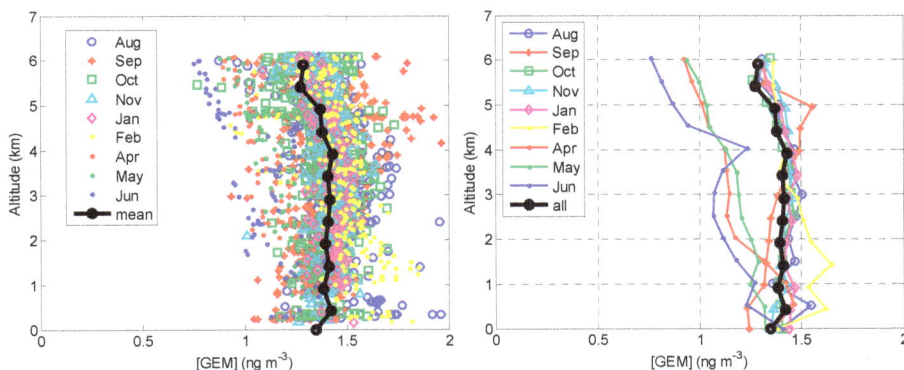

Figure 3. Vertical profile of GEM concentrations: all data points with 2.5 min intervals (**Left**) and mean vertical profile for each month of flights (**Right**). The black line represents mean GEM levels in 0.5-km altitude bins.

2.2.2. GOM

In contrast to GEM, vertical profiles of GOM show large variations (Figure 4). Each KCl denuder collected GOM over a block of altitude over a 30-min time period. With the exception of January 2013, which showed little vertical variation, GOM concentration maxima were consistently measured between 2 and 4.5 km (Figure 5).

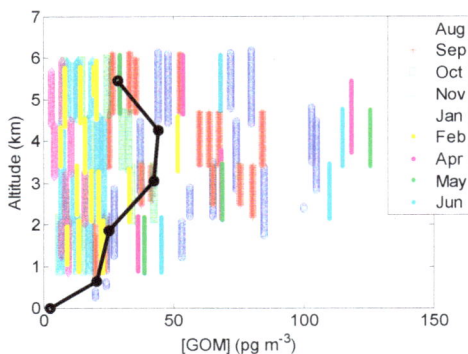

Figure 4. Vertical profile of GOM concentrations. All GOM measurements had a duration of 30 min. Each symbol line represents the GOM concentration of a 30-min sample over the corresponding altitude range. The linked black circles represent the mean vertical profile for all measurements.

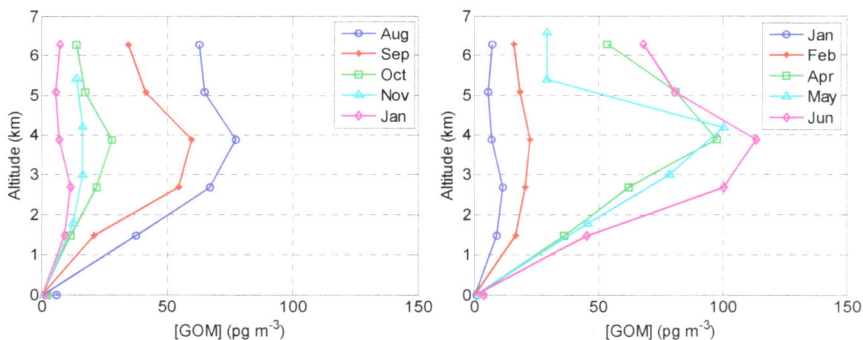

Figure 5. (**Left**) Mean vertical profile of GOM for each month of flights from August 2012 to January 2013 (**Right**) mean vertical profile of GOM for each month of flights from January 2013 to June 2013.

GOM showed a strong seasonal variation. Overall, GOM levels continuously dropped from August to January, were minimum in January and increased from February to June (Figure 5). We observed that GOM concentration at an altitude of ~2–4 km was 10–30-times higher than GOM concentrations in the near-surface air (shown at an altitude of 0 km in Figure 5) and that the ground measurements of GOM did not correlate with the GOM above the boundary layer ($r^2 = 0.13$; detailed sample data not shown). The airborne GOM concentration showed a strong seasonality with "winter" minima in November to February (we did not conduct measurements in December) and "summer" maxima in April to August (we did not conduct measurements in July and only one flight was performed each month in April, May and June 2013). The spring/summer maximum observed in this study is consistent with other measurements made in the southeastern United States (e.g., [12]). GOM during the "summer" (April to August) showed lower concentrations below ~2 km, which is often the summertime midday boundary layer height.

2.2.3. PBM

Measurements of PBM began in October 2012, two months after the study started. Similar to GOM, each University Research Glassware (URG) filter tube downstream of a KCl denuder collected PBM over a block of altitude over a 30-min time period. The URG filter tube we used was essentially the same as the regenerable particulate filter (RPF) used in the commercial Tekran 1135 instrument. The slight difference of our filter tube was in the inlet/outlet connectors in order to fit the aircraft configuration. No PBM measurements were conducted at the ground level, due to resource constraints. As shown in Figure 6, PBM showed "winter" minima in January and February and "summer" maxima in June through October. Little vertical variations were noted in the winter months; however, during summer months, a significant peak occurred in the range of 3–4.5 km with lower PBM at higher altitudes.

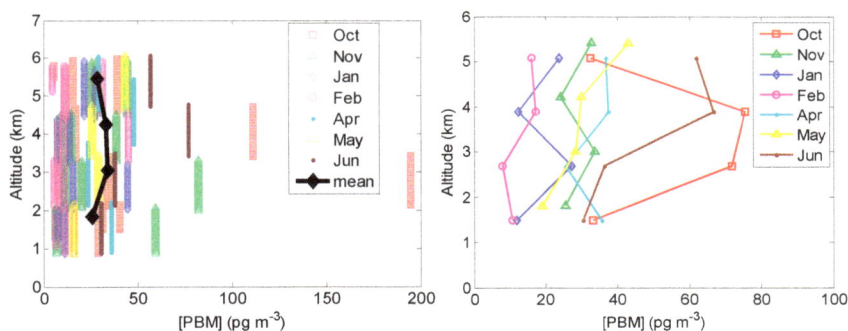

Figure 6. Vertical profile of PBM concentrations. (**Left**) All PBM measurements with 30-min intervals. Each symbol line represents the PBM concentration of a 30-min sample over the corresponding altitude range. The linked black diamonds represent the mean vertical profile of all measurements; (**Right**) the mean vertical profile for each month of flights from October 2012 to June 2013.

225

Surprisingly, our results from a single nighttime flight (11 pm–2 am local time) on 18 October 2012, did not differ significantly from our mid-day measurements earlier that same day. Due to logistical difficulties and the lack of differences in Hg concentrations between the midday and midnight flights on 18 October 2012, we did not conduct any additional nighttime flights.

2.3. Other Supporting Measurements

Figure 7 shows vertical profiles of the ancillary measurements of ozone (O_3), sulfur dioxide (SO_2) and condensation nuclei (CN). O_3 concentrations were relatively low in the boundary layer and relatively constant in the free troposphere. In general, ozone was elevated during the summer months with peak values in August and September, as would be expected. SO_2 was occasionally elevated in the boundary layer, but was near-zero above the boundary layer. Condensation nuclei were generally low with episodic peaks within the boundary layer and at 5–6 km. Overall, the vertical profiles of GEM and GOM were not significantly correlated with ozone ($r^2 = 0.01$ and sample number N = 4,371 for GEM and $r^2 = 0.06$ and N = 107 for GOM), sulfur dioxide ($r^2 = 0.002$ and N = 3,042 for GEM and $r^2 = 0.006$ and N = 107 for GOM) or condensation nuclei ($r^2 = 0.002$ and N = 3,874 for GEM and $r^2 = 0.02$ and N = 107 for GOM) with all p-values greater than 0.1.

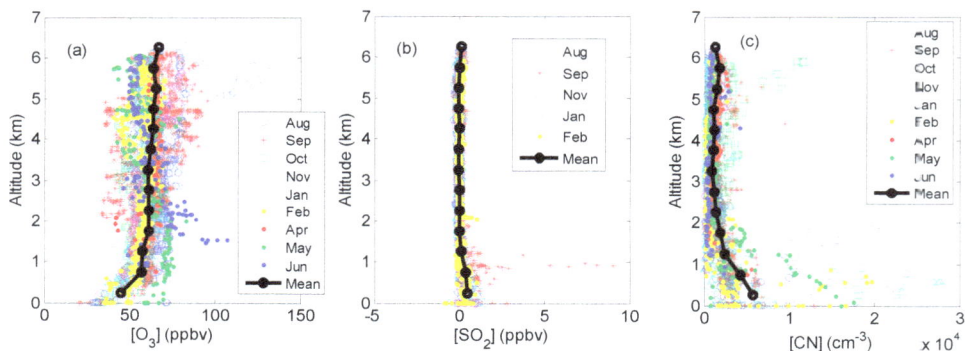

Figure 7. Vertical profiles of ancillary measurements of ozone (**a**), sulfur dioxide (**b**), and condensation nuclei (**c**) for the entire study. Individual symbols represent 1-min data points, and the linked black circles are averages in 0.5-km altitude bins.

2.4. Oxidation of GEM via Hydroxyl Radical

Major uncertainties in current atmospheric mercury models arise from calculated GEM oxidation rates via ozone, hydroxyl radical (OH) and Br [13]. The hydroxyl radical (OH) is a major oxidizing agent in the troposphere and, with ozone, was the first oxidizer postulated for GEM in the troposphere [14]. Model estimates of the GEM lifetime with respect to OH oxidation [15,16] were approximately 120 days in the troposphere. The OH oxidation mechanism then fell out of favor when [17] and [18,19] concluded that the product of GEM + OH → Hg(OH) was too shorted-lived to be atmospherically relevant. Calvert and Lindberg [20] further suggested that OH oxidation of GEM might be too slow to account for the observed atmospheric lifetime of GEM. A review by Hynes *et al.* [21] determined that the OH initiated oxidation of GEM was unlikely, but that more direct observations of HgOH lifetime and reactivity were needed to resolve its potential atmospheric importance. In spite of these arguments against OH, current atmospheric Hg models (e.g., [22,23]) include ozone, OH and Br as oxidizing agents. Holmes *et al.* [13] concluded that Br is likely more influential over the ocean and the polar regions, but that OH and/or ozone may be more significant over regions, such as the southeastern continental USA where Br levels are low.

To a first approximation, OH can be considered to be a product of sunlight, ozone and water vapor. The primary tropospheric production of OH at temperate latitudes is:

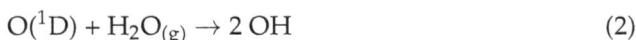

$$O_3 + h\nu \rightarrow O_2 + O(^1D) \tag{1}$$

$$O(^1D) + H_2O_{(g)} \rightarrow 2\,OH \tag{2}$$

In the first reaction, the formed $O(^1D)$ can be quenched by collisions with molecular oxygen and nitrogen to the lower-energy $O(^3P)$ state while a smaller fraction (1%–20%) reacts with water vapor, producing OH via the second reaction. The primary loss mechanisms for OH at temperate latitudes are the oxidation of CO and CH_4, giving OH a tropospheric lifetime of 0.1 to 1.0 s. With this short of a lifetime, OH concentrations are determined by local chemical processes rather than transport [24,25].

Approximate, modeled OH profiles are shown in Figure 8, adapted from Fishman and Crutzen [24], for average [OH] concentrations at solar-noon for our study location on January 1, April 1, July 1 and October 1. The Fishman and Crutzen [24] model is a quasi-steady state photochemical numerical simulation developed to calculate the tropospheric distribution of OH. The OH concentrations are highest in the summer (July) and lowest in the winter (January). The modeled OH vertical gradient over our profile altitudes (0–6 km) was minimal in winter (January) and maximum (decreasing with height) in summer (July), roughly consistent with our observed gradients of GOM and PBM in winter and above 4 km in summer. We speculate that below ~3 km, removal processes, related to deep summertime boundary layer entrainment, downward mixing and surface sinks, deplete GOM and PBM at a rate exceeding their production via oxidation. In additional to the seasonality of GOM concentration, its monthly profile averages were strongly correlated ($r^2 = 0.83$; maximum OH *versus* averaged GOM concentration to our modeled hydroxyl radical concentrations, as were maximum OH *versus* average PBM ($r^2 = 0.67$). In addition, for the months when both GOM and PBM were measured, the total oxidized mercury (averaged monthly peak GOM + PBM) and maximum OH showed a correlation of $r^2 = 0.94$ (Figure 8). The summertime decrease of GOM concentrations at altitudes below ~3 km could be due to entrainment into the atmospheric boundary layer followed by relatively rapid dry or wet deposition.

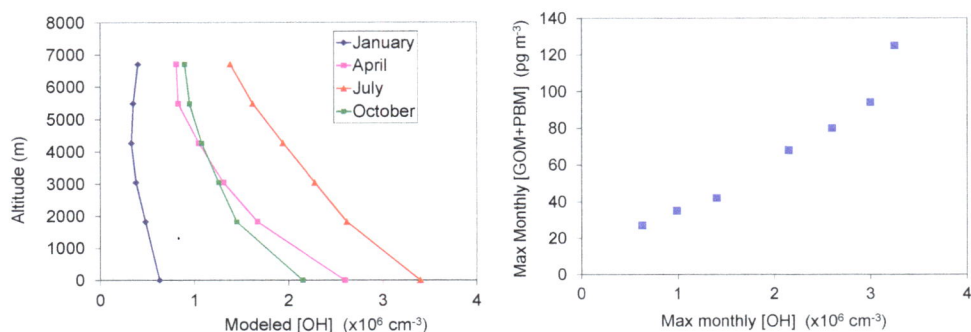

Figure 8. (**Left**) Modeled OH at solar noon over the study location 1 January, 1 April, 1 July and 1 October (adapted from Fishman and Crutzen [24]). (**Right**) peak GOM + PBM from the monthly mean profiles from the months when both were collected *versus* the monthly modeled OH maxima.

The significant correlation between maximum monthly GOM + PBM and our modeled maximum monthly OH (Figure 8) certainly does not prove that OH plays a major role in GEM oxidation, as correlation does not prove causation. As noted above, Hynes *et al.* [21] and others have argued against OH being a significant oxidizer of GEM. Other explanations might include cloud processes, typically at the top of the summertime boundary layer (~1–1.5 km), or GEM oxidation by halogens, none of which were measured or modeled in this study.

2.5. HYSPLIT back Trajectory Analysis

Two examples of back trajectories for GOM concentration >70 pg·m^{-3} and PBM concentration >30 pg·m^{-3} are shown in the left and middle panels of Figure 9, respectively. We found that back trajectories for GOM concentration >70 pg·m^{-3} or PBM concentration >30 pg·m^{-3} were largely associated with air masses coming from west/northwest, while samples with lower GOM and PBM were generally associated with winds from a wider range of directions. It can also be seen in Figure 9 that the trajectories for different starting altitudes are relative flat, suggesting that there was little downward or upward vertical mixing during those period with enhanced GOM and PBM concentrations. A detailed source-receptor analysis is beyond the scope of this paper. Additional back-trajectory and/or forward dispersion modeling will be necessary to determine the extent to which local/regional sources may have contributed to the levels of mercury seen.

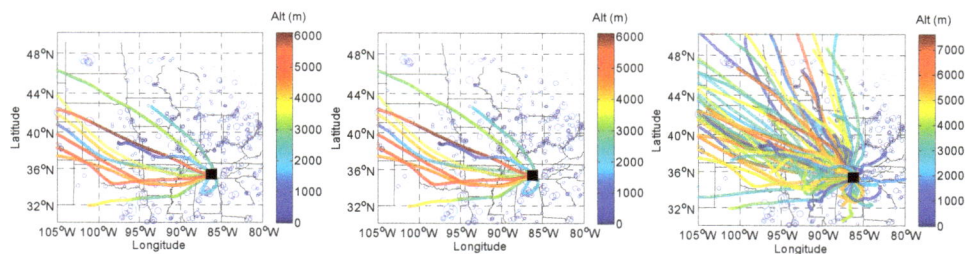

Figure 9. HYSPLIT two-day back trajectories during the flight hours with a starting location of Tullahoma, TN, and starting altitudes at the mid-point of each altitude blocks where airborne GOM and PBM samples were collected. (**Left**) Back trajectories for samples with measured GOM concentration >70 pg·m^{-3}; (**Middle**) back trajectories for samples with measured PBM concentration >30 pg·m^{-3}; (**Right**) back trajectories for all GOM/PBM samples. Blue circles show the locations of large mercury point source emissions in the region, based on the U.S. EPA's 2011 National Emission Inventory. The size of each circle is proportional to the emission rate. The black square at the end of trajectories in each plot represents the flight area.

3. Experimental Section

3.1. Study Area

The study was conducted over an area near Tullahoma (35°23′N, 86°14′W), TN, USA, which is located about 110 km southeast of Nashville, TN, USA. To simplify the flight track and to easily accommodate the air traffic control in this region, we requested a square flight area centered on Tullahoma with a length of ~50 km for each side. The U.S. Environmental Protection Agency (EPA) 2011 National Emission Inventory (NEI) shows that there are no significant anthropogenic mercury emission sources within the flight area (within ~100 km) (Figure 10). However, regionally, there are many Hg point sources, mainly power plants and industrial facilities.

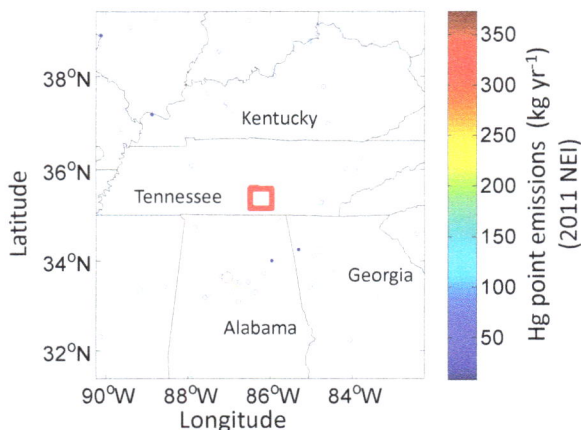

Figure 10. Regional anthropogenic mercury point emissions near Tullahoma, TN, USA, as indicated by the open circles with total mercury emission rates greater than 10 kg Hg·yr^{-1}, based on the U.S. EPA's 2011 National Emissions Inventory (NEI). The size of each circle is proportional to the emission amount. The red square in the middle indicates the flight region.

3.2. Measurements

3.2.1. Aircraft Measurements

The University of Tennessee Space Institute (UTSI) Piper Navajo airborne science research aircraft (N11UT) was used in this study to conduct airborne vertical profiling of speciated mercury and other pollutants at altitudes up to ~6 km above ground level in the Tullahoma region. The aircraft was based at Tullahoma Regional Airport (35°23′N, 86°14′W). Airborne measurements of GEM, GOM, PBM, ozone (O$_3$), sulfur dioxide (SO$_2$), condensation nuclei (CN) and meteorological parameters (temperature, pressure, dew point and solar radiation) were made (Figure 11). The study was conducted from August 2012 to June 2013, with typically one week of flights during each month (typically 3–5 flights each month) to characterize seasonality in mercury concentrations. The flight dates are listed in Table 2. Flights were conducted away from clouds in order to minimize potential scavenging of GOM and PBM into cloud drops and to avoid wetting the walls of the forward-facing inlet, with possible attendant losses. On the two occasions when the airplane flew through thin cloud layers during descent, GEM concentrations were not noticeably affected. GOM and PBM samples were conducted through these cloud layers; however, the duration of the flight through the thin cloud layers was only a few minutes, and the impacts appeared to be insignificant.

An advantage of using a twin-engine aircraft with outboard engines is that the forward fuselage area is free of exhaust contamination. Here, we extended the sampling inlets through the fuselage near the middle window on the starboard side. Sample air for the mercury speciation was brought into the onboard mercury system through a forward-facing 30 cm-long, 2 cm-diameter Teflon-coated aluminum tube protruding 15 cm into the free stream flow, with a volumetric flow rate of 20 $L \cdot m^{-1}$. Due to its very short length (30 cm), the tube was left unheated. Sample air for the Tekran 2537, ozone and SO_2 instruments were obtained through a rear-facing Teflon tube protruding 8 cm into the free stream. A rear-facing inlet instead of a forward-facing inlet was used for the Tekran 2537 to minimize the effects of pressure changes due to speed changes during flight. Another advantage of using different inlets for GEM and GOM/PBM was to have continuous GEM measurement during flight because the air flow was blocked intentionally by the valves shown in Figure 11 during some ascent/descent periods when all valves were closed. For the GOM/PBM sampling, a flow of 20 $L \cdot m^{-1}$ (*i.e.*, $2 \times 10 \ L \cdot m^{-1}$) was needed, so a forward-facing inlet was used, so the ram air could assist in meeting the flow requirements. Sample air for the CN counter was obtained through a rear-facing stainless steel tube (conductive, to avoid static charge buildup and loss of aerosols on the inlet. Tube lengths were minimized to avoid adsorptive loses. During the campaign, the Hg inlet tube was removed and periodically rinsed with DI water, which was subsequently analyzed for total mercury. The total mercury quantity in the rinsed water was found to be negligible. We concluded that oxidized Hg was not trapped by this short inlet tube.

Table 2. Measurement flight months and dates.

Month	# of Flights	Dates
August, 2012	5	3,6,7,8 and 9 August
September, 2012	5	10,11,12,13 and 14 September
October, 2012	5	15,16,17,18 (day), 18 (night) October
November, 2012	4	13,14, 15 and 16 November
January, 2013	2	18 and 31 January
February, 2013	4	1,4,5 and 6 February
April, 2013	1	10 April
May, 2013	1	14 May
June, 2013	1	4 June
Total	28 Flights	

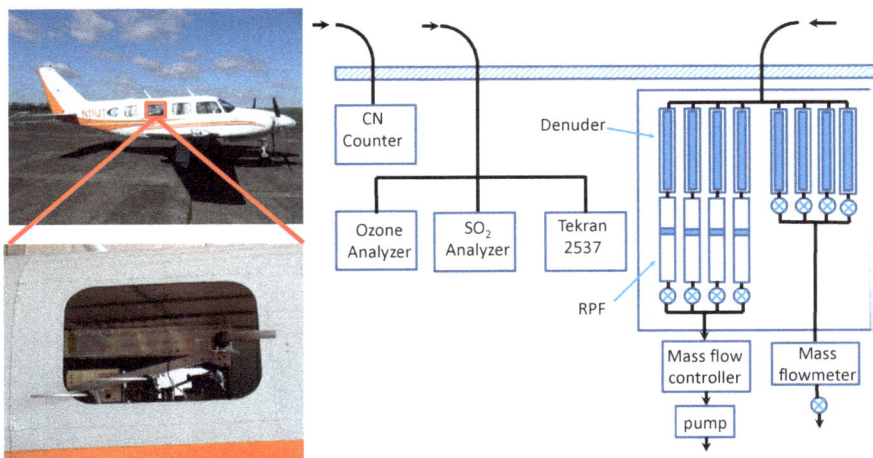

Figure 11. Pictures showing the University of Tennessee Space Institute (UTSI) Navajo airplane, sample inlets through a window and the instrumentation inside the airplane. On top of the instrument rack is a box assembled with 8 denuders and 4 regenerable particle filters (RPFs) used to sample GOM and PBM, respectively.

GEM was measured with an on-board Tekran ambient mercury vapor analyzer (Model 2537) based on cold vapor atomic fluorescence spectrophotometry (CVAFS) with a pressure controller on the cell vent to maintain a constant detection cell pressure (~780 Torr) at different altitudes. The instrument was operated with a sample cycle time of 2.5 min with a detection limit of about 0.1 ng m^3. Ozone was measured with a UV photometer constructed from components of a commercial detector (Model 49, Thermo Environmental Corporation, now Thermo Fisher Scientific, Waltham, MA, USA) mated to custom electronics for enhanced stability and response speed [26,27]. Sulfur dioxide was measured with a similarly modified pulsed fluorescence detector (Model 43s, Thermo Electron Corporation, Waltham, MA) optimized to minimize interferences posed by aromatic hydrocarbons and zeroed periodically with a carbonate-impregnated cellulose filter [28]. The number concentration of particles ranging in size from approximately 0.01 to 3 μm was measured with a TSI condensation nucleus counter (Model 3760). The particles are detected by condensing n-butanol on the particles, which allow them to grow to a size to be detected and counted by a laser-diode optical detector. All data were recorded at a 1-Hz frequency. The Tekran 2537 analyzer was calibrated using its internal Hg permeation source before each flight and periodically checked with an Hg calibrator (Tekran 2505) using manual injections. The O$_3$ detector was calibrated before and after the field campaign with a primary ozone standard. The SO$_2$ detector was calibrated weekly at concentrations ranging from 0–40 ppbv by dynamic dilution

232

of a NIST-traceable compressed gas standard. No direct calibration for the CN counter is required, because particle pulses are well above the electronics noise level, and each pulse corresponds to exactly one particle. The detection limits were 1 ppbv for O_3 and 0.07 ppb for SO_2. The uncertainty for 1-min average concentrations is estimated to be $\pm$ (2 ppbv + 2% of reported concentration) for O_3 and $\pm$ (0.07 ppb + 6% of reported concentration) for SO_2. GOM and PBM samples were collected on each flight using eight potassium chloride (KCl)-coated University Research Glassware (URG®) Corporation quartz denuders and four URG® regenerable particulate filters (RPFs), respectively, similar to those used in a Tekran Model 1130 Speciation Unit and a Model 1135 Particulate Mercury Unit, respectively. The volumetric sample flow rate through each denuder and RPF was maintained at 10 L·min^{-1} to ensure a constant residence time in the denuder. Each denuder pair/RPF assembly sampled air for about 30 min through a given altitude block (Figure 2). After each flight, denuder and RPF samples were immediately analyzed by heating them in a temperature-controlled tube furnace (set at 500 °C for denuder analysis and 800 °C for RPF analysis) and measuring the resulting, liberated mercury with a Tekran Model 2537 ambient mercury vapor analyzer. Concentrations of GOM and PBM were calculated based on the integrated elemental mercury driven off during heating and total sample volume corrected to standard conditions (*i.e.*, 0 °C and 1 atm). Field blank denuders (for GOM) and RPFs (for PBM) were deployed by installing them exactly the same as the sample denuders and RPFs, but with no air flow allowed. Both GOM and PBM blanks were very low and close to the detection limit of the Tekran 2537. During the normal sampling, denuders and RPFs were typically loaded with mercury amounts at least an order of magnitude greater than the blank levels. The average blank Hg level was about 0.8 pg for both GOM and PBM samples, and the detection limits were 2–3 pg·m^{-3}. The GOM concentration differences between the replicate denuders collected for all samples at all altitudes averaged 13% with a standard deviation of 8.6%.

There are recent concerns that KCl denuders significantly underestimate GOM, particularly in the presence of high ozone [29,30]. This will be discussed further in the Conclusions section. Another bias inherent in the collected dataset is a "sunny and clear" weather bias. Due to regulatory restrictions and safety concerns, the aircraft was not operated in rain, poor visibility or anytime there were significant cloud layers below about 7 km. Particularly in the summer months, this precluded many potential flight days.

3.2.2. Ground-Based Measurements

Measurements of GEM and GOM at a ground site at the airport were also made during the flight periods utilizing a Tekran 2537/1130 system (Toronto, ON, Canada) with the inlet ~1.3 m above the airport tarmac surface. The system was operated according to National Atmospheric Deposition Program Atmospheric's (NADP) Mercury Network (AMNet) protocol [31], except a shorter sample/analysis period (2.5 min) was used. The system was located next to a hangar where the aircraft parked. The system did not start to sample the ambient air until the aircraft was in the air, and there were few other airplanes around the hanger that were operated during ground-level sampling, so the risk of the sampling airplane exhaust was eliminated. A particulate mercury sample unit was not available for the ground-based measurements, due to resource constraints.

3.3. HYSPLIT Trajectory Model

Five-day back trajectory calculations were conducted for each high GOM and PBM event to establish the transport history of the associated air masses. The back trajectories were calculated using a PC version of the NOAA Hybrid Single-Particle Lagrangian Integrated Trajectory model (HYSPLIT, v4.9) [32] and the Eta Data Assimilation System (EDAS40) archive, having a horizontal resolution of 40 km × 40 km covering the continental United States, and 3-h time resolution. Trajectories were initialized from Tullahoma at 5 different altitudes (100 m, 1800 m, 3000 m, 4200 m and 5400 m) for the hours when the flights were conducted. These altitudes represent the surface and the typical middle altitudes of four altitude blocks where airborne GOM and PBM samples were collected.

4. Conclusions

Vertical profiling flights primarily for mercury species from the surface to ~6 km were conducted between August 2012 and June 2013, over central Tennessee, USA. Profiles were predominantly characterized by high spring/summer concentrations of gaseous oxidized mercury (GOM) up to 126 $pg \cdot m^{-3}$ (mean: 46 ± 30 pg m^{-3} and range: ~6–126 $pg \cdot m^{-3}$) and low fall/winter concentrations, ~10–20 $pg \cdot m^{-3}$. The predominant vertical variation was low concentrations of GOM at the surface (1–5 $pg \cdot m^{-3}$), peak concentrations at 2–4 km and decreasing concentrations extending to the highest altitude (6.1 km) at which measurements were conducted. The profiles of particulate bound mercury (PBM) showed some of the same general variations as observed for GOM, but were far less pronounced. The vertical and seasonal variation for gaseous elemental mercury (GEM) varied the least with concentrations ~1.4 $ng \cdot m^{-3} \pm 10\%$ for the entire vertical range and annual cycle.

Comparing our summer GOM profiles (Figure 5) to those airborne GOM measurements previously conducted by Swartzendruber *et al.* [1] (Supplemental Materials Figure S2) in July and August over Washington State, DC, USA, we see the same variation of low GOM concentrations in the boundary layer (<1–2 km), peak GOM concentrations in the 2–4 km range and decreasing GOM concentrations above ~4 km. The technique used in Swartzendruber *et al.* [1] was based on the difference method (TGM-GEM), which they determined to be well correlated to the denuder method with a nearly consistent factor of two difference in concentrations (with KCl denuders showing less GOM concentrations). Over their five flights in July (one flight) and August (four flights), the maximum mean GOM concentrations at 2–4 km altitude were ~150 and 75 $pg \cdot m^{-3}$ for the difference and denuder methods, respectively [1]. This latter concentration is remarkably close to our (denuder) mean August (four flights) GOM concentration of 73 $pg \cdot m^{-3}$ at 2–4 km.

For future measurement of oxidized mercury Gustin *et al.* [30] recommends total pyrolyzer methods over the use of KCl-coated denuders. We could not take the suggestions from this 2013 publication, because our measurement began in 2012. There are current concerns that the KCl denuders used in this study may significantly under-collect GOM, especially in the presence of enhanced ozone [29,30]. GOM collections also suffer from a lack of calibration standards [33]. However, there are challenges with both the denuder and total pyrolyzer techniques. Despite these concerns, KCl denuders are currently utilized by the Tekran model 1130 and by the National Atmospheric Deposition Program Atmospheric Mercury Network (AMNet). In light of the above, the GOM concentrations reported here may be biased low, particularly in the summer months, when ozone is elevated. We feel that this potential bias does not significantly affect our main findings regarding the vertical profiles of the Hg species profiles and their seasonality. Swartzendruber *et al.* [1] utilized KCl denuders and a pyrolyzer unit concurrently on their aircraft and determined a roughly consistent factor of ~2 difference in concentrations, with KCl denuders showing less GOM. As presented, our GOM results match well with Swartzendruber *et al.*'s [1] KCl denuder GOM results. If we doubled our oxidized mercury reported results, then these revised results would closely match the pyrolyzer (TGM-GEM) results of Swartzendruber *et al.* [1]. Results from CTM-Hg, GEOS-Chem and GRAHM are consistent with our observations up to 4 km, predicting low GOM concentrations in the near surface air with enhancements beginning above the boundary layer and concentrations in the range of ~30 to 120 $pg \cdot m^{-3}$ at an altitude of 3 km [7]. However, none of these models predict our and Swartzendruber's *et al.*'s [1] observed decrease in GOM concentrations above 4 km. All of these models show a positive (increasing) GOM concentration gradient at 4–5 km. Here, we find these modeled results of the increased oxidation rate above ~4 km to be inconsistent with our "summer" GOM profiles and the previous mean summer GOM concentration profile of Swartzendruber *et al.* [1]. It is possible that these current models overestimate the GEM oxidation in the 4–6 km range.

Our GOM concentration profiles are consistent with other indirect measurements of free-tropospheric GOM. Greatly elevated GOM concentrations in the free troposphere relative to normal ground-based measurements have been reported from mountain peak measurement sites under special conditions where free tropospheric air has been rapidly mixed downward [4,5].

Our measurements over Tullahoma showed generally that GOM concentrations peaked at an altitude of ~2–4 km, where it was 10–30-times higher than GOM concentrations in the near surface air. GOM and PBM concentrations showed their lowest concentrations at the lowest altitudes where boundary layer entrainment leads to surface deposition. Ground-based GOM concentration was not significantly correlated (r^2 = 0.13) to the maximum GOM concentration in the profile, making assessment of the GOM column difficult from just ground-based measurements. GOM concentrations showed a strong seasonality with "winter" minima in November to February and "summer" maxima in April to August. PBM concentration, similar to GOM concentration, showed "winter" minima in January and February.

We find that some current models (e.g., CTM-Hg, GEOS-Chem and GRAHM) [11] appear to "miss" the summer decrease in GOM concentration above 4 km observed here and by Swartzendruber *et al.* [1]. We suggest that possibly these current models overestimate the GEM oxidation rate in the 4–6 km range.

Here, we have presented for the first time a set of speciated mercury profiles at a single location over the course of a year. Users of these data are cautioned that these measurements are biased towards "clear sunny days" due to safety concerns of flying in the rain, poor visibility or cloud layers below 7 km.

Acknowledgments: This study was funded by the National Oceanic and Atmospheric Administration (NOAA) (Project #: NA09OAR4600198 and NA10OAR4600209). We thank the University of Tennessee Space Institute flight crew, Devon Simmons, Gregory Heatherly, Samuel Williams and Jacob Bowman, for their dedicated work to make the airborne measurements successful. Support for this research was also partially provided by the Cooperative Institute for Climate and Satellites agreement funded by NOAA's Office of Oceanic and Atmospheric Research under a NOAA Cooperative Agreement.

Author Contributions: Steve Brooks and Xinrong Ren wrote the manuscript and performed much of the data analysis. Steve Brooks, Xinrong Ren, Winston T. Luke, Paul Kelley, Mark Cohen, Richard Artz, Anthony Hynes, William Landing and Borja Martos were involved in the planning, preparation and data collection in the field deployment. Mark Cohen helped with HYSPLIT trajectory simulations and provided scientific insight and editing.

Conflicts of Interest: The authors declare no conflict of interest.

References

1. Swartzendruber, P.C.; Jaffe, D.A.; Finley, B. Development and first results of an aircraft-based, high time resolution technique for gaseous elemental and reactive (oxidized) gaseous mercury. *Environ. Sci. Technol.* **2009**, *43*, 7484–7489.

2. Murphy, D.M.; Hudson, P.K.; Thomson, D.S.; Sheridan, P.J.; Wilson, J.C. Observations of mercury-containing aerosols. *Environ. Sci. Technol.* **2006**, *40*, 3163–3167.

3. Lyman, S.N.; Jaffe, D.A. Formation and fate of oxidized mercury in the upper troposphere and lower stratosphere. *Nat. Geosci.* **2012**, *5*, 114–117.

4.   Fain, X.; Obrist, D.; Hallar, A.G.; Mccubbin, I.; Rahn, T. High levels of reactive gaseous mercury observed at a high elevation research laboratory in the Rocky Mountains. *Atmos. Chem. Phys.* **2009**, *9*, 8049–8060.

5.   Swartzendruber, P.C.; Jaffe, D.A.; Prestbo, E.M.; Weiss-Penzias, P.; Selin, N.E.; Park, R.; Jacob, D.J.; Strode, S.; Jaegle, L. Observations of reactive gaseous mercury in the free troposphere at the Mount Bachelor Observatory. *J. Geophys. Res.: Atmos.* **2006**.

6.   Gay, D.A.; Schmeltz, D.; Prestbo, E.; Olson, M.; Sharac, T.; Tordon, R. The atmospheric mercury network: Measurement and initial examination of an ongoing atmospheric mercury record across North America. *Atmos. Chem. Phys.* **2013**, *13*, 11339–11349.

7.   Bullock, O.R.; Atkinson, D.; Braverman, T.; Civerolo, K.; Dastoor, A.; Davignon, D.; Ku, J.Y.; Lohman, K.; Myers, T.C.; Park, R.J.; *et al.* The North American Mercury Model Intercomparison Study (NAMMIS): Study description and model-to-model comparisons. *J. Geophys. Res.: Atmos.* **2008**.

8.   Seigneur, C.; Karamchandani, P.; Lohman, K.; Vijayaraghavan, K.; Shia, R.L. Multiscale modeling of the atmospheric fate and transport of mercury. *J. Geophys. Res.: Atmos.* **2001**, *106*, 27795–27809.

9.   Selin, N.E.; Jacob, D.J.; Yantosca, R.M.; Strode, S.; Jaegle, L.; Sunderland, E.M. Global 3-D land-ocean-atmosphere model for mercury: Present-day *versus* preindustrial cycles and anthropogenic enrichment factors for deposition. *Glob. Biogeochem. Cycles* **2008**.

10.   Dastoor, A.P.; Larocque, Y. Global circulation of atmospheric mercury: A modelling study. *Atmos. Environ.* **2004**, *38*, 147–161.

11.   Holmes, C.D.; Jacob, D.J.; Corbitt, E.S.; Mao, J.; Yang, X.; Talbot, R.; Slemr, F. Global atmospheric model for mercury including oxidation by bromine atoms. *Atmos. Chem. Phys.* **2010**, *10*, 12037–12057.

12.   Nair, U.S.; Wu, Y.; Justin, W.; Jansen, J.; Edgerton, E.S. Diurnal and seasonal variation of mercury species at coastal-suburban, urban, and rural sites in the southeastern United States. *Atmos. Environ.* **2012**, *47*, 499–508.

13.   Holmes, C.D.; Jacob, D.J.; Yang, X. Global lifetime of elemental mercury against oxidation by atomic bromine in the free troposphere. *Geophys. Res. Lett.* **2006**.

14.   Lin, C.J.; Pongprueksa, P.; Lindberg, S.E.; Pehkonen, S.O.; Byun, D.; Jang, C. Scientific uncertainties in atmospheric mercury models I: Model science evaluation. *Atmos. Environ.* **2006**, *40*, 2911–2928.

15.   Sommar, J.; Gardfeldt, K.; Stromberg, D.; Feng, X.B. A kinetic study of the gas-phase reaction between the hydroxyl radical and atomic mercury. *Atmos. Environ.* **2001**, *35*, 3049–3054.

16.   Pal, B.; Ariya, P.A. Gas-phase HO center dot-Initiated reactions of elemental mercury: Kinetics, product studies, and atmospheric implications. *Environ. Sci. Technol.* **2004**, *38*, 5555–5566.

17.   Tossell, J.A. Calculation of the energetics for oxidation of gas-phase elemental Hg by Br and BrO. *J. Phys. Chem. A* **2003**, *107*, 7804–7808.

18.   Goodsite, M.E.; Plane, J.M.C.; Skov, H. A theoretical study of the oxidation of Hg^0 to $HgBr_2$ in the troposphere. *Environ. Sci. Technol.* **2004**, *38*, 1772–1776.

19. Goodsite, M.E.; Plane, J.M.C.; Skov, H. Correction to a theoretical study of the oxidation of Hg^0 to $HgBr_2$ in the troposphere. *Environ. Sci. Technol.* **2012**, *46*, 5262–5262.

20. Calvert, J.G.; Lindberg, S.E. Mechanisms of mercury removal by O-3 and OH in the atmosphere. *Atmos. Environ.* **2005**, *39*, 3355–3367.

21. Hynes, A.J.; Donohoue, D.L.; Goodsite, M.E.; Hedgecock, I.M. Our current understanding of major chemical and physical processes affecting mercury dynamics in the atmosphere and at the air-water/terrestrial interfaces. In *Mercury Fate and Transport in the Global Atmosphere: Emissions, Measurements and Models*; Pirrone, N., Mason, R.P., Eds.; Springer: Berlin, Germany, 2009; pp. 427–457.

22. Seigneur, C.; Lohman, K. Effect of bromine chemistry on the atmospheric mercury cycle. *J. Geophys. Res.: Atmos.* **2008**.

23. Dastoor, A.P.; Davignon, D.; Theys, N.; Van Roozendael, M.; Steffen, A.; Ariya, P.A. Modeling dynamic exchange of gaseous elemental mercury at polar sunrise. *Environ. Sci. Technol.* **2008**, *42*, 5183–5188.

24. Fishman, J.; Crutzen, P. *The Distribution of the Hydroxyl Radical in the Troposphere*; Department of Atmospheric Science. Colorado State University: Fort Collins, CO, USA, 1978.

25. Berresheim, H.; Plass-Dulmer, C.; Elste, T.; Mihalopoulos, N.; Rohrer, F. OH in the coastal boundary layer of Crete during MINOS: Measurements and relationship with ozone photolysis. *Atmos. Chem. Phys.* **2003**, *3*, 639–649.

26. Luke, W.T.; Arnold, J.R.; Watson, T.B.; Dasgupta, P.K.; Li, J.Z.; Kronmiller, K.; Hartsell, B.E.; Tamanini, T.; Lopez, C.; King, C. The NOAA Twin Otter and its role in BRACE: A comparison of aircraft and surface trace gas measurements. *Atmos. Environ.* **2007**, *41*, 4190–4209.

27. Luke, W.T.; Kelley, P.; Lefer, B.L.; Flynn, J.; Rappengluck, B.; Leuchner, M.; Dibb, J.E.; Ziemba, L.D.; Anderson, C.H.; Buhr, M. Measurements of primary trace gases and NOy composition in Houston, Texas. *Atmos. Environ.* **2010**, *44*, 4068–4080.

28. Luke, W.T. Evaluation of a commercial pulsed fluorescence detector for the measurement of low-level SO_2 concentrations during the gas-phase sulfur intercomparison experiment. *J. Geophys. Res.: Atmos.* **1997**, *102*, 16255–16265.

29. Lyman, S.N.; Jaffe, D.A.; Gustin, M.S. Release of mercury halides from KCl denuders in the presence of ozone. *Atmos. Chem. Phys.* **2010**, *10*, 8197–8204.

30. Gustin, M.S.; Huang, J.Y.; Miller, M.B.; Peterson, C.; Jaffe, D.A.; Ambrose, J.; Finley, B.D.; Lyman, S.N.; Call, K.; Talbot, R.; *et al.* Do we understand what the mercury speciation instruments are actually measuring? Results of RAMIX. *Environ. Sci. Technol.* **2013**, *47*, 7295–7306.

31. Ren, X.; Luke, W.T.; Kelley, P.; Cohen, M.; Ngan, F.; Artz, R.; Walker, J.; Brooks, S.; Moore, C.; Swartzendruber, P.; *et al.* Mercury speciation at a coastal site in the northern Gulf of Mexico: Results from the Grand Bay intensive studies in summer 2010 and spring 2011. *Atmosphere* **2014**, *5*, 230–251.

32. Draxler, R.R.; Rolph, G.D. *HYSPLIT (HYbrid Single-Particle Lagrangian Integrated Trajectory) Model*; NOAA Air Resources Laboratory: College Park, MD, USA, 2014.

33. Steffen, A.; Scherz, T.; Olson, M.; Gay, D.; Blanchard, P. A comparison of data quality control protocols for atmospheric mercury speciation measurements. *J. Environ. Monit.* **2012**, *14*, 752–765.

239

Mercury Plumes in the Global Upper Troposphere Observed during Flights with the CARIBIC Observatory from May 2005 until June 2013

Franz Slemr, Andreas Weigelt, Ralf Ebinghaus, Carl Brenninkmeijer,
Angela Baker, Tanja Schuck , Armin Rauthe-Schöch, Hella Riede,
Emma Leedham, Markus Hermann, Peter van Velthoven, David Oram,
Debbie O'Sullivan, Christoph Dyroff, Andreas Zahn and Helmut Ziereis

Abstract: Tropospheric sections of flights with the CARIBIC (Civil Aircraft for Regular Investigation of the Atmosphere Based on an Instrumented Container) observatory from May 2005 until June 2013, are investigated for the occurrence of plumes with elevated Hg concentrations. Additional information on CO, CO_2, CH_4, NOy, O_3, hydrocarbons, halocarbons, acetone and acetonitrile enable us to attribute the plumes to biomass burning, urban/industrial sources or a mixture of both. Altogether, 98 pollution plumes with elevated Hg concentrations and CO mixing ratios were encountered, and the Hg/CO emission ratios for 49 of them could be calculated. Most of the plumes were found over East Asia, in the African equatorial region, over South America and over Pakistan and India. The plumes encountered over equatorial Africa and over South America originate predominantly from biomass burning, as evidenced by the low Hg/CO emission ratios and elevated mixing ratios of acetonitrile, CH_3Cl and particle concentrations. The backward trajectories point to the regions around the Rift Valley and the Amazon Basin, with its outskirts, as the source areas. The plumes encountered over East Asia and over Pakistan and India are predominantly of urban/industrial origin, sometimes mixed with products of biomass/biofuel burning. Backward trajectories point mostly to source areas in China and northern India. The Hg/CO_2 and Hg/CH_4 emission ratios for several plumes are also presented and discussed.

Reprinted from *Atmosphere*. Cite as: Slemr, F.; Weigelt, A.; Ebinghaus, R.; Brenninkmeijer, C.; Baker, A.; Schuck, T.; Rauthe-Schöch, A.; Riede, H.; Leedham, E.; Hermann, M.; van Velthoven, P.; Oram, D.; O'Sullivan, D.; Dyroff, C.; Zahn, A.; Ziereis, H. Mercury Plumes in the Global Upper Troposphere Observed during Flights with the CARIBIC Observatory from May 2005 until June 2013. *Atmosphere* **2014**, *5*, 342–369.

1. Introduction

Mercury (Hg) is emitted by natural and anthropogenic processes, and because of its rather long atmospheric lifetime of one year, it can be transported over long distances [1,2]. After oxidation and deposition, part of it can be transformed to highly neurotoxic methyl mercury. The latter is then bio-accumulated in the aquatic food web and may harm both human populations and fauna, which are dependent on fish [3,4]. Emissions from different natural and anthropogenic processes, such as volcanic emissions, emission from the oceans, from soils, coal and biomass burning, as well as many other anthropogenic activities, have been estimated, and spatially and temporally resolved emission inventories have been calculated from the emission factors obtained in these studies (e.g., [5–15]). Despite all these efforts, the emission estimates are still quite uncertain, especially those related to natural sources and anthropogenic emissions in rapidly developing countries in East and South-East Asia [9,13,16,17]. Thus, more data on mercury emissions are required, and the existing inventories need to be evaluated by measurements.

Direct measurements of emissions by techniques, such as a mass balance technique or using an artificially emitted tracer substance [18], are complex and expensive. The mass balance technique measures the fluxes in and out of a chosen source area and calculates the emissions as a difference between them. For a middle-sized city, it requires the use of several aircraft equipped with precise chemical and meteorological instrumentation to resolve small differences of large fluxes. Alternatively, an artificial tracer, such as SF_6, is emitted in an area under investigation and the emission of the target substance is calculated from the known emission of the artificial tracer and the correlations of the target substance concentrations with those of the tracer. Both techniques have been successfully used to determine emissions of CO and NOy of a middle-sized city [18], but they can hardly be scaled up to larger areas. A determination of emission ratios of two substances from their concentrations in the plumes even of large areas is experimentally much more amenable [18–20]. Consequently, emission ratios are promising to be the most practicable way to evaluate the consistency of an emission inventory of one substance with an inventory of another substance [21]. Emission ratios can also be used to constrain the lesser known emissions of one substance using the better known emissions of another substance [21,22].

The CARIBIC (Civil Aircraft for Regular Investigation of the Atmosphere Based on an Instrumented Container) observatory is a long-term project aimed at the monitoring of atmospheric composition and its changes by using an instrumented freight container flown on-board a passenger aircraft during intercontinental flights [23]. It started in 1997 and, apart from an interruption between 2002 and 2004, has been operational continuously over more than 15 years. Despite cruising most of the time at altitudes from 10 to 12 km, plumes of polluted air lifted mostly

241

by convection or warm conveyor belts [24,25] are frequently encountered in the tropospheric sections of the flights.

Here, we report on plumes with elevated mercury concentrations observed during CARIBIC flights since May 2005, when a mercury instrument was installed until June 2013. Rich ancillary data on other gases, such as carbon monoxide (CO), carbon dioxide (CO_2), methane (CH_4), total reactive nitrogen (NOy), hydrocarbons, halocarbons and on atmospheric aerosol, enable a detailed characterization of the plumes, its attribution to the emission processes and, in combination with backward trajectories, an approximate localization of the emissions. Correlations of Hg with CO, CO_2 and CH_4 provide Hg/CO, Hg/CO_2 and Hg/CH_4 emission ratios, which may help to constrain the estimates of mercury emissions using the CO, CO_2 and CH_4 emission inventories [21].

2. Experimental Section

Since December 2004, a new CARIBIC measurement container [23] on-board an Airbus 340–600 of Lufthansa has been flown monthly on transcontinental flights. The corresponding routes (until June 2013) are shown in Figure 1, and the complete list of flights can be found at www.caribic-atmospheric.com. Typically, a sequence of 4 consecutive intercontinental flights is executed every month. The modified freight container (gross weight: 1.5 metric tons) holds 15 automated analysers for the *in situ* measurements of mercury concentrations and mixing ratios of CO, O_3, NO, NOy, CO_2, total (including cloud droplets) and gaseous water, oxygenated organic compounds and concentrations of fine particles (three counters for particles with diameters >4 nm, >12 nm and >18 nm, all up to 2 μm), as well as one optical particle counter for particles with diameters of 130–900 nm. In addition, air and aerosol samples are taken in flight and subsequently analysed in the laboratory for greenhouse gases, halocarbons, non-methane hydrocarbons (NMHCs) and particle elemental composition and morphology, respectively [23]. In May 2010, several instruments were upgraded and new instruments were added. In the context of this paper, the most important change was the addition of a whole air sampler with a capacity of 88 samples and of an instrument for continuous measurements of CH_4 (Fast Greenhouse Gas Analyzer, Los Gatos Research, [26]). With the new whole air sampler, the measurement frequency of greenhouse gases [27] and hydrocarbons [28] could be increased from 28 to 116 measurements per flight sequence. Halocarbon measurements (except for CH_3Cl, which can be determined by both hydrocarbon and halocarbon analytical methods) were unaffected, due to the limited volume of air available for analysis in the new sampler.

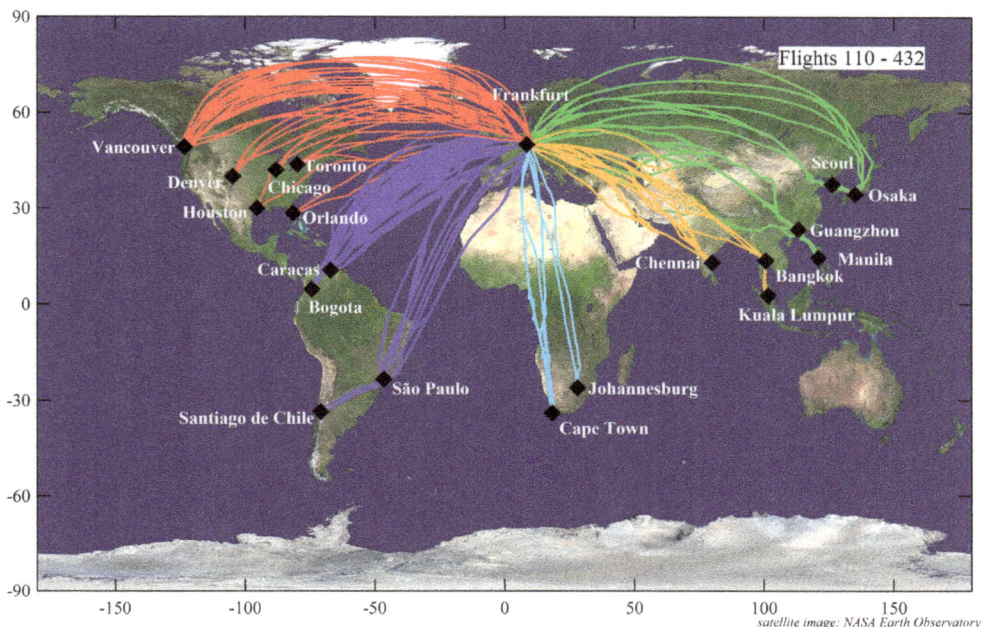

Figure 1. The tracks of 328 CARIBIC (Civil Aircraft for Regular Investigation of the Atmosphere Based on an Instrumented Container) flights from May 2005 until June 2013. The colours denote the classification of destination airports used in this paper: green, East Asia; yellow, South Asia; light blue, Africa; dark blue, South America; red, North America.

The air and aerosol inlet system and instrument tubing are described in detail by Brenninkmeijer *et al.* [23] and Slemr *et al.* [29]. Briefly, the trace gas probe consists of a 3-cm inner diameter diffuser tube with a forward facing inlet orifice of 14 mm in diameter and an outlet orifice of 12 mm in diameter, providing an effective ram pressure of about 90–170 hPa, depending on cruising altitude and speed. This ram pressure forces about 100 L/min of ambient air through a PFA tubing heated to 40 °C (a 3 m-long, 16-mm ID PFA-lined tube connecting the inlet and the container and 1.5 m-long, 16-mm ID PFA tubing within the container to the instrument manifold). The sample air for the mercury analyser is taken at a flow rate of 0.5 L(STP, *i.e.*, at standard temperature of 273.15 K and pressure of 1013.25 hPa)/min from the manifold using 4-mm ID PFA tubing heated by the energy dissipated in the container to ~30 °C. The arrangement similar to that described by Talbot *et al.* [30] was optimized to transmit highly surface reactive HNO$_3$ [31] and can thus be presumed to facilitate the transfer of gaseous oxidized mercury (GOM), as well [30]. The large flow through the trace gas diffuser inlet tube of more than 2000 L/min and perpendicular sampling at much smaller flow rates of

243

about 100 L/min discriminate against particles larger than about one micrometre in diameter (50% aspiration efficiency [32]). Consequently, all smaller particles and, thus, the major fraction of particle mass in the upper troposphere will be transported to the manifold in the container.

The mercury instrument, which is based on an automated dual channel, single-amalgamation, cold vapour atomic fluorescence analyser (Tekran-Analyzer Model 2537 A, Tekran Inc., Toronto, ON, Canada), is described by Slemr et al. [29]. The instrument features two gold cartridges. While one is adsorbing mercury during a sampling period, the other is being thermally desorbed using argon as a carrier gas. Hg is detected using cold vapour atomic fluorescence spectroscopy (CVAFS). The functions of the cartridges are then interchanged, allowing continuous sampling of the incoming air stream. A 45-mm diameter PTFE pre-filter (pore size 0.2 µm) protects the sampling cartridges against contamination by particles that pass through the inlet system. The 0.5 L(STP)/min of air sample, typically at 200–300 hPa, is compressed to about 500 hPa, needed to operate the instrument with its internal pump. Extensive laboratory tests of this PTFE diaphragm pump (KNF-Neuberger, Model N89KTDC) did not show either any contamination of the system with Hg or Hg losses. To avoid the contamination of the instruments and of the tubing connecting the sampling manifold with the instruments during ascents and descents in polluted areas near airports, the sampling pumps are activated only at an ambient pressure below 500 hPa. Consequently, no measurements below an altitude of about 5 km were made.

Initially, the instrument was operated with a gas mixture of 0.25% CO_2 in argon, which also is used for the operation of the CO instrument. Because the addition of CO_2 to argon reduced the sensitivity of the fluorescence detector by ~35%, the instrument was run initially with a 15-min sampling time (corresponding to a ~225 km-flying distance) until March 2006 (Flight 145) and with 10 min until June 2007 (Flight 197). Since August 2007, the CO_2 has been removed from the gas mixture using a tube filled with an X10 molecular sieve. The corresponding sensitivity gain enabled us to reduce the sampling interval to 5 min (corresponding to a ~75-km flying distance). The instrument is calibrated after every other month in the laboratory by ~48 h of parallel operation to a well-calibrated identical instrument. A detection limit of ~0.1 ng·Hg·m^{-3} and a reproducibility of about 0.05 ng·Hg·m^{-3} is achieved at our operating conditions. To improve the detection limit and reproducibility of the measurements, we returned to 10-min sampling in August 2011 (Flight 349). For this paper, the data from May 2005 to June 2013, were analysed. All mercury concentrations are reported in ng·Hg·m^{-3} (STP, i.e., at 1013.25 hPa and 273.15 K).

Speciation experiments on-board the CARIBIC container, where gaseous oxidized mercury (GOM) was removed in the instrument using a KCl or soda lime trap upstream of one of the gold cartridges, showed qualitatively that GOM (essentially Hg^{2+}) is transmitted through the inlet system to the instrument and will be measured. A demonstration of a quantitative transmission would require capabilities to prepare GOM test mixtures at high flow rates and to replicate the flight conditions (*i.e.*, $-50\,°C$, $900\ km·h^{-1}$), which is beyond the constraints imposed by a commercial airliner. Temme *et al.* [33] found the GOM transmission to be quantitative at conditions similar to those in the upper troposphere, *i.e.*, low temperatures and dry air, which allows us to assume the same for our system. A definitive verification of this assumption has to wait for an in-flight intercomparison with a research aircraft with proven speciation capabilities. Newer data on the gas-particle partitioning of atmospheric Hg^{2+} [34,35] suggest that particle bound mercury (PBM, also mostly Hg^{2+}) sampled near the tropopause at temperatures of $\sim-50\,°C$ will evaporate when warmed up to $\sim+40\,°C$ during the transport in the sampling tubing to the instrument. PBM on particles that make it into the trace gas inlet will thus be most likely also measured. Consequently, the CARIBIC measurements approximate the total mercury concentration in the troposphere. We note that even if GOM concentrations in the upper troposphere represent more than 1% or less of total gaseous mercury concentrations typically found in the boundary layer [36,37], its non-quantitative transmission by our inlet system would not substantially influence the results presented in this paper.

The plumes with elevated Hg concentrations showed, apart from a few exceptions mentioned in Section 3.1, also elevated CO and, sometimes, CO_2 and CH_4 mixing ratios. For these plumes, the Hg/CO, Hg/CO_2 and Hg/CH_4 emission ratios were calculated by bivariate least-squares correlations of Hg with CO, CO_2 and CH_4 [38], respectively, which take into account the uncertainties in both variables. For these correlations, the continuously measured CO, CO_2 and CH_4 were averaged over the Hg sampling interval. The uncertainties of Hg, CO, CO_2 and CH_4 measurements were set to $0.05\ pg·m^{-3}$, 1 ppb, 0.05 ppm and 3 ppb, respectively. The underlying assumptions in the emission ratio calculations are: (1) that none of the correlated substances are lost during the transport from the source to the point of encounter by chemical reactions; (2) that the emission ratios are nearly constant during the observation interval; and (3) that the plume is embedded in a homogeneous air mass, *i.e.*, that the Hg concentration and CO, CO_2 or CH_4 mixing ratios before and after the plume are nearly the same [20]. Assumption (1) is fulfilled, as the transport times (ranging from a few days to about one week) are much shorter than the atmospheric lifetime of our target compounds (CO has the shortest lifetime of ~2 months, with a local lifetime in the tropics being several weeks). Assumptions (2) and (3) are probably fulfilled for a majority of smaller plumes. The large plumes stretching over thousands of kilometres north and south of the intertropical convergence zone (ITCZ) during the flights to South Africa are superimposed on a north-south Hg gradient, which violates Assumption (3). Sometimes, large overlapping plumes with

245

different sources in different areas are sampled, also violating the Assumption (2). Consequently, even statistically significant Hg *vs.* CO, CO_2 and CH_4 correlations may sometimes provide biased Hg/CO, Hg/CO_2 or Hg/CH_4 emission ratios in the case of the African flights.

Meteorological analyses for all CARIBIC flights are provided by KNMI (Royal Netherlands Meteorological Institute) at http://www.knmi.nl/samenw/CARIBIC. Trajectories were calculated at 3-min intervals along the flight track for each flight with the KNMI trajectory model, TRAJKS [39], using data from ECMWF (European Centre for Middle Weather Forecast) data.

3. Results and Discussion

3.1. Overview

Compared to the search for plumes at a ground station [21], the processing of the CARIBIC data is complicated by the variability of the data over large distances in the upper troposphere and by the frequent changes of tropospheric and stratospheric air masses. Figure 2 shows an overview of the data from Flight 158 from Frankfurt to Guangzhou on 31 July and 1 August 2006. The aircraft flew in the troposphere until ~23:00 UTC and then in the stratosphere until about 1:45 UTC on 1 August. The stratospheric section is evidenced by the high potential vorticity and O_3 mixing ratio, as well as the low CO mixing ratio shown in the upper two panels of the data time series. In the tropospheric section after about 1:45 UTC on 1 August 2006, mercury background concentrations vary between about 1.25 and 1.35 ng·m^{-3} (the second panel from the top of the data time series). Three events with elevated Hg concentrations, denoted as A, B and C, with peak Hg concentrations of 1.55, 1.5 and 2.3 ng·m^{-3} are observed at about 2:40, 4:50 and 6:00 UTC, respectively, on 1 August 2006. All events are accompanied by elevated CO (the second panel from the top), NOy, H_2O (middle panel) and acetone (bottom panel). We base our approach on the visual inspection of the data overview plots of each flight for the coincident occurrence of elevated Hg concentrations with elevated CO mixing ratios. Plumes identified in this way are cross-checked using variations of other tracers for anthropogenic emissions, such as NOy (middle panel), acetone (bottom panel), CH_4 (the second panel from the bottom), non-methane hydrocarbons (not available for this flight) and halocarbons (the second panel from the bottom). Humidity and cloud water content (determined as the difference between total water content and the water vapour mixing ratio) as tracers for convective processes are also sometimes useful.

An event within the stratospheric section of the flight at ~23:25 UTC (marked as D) has a similar characteristics as Events A, B and C and can easily be mistaken for a plume. The only pronounced difference is that the maximum of Hg concentration (and CO, acetone and H_2O mixing ratios) in Event D coincides with dips in potential vorticity and O_3 mixing ratios, both at higher levels characteristic for the lower stratosphere. Such dips in potential vorticity and O_3 indicate a crossing of a filament of tropospheric air in the stratosphere. The variation of all mentioned species during such crossing results from their strong gradients above the tropopause [29]. Events of this type are thus not related to surface emissions and have to be eliminated from further consideration. Consequently, only events embedded in air with a potential vorticity of less than 1.5 PVU (1 PVU = $10^{-6} \cdot m^2 \cdot K \cdot kg^{-1} \cdot s^{-1}$) and/or less than 150 ppb O_3 were considered.

Because we rely mostly on Hg and CO as plume tracers, only those processes that emit Hg and CO, such as biomass burning, will be detected. This includes also collocated emissions of Hg and CO, CO_2 or CH_4, which applies for most of the urban and industrial emissions. However, emissions from mining and smelting, which emit hardly any CO, CO_2 or CH_4, will not be detected (unless collocated with other CO, CO_2 or CH_4 sources), because a suitable specific tracer for these processes, such as SO_2, is not on the otherwise comprehensive list of CARIBIC *in situ* measurements. Lacking *in situ* SO_2 measurements also prevents the direct detection and identification of volcanic Hg emissions. One such SO_2 plume was detected during the descent to Frankfurt airport on 15 August 2008, using remote SO_2 sensing by a nadir looking differential optical absorption system (DOAS) on-board CARIBIC and remote sensing satellite [40]. Quantitative evaluation of this plume in terms of Hg emission, however, was not possible, because the DOAS measurement does not provide *in situ* SO_2 concentrations and the elevated Hg concentrations were documented by only one Hg measurement point. Among the substances measured in the whole air samples are tracers for marine emissions, such as short-lived bromine and iodine containing halocarbons, but the low sampling frequency (28 samples taken over four intercontinental flights) makes them unsuitable as tracers for the detection of the Hg plumes of marine origin. ^{222}Rn as a tracer for terrestrial Hg emissions [22] is also not being measured on-board CARIBIC. Consequently, in this study, we can only distinguish between Hg plumes of biomass burning or urban/industry origin.

The short encounter with the Kasatochi volcano plume (~5 min) in 2008 during the descent to Frankfurt airport [40] also illustrates some practical limits of our approach. The most severe limitation is given by the low temporal resolution of the Hg measurements of 5–15 min corresponding to a ~75–225 km flight distance. Statistically significant Hg *vs.* CO, CO_2 and CH_4 correlations require at least three measurements. Thus, only plumes larger than ~300 km can be captured by our approach, which also means that many plumes encountered during the short aircraft ascents and descents cannot be resolved. The CARIBIC measurements start and stop at ~500 hPa to prevent the contamination of the CARIBIC system by polluted air in the boundary layer near the airports [23]. Consequently, information for the boundary layer, the most polluted part of the troposphere, is missing in our data set. In addition, the Hg, CO, CO_2 and CH_4 data from ascents and descents through quasi-horizontal layers in the troposphere are likely to violate the assumption of a plume being embedded in a homogeneous air mass on which the Hg vs. CO, CO_2 and CH_4 correlations are based.

Although almost all of the Hg plumes were accompanied by elevated CO mixing ratios, seven of them were observed during flight sections with nearly constant CO mixing ratios. All of these plumes were encountered over the equatorial Atlantic Ocean between $0°$ and $15°$ N during the flights to South American destinations. They were embedded in background mercury concentrations varying between ~0.95 and 1.3 ng·m^{-3}, and the elevation above the background (ΔHg) varied between ~0.25 and 0.45 ng·m^{-3}. In these events, elevated Hg concentrations were always accompanied by elevated humidity, frequently also with clouds and elevated NOy, as well as with low O_3 mixing ratios; frequently, only around 30 ppb. Such low O_3 mixing ratios are typical for the marine boundary layer over the equatorial Atlantic Ocean [41]. Backward trajectories reveal contact with the equatorial Atlantic Ocean surface. Satellite images of cloud cover indicate that these events are due to the convection of the air masses from the marine boundary layer at the ITCZ. Elevated Hg concentration in these events encountered in the upper free troposphere can point to emissions of mercury by ocean [42], but we lack highly resolved tracer data for air from the marine boundary layer to quantitatively describe them.

Despite these caveats and limitations, 98 encounters with plumes with elevated CO mixing ratios and simultaneously elevated Hg concentrations were counted during 309 CARIBIC flights with valid Hg measurements between May 2005 and June 2013. Taking into account the number of flights to the respective regions listed in Table 1, the probability of plume encounters was highest during the flights to South Africa with 85% of the flights. The second highest probability of plume occurrence was over East Asia with 46%, followed by flights to South Asia with 26%, South America with 20% and North America with 17%. The low frequency of plume encounters over North America is partly due to the high northern latitude of the flight routes of these flights, which, at usual flight altitudes of 10–12 km, results in a high proportion of stratospheric sections with a potential vorticity >1.5 PVU. However, the high frequency of plume encounters during the flights to Osaka and Seoul with flight routes at similarly high northern latitudes shows that this bias alone cannot explain the low frequency of plume encounters over Europe and North America.

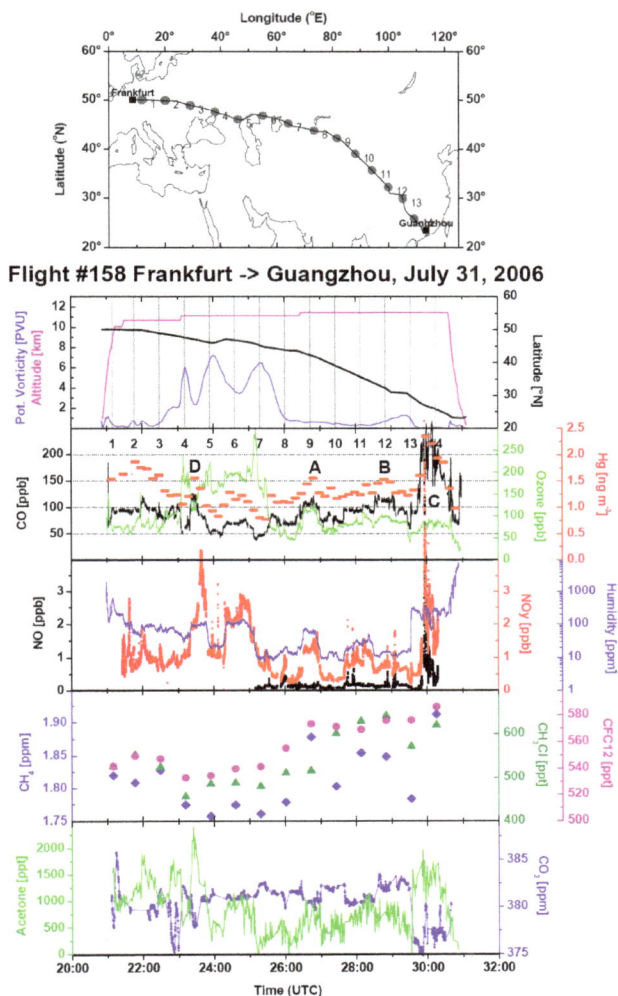

Figure 2. Overview of the data from Flight 158 from Frankfurt to Guangzhou on 31 July 2006. (**Top**) Flight track and the locations of whole air samples. Time series plots are below: (**uppermost panel**): flight altitude (magenta) and latitude (black), potential vorticity (blue), sampling intervals (grey bars); (**second panel from top**): mixing ratios of CO (black), O_3 (green) and Hg concentrations (red); (**middle panel**): mixing ratios of NO (black), NOy (red) and total water content (blue); (**second panel from bottom**): mixing ratios of CH_4 (blue), CH_3Cl (olive green) and CFC12 (CCl_2F_2, magenta)) in whole air samples; (**bottom panel**): mixing ratios of acetone (green) and CO_2 (blue). The three identified plumes are marked with A, B and C in the second panel from top. Another event, due to a crossing of a filament of tropospheric air within the lower stratosphere, is marked with D. Although similar to Events A, B and C, this event has no relation to surface emissions (see the text).

249

For 56 plume encounters (out of 98), the Hg *vs.* CO correlations were statistically significant at a confidence level of at least 95%. For these correlations, data from the same plume encountered twice in the vicinity of an airport were combined, e.g., for a plume near Guangzhou encountered during the forward and return flight to Manila (Flights 203 and 204) or a plume encountered near São Paulo during Flights 123 (Frankfurt→São Paulo), 124 (São Paulo→Santiago de Chile) and 125 (Santiago de Chile→São Paulo). Seven extremely high Hg/CO slopes of 12.9–23.7 $pg \cdot m^{-3} \cdot ppb^{-1}$ were connected with physically unrealistic negative intercepts and are thus eliminated from the data set, leaving 49 plume encounters with valid Hg/CO slopes. The geographic distribution of these plumes is shown in Figure 3.

Table 1. Overview of encounters with plumes with elevated Hg concentrations and CO mixing ratios.

Destination Airport	Number of Flights	Number of Plume Encounters	Number of Plumes With Significant Hg *vs.* CO Correlations	Median and Range of Hg/CO Emission Ratios ($pg \cdot m^{-3} \cdot ppb^{-1}$)
South Africa	13	11	4	2.9 (2.2–7.5)
East Asia	101	46	31	8.2 (2.3–16.6)
South Asia	57	15	6	7.4 (5.0–10.0)
South America	90	18	5	1.3 (1.1–3.3)
North America	48	8	2	9.2 (6.9 and 11.4)

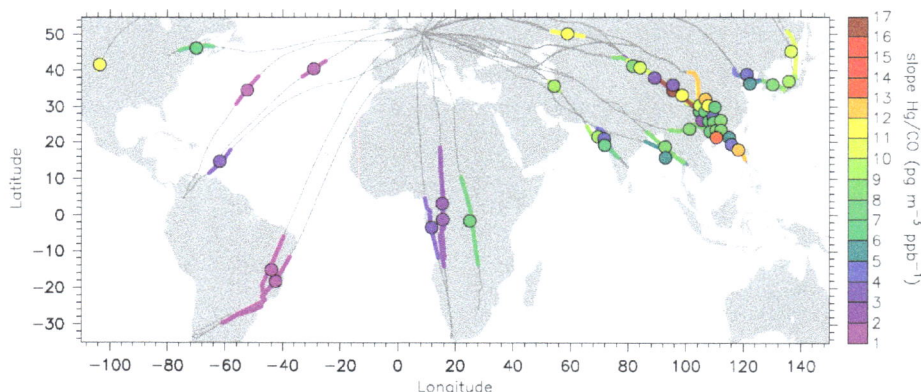

Figure 3. Geographic distribution and the extension of the plumes with statistically significant Hg *vs.* CO correlations. The magnitude of Hg/CO emission ratios in $pg \cdot m^{-3} \cdot ppb^{-1}$ is colour coded.

Most of the plumes with statistically significant Hg *vs.* CO correlations (31 plumes) were encountered over the East Asian region, and these are listed in

Table S1 (Supplementary Information). Table S2 lists 18 plumes with statistically significant Hg *vs.* CO correlations for all other regions. Relative to the number of flights, the frequency of plumes with statistically significant Hg/CO correlations is the highest for the South African and East Asian flights, with each being 31%, followed by flights to South Asia with 11%, South America with 6% and North America with 4%. The high occurrence of plumes during the flights to South Africa in which Hg does not correlate with CO is caused by their large extension over a few thousands of km, changing the Hg and CO background from north to south hemispheric concentrations and the inhomogeneity of the plumes. This will be discussed later in Section 3.3.

The colour code of Figure 3 reveals a pronounced difference between the Hg/CO emission ratios in different regions. The Hg/CO emission ratios for plumes encountered over East Asia range from 2.3 to 16.6 $pg \cdot m^{-3} \cdot ppb^{-1}$ (median 8.2 $pg \cdot m^{-3} \cdot ppb^{-1}$) and are similar to those over South Asia, ranging from 5.0 to 10.0 $pg \cdot m^{-3} \cdot ppb^{-1}$ (median 7.4 $pg \cdot m^{-3} \cdot ppb^{-1}$). On the contrary, the Hg/CO emission ratios for plumes observed during the flights to South America and equatorial Africa range from 1.1 to 7.4 $pg \cdot m^{-3} \cdot ppb^{-1}$, with a median value of 1.8 $pg \cdot m^{-3} \cdot ppb^{-1}$ (a median of 1.3 $pg \cdot m^{-3} \cdot ppb^{-1}$ for South America and 2.9 $pg \cdot m^{-3} \cdot ppb^{-1}$ for equatorial Africa). These plumes include also two plumes encountered over the Atlantic Ocean, which, based on backward trajectories, can be attributed to forest fires in the USA, as will be discussed in Section 3.4. The range and median of Hg/CO emission ratios observed over South America and equatorial Africa is similar to the one for plumes observed at Cape Point in South Africa [21]. A compilation of previously published Hg/CO emission ratios reported for different processes and regions [29] shows that biomass burning is characterized by ratios below 2 $pg \cdot m^{-3} \cdot ppb^{-1}$ [43], whereas the ratios for urban/industrial emissions tend to be around 6 $pg \cdot m^{-3} \cdot ppb^{-1}$ and higher. Applying these criteria to the Hg/CO emission ratios shown in Figure 3 thus leads to the conclusion that the plumes encountered during the flights to South America and equatorial Africa originate predominantly from biomass burning (see also [44]), whereas the plumes observed over East Asia, South Asia and North America are of industrial/urban or mixed origin. This preliminary classification is supported by the detailed discussion in Sections 3.2–3.4.

CO_2 emissions are better known than those of CO, and thus, the Hg/CO_2 emission ratios have a potential to provide a more accurate estimate of Hg emissions [21]. Unfortunately, CO_2 data were available for only 69 (out of 98) of the encountered plumes, of which nine were too narrow for Hg *vs.* CO_2 correlations. Statistically significant correlations of Hg *vs.* CO_2 were only found for the 17 plumes listed in Table S3. These correlations will be discussed in Section 3.5.

Mercury was also found to correlate frequently with methane at Cape Point, which proved to be useful for constraining the mercury emissions in South Africa [21]. Methane data were available only for flights since October 2010, and altogether, 26 correlations of Hg $vs.$ CH_4 could be calculated, of which, only the six listed in Table S4 were statistically significant. These will be discussed in Section 3.6.

3.2. Plumes Encountered during the East Asian Flights

The East Asian plumes were encountered during the flights from Frankfurt to Manila with a stopover at Guangzhou and during the flights to Osaka and Seoul. Apart from a few plumes over Central Asia, most of them were encountered within a ~2000-km distance from the airports at Guangzhou, Osaka and Seoul. Data from one of the flights have already been shown in Figure 2. The Hg/CO emission ratios were 8.8 ± 1.5, 11.3 ± 2.0 and 7.49 ± 1.0 pg·m^{-3}·ppb^{-1} for Events A, B and C, respectively. Air sample analyses (the second panel from the bottom) show high CH_4 in Sample 9 (Event A, sample numbers are marked in the uppermost panel), 11 and 12 (Event B), and the highest level in Sample 14 (Event C). Acetonitrile as a tracer for biomass burning is not available for this flight section, but elevated CH_3Cl mixing ratios (dark green triangles in the second panel from the bottom) indicate some influence of biomass/biofuel burning in Samples 10, 11, 12 and 14, but not in Sample 9. Backward trajectories for the preceding eight days for Samples 9, 11 and 12 in Figure 4 all show that transport took place at a higher altitude (<250 hPa). Notwithstanding, elevated H_2O mixing ratios and satellite cloud images (not shown) indicate convection in the area above the Persian Gulf and the Gulf of Oman for Samples 9 and 11 and above the Iberian Peninsula for Sample 12. Backward trajectories for Sample 14 had a surface contact over Sichuan, China. Fire maps (not shown) show fire counts in Oman and the Indus River Valley, which are reached by the trajectories of Sample 11, but not of 9. The biomass burning contribution for Sample 12 probably originates from fire activity in north-western Spain. No fire counts were reported for Sichuan during this period, but the observed influence of biomass burning could originate from biofuel use.

The composition and origin of the plumes over southern China and the Philippines (around the airport of Guangzhou, east of 103° E and south of 32° N) were analysed in detail by Lai et $al.$ [45]. According to their chemical signatures, the 51 identified high CO events were attributed to biomass/biofuel burning, anthropogenic emissions or a mixture of both. The backward trajectories pointed to three source areas, namely: southern China, the Indochinese Peninsula and the Philippines/Indonesia. The emissions from southern China were found to be dominated by urban/industrial emissions, while emissions from biomass/biofuel burning contributed substantially to plumes from the Indochinese Peninsula. Mixed emissions were attributed to plumes originating from the Philippines/Indonesia.

ECMWF/KNMI/TRAJKS: BW MA_20060731_FRA_CAN_158_W09_TR
sample date: Tuesday 01 August 2006; 03 UTC
end date: Monday 24 July 2006; 03 UTC

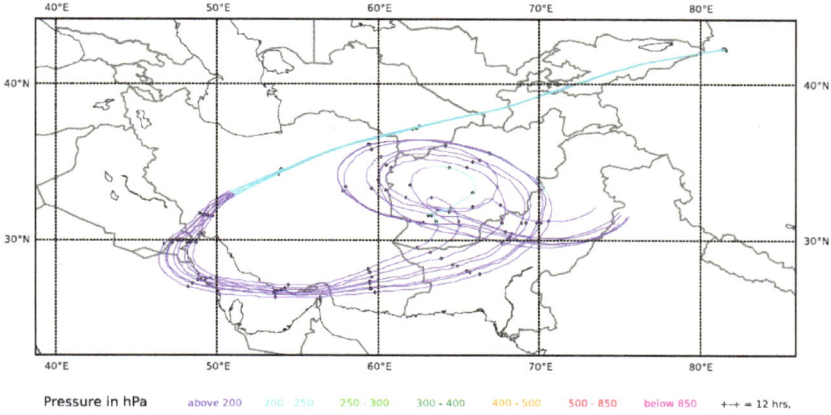

Pressure in hPa   above 200   200 - 250   250 - 300   300 - 400   400 - 500   500 - 850   below 850   +-+ = 12 hrs,

(a)

ECMWF/KNMI/TRAJKS: BW MA_20060731_FRA_CAN_158_W11_TR
sample date: Tuesday 01 August 2006; 05 UTC
end date: Monday 24 July 2006; 05 UTC

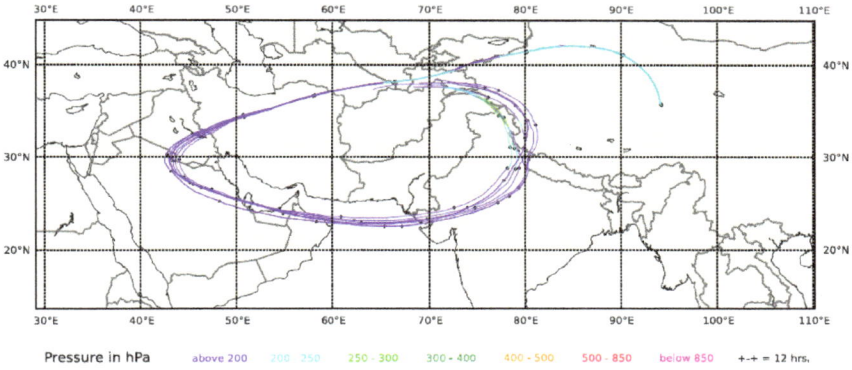

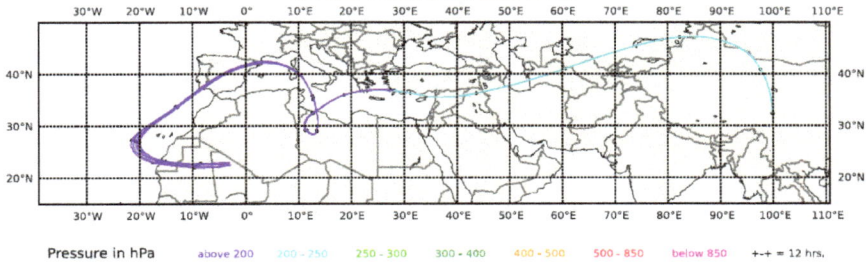

Pressure in hPa   above 200   200 - 250   250 - 300   300 - 400   400 - 500   500 - 850   below 850   +-+ = 12 hrs,

(b)

ECMWF/KNMI/TRAJKS: BW MA_20060731_FRA_CAN_158_W12_TR
sample date: Tuesday 01 August 2006; 05 UTC
end date: Monday 24 July 2006; 05 UTC

Pressure in hPa   above 200   200 - 250   250 - 300   300 - 400   400 - 500   500 - 850   below 850   +-+ = 12 hrs,

(c)

Figure 4. *Cont.*

253

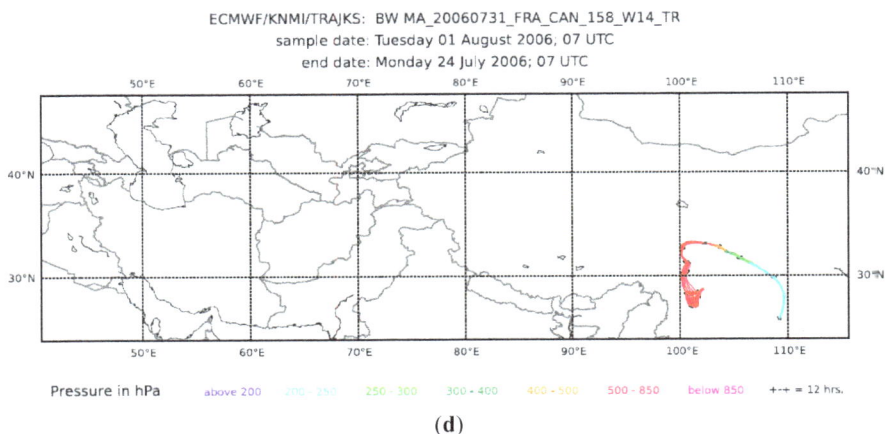

ECMWF/KNMI/TRAJKS: BW MA_20060731_FRA_CAN_158_W14_TR
sample date: Tuesday 01 August 2006; 07 UTC
end date: Monday 24 July 2006; 07 UTC

(d)

Figure 4. Eight-day backward trajectories for whole air Samples 9 (**a**), 11 (**b**), 12 (**c**) and 14 (**d**) taken during Flight 158.

A dense plume observed during Flight 300 from Osaka to Frankfurt on 24 June 2010, covering a distance of about 1000 km over the Korean Peninsula and Yellow Sea, was characterized by CO mixing ratios of ~240 ppb, Hg concentrations of 2.25 ng·m^{-3} and an Hg/CO emission ratio of 4.4 pg·m^{-3}·ppb^{-1}. High mixing ratios of biomass burning tracers, such as CH_3Cl, and of tracers of anthropogenic origin, such as SF_6, together with elevated levels of pollutants, which may originate from both biomass burning and anthropogenic processes, such as ethyne and propane, point to the mixed origin of this plume, both from anthropogenic processes and from biomass burning. Backward trajectories and a map of fire counts for 20–24 June 2010 (not shown, FIRMS (Fire Information for Resource Management System) web fire mapper http://firefly.geog.umd.edu, accessed on 28 October 2010), indicate that the biomass burning component originated most likely from a region with a high burning density in Shandong, Henan, Shanxi and Hebei provinces of China and possibly from some fires in southern Siberia. The anthropogenic component most likely originated from the Chinese provinces mentioned above.

254

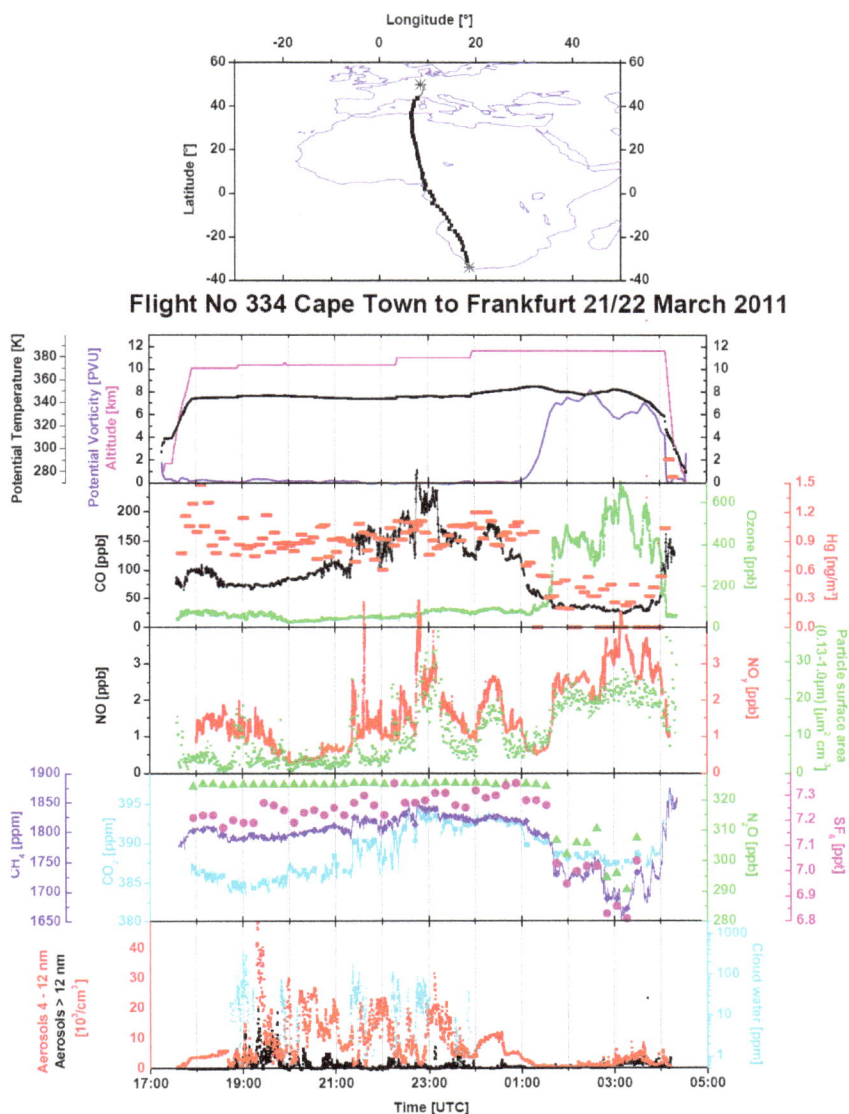

Figure 5. The overview of the data from Flight 334 from Cape Town to Frankfurt on 21/22 March 2011. The parameters displayed here are similar to those in Figure 2. (**Middle**) The time series plots additionally show the particle surface area concentrations (green); (**second panel from the bottom**) mixing ratios of SF_6 (magenta) and N_2O (green) in whole air samples, as well as continuously measured mixing ratios of CH_4 (dark blue) and CO_2 (light blue); (**bottom**) cloud water content (light blue) and concentrations of particles within the 4–12 nm size range (red) and larger than 12 nm (black).

255

The plumes observed near Osaka during two flights, 331 and 332, on 26/27 February 2011, are characterised by similar CO mixing ratios as in June 2010, but higher Hg concentrations of ~2.7 ng·m^{-3}, resulting in the higher Hg/CO emission ratios of 8.4 and 10.0 pg·m^{-3}·ppb^{-1}, which point to urban/industrial origin. Air samples taken within the plume had elevated mixing ratios of SF$_6$ (~7.6 ppt), ethyne (~500 ppt), CH$_4$ (~1870 ppb), CO$_2$ (~397 ppm) and several other hydrocarbons, documenting the urban/industrial component of the plumes. Biomass burning also contributed to this plume, as evidenced by high CH$_3$Cl mixing ratios of ~700 ppt. However, the high Hg/CO slope of ~9 pg·m^{-3}·ppb^{-1} and the high SF$_6$ mixing ratios imply urban/industrial origin to be dominating. Backward trajectories for the plume observed during Flight 331 show a fast transport from the west with surface contact over northern India, southern Pakistan and southern Iran (within ~3 days) and a transport at high altitudes afterwards. Backward trajectories for the plume observed during Flight 332 are similar, but because of their lower altitude contributions from sources in China, cannot be ruled out. However, based on the trajectories from Flight 331 and the low probability of convection in February, we deem northern India and southern Pakistan to be the major source of the observed plumes.

A narrow plume observed during the descent to Seoul during Flight 383 on 29 March 2012, is characterised by very high CO mixing ratios of up to 357 ppb, an Hg concentration of up to 2.49 ng·m^{-3} and an Hg/CO emission ratio of 5.6 pg·m^{-3}·ppb^{-1}. Apart from high NOy and CO$_2$, there are no other measurements available to characterise this plume. The medium Hg/CO slope indicates a mixture of emissions from biomass burning and industrial/urban emissions. The backward trajectories are changing during the descent and point to northern China or/and southern Siberia as possible source areas.

Several plumes were also encountered over central Asia on the way to Guangzhou and back, especially during Flights 158, 161 (both flights were on 1 August 2006), 166 and 169 (both flights were on October 20, 2006). The events during the outward bound flight on October 20 at 3:53–4:53 UTC and during the return flight on the same day at 17:49–20:29 UTC have very similar Hg/CO emission ratios of 3.5 ± 0.7 and 3.2 ± 1.1 pg·m^{-3}·ppb^{-1}, respectively, and originate, with little doubt, from one and the same plume. This plume is analysed in detail by Baker *et al.* [46]. The low Hg/CO emission ratio and elevated acetonitrile, CH$_3$Cl and CH$_3$Br mixing ratios suggest that the plume originates partly from biomass/biofuel burning, whereas elevated mixing ratios of C$_2$Cl$_4$ and toluene, which are used as solvents, indicate anthropogenic contributions. Backward trajectories for the samples taken in this plume pass over Afghanistan, Pakistan and a region of Northern India, where extensive fire activity was recorded during 17–29 October 2006 [46]. Another plume with an Hg/CO emission ratio of 11.2 ± 3.4 pg·m^{-3}·ppb^{-1} and very high acetone mixing ratios was encountered during Flight 161 on 1 August 2006, at

22:07–23:07 UTC. No air samples were taken during this event. Backward trajectories pass at high altitude (<300 hPa) over the Black Sea partly to northern Europe and partly to the European Mediterranean coast. Elevated H_2O mixing ratios and satellite images indicate some convective activity west of the Black Sea. Extensive fire activity was recorded during this time for a broad area along the northern coast of the Black Sea and for the Mediterranean coast.

3.3. Plumes over Africa

Data are available for a total of 13 flights en route between Frankfurt and Cape Town or Johannesburg. As the South African airports were served by Lufthansa Airbus A340-600 aircraft only during the winter flight schedule, the data cover the months from November to March (March 2009–March 2011) encompassing the austral summer. Elevated CO mixing ratios over equatorial Africa were encountered during all flights with peak values varying between ~120 and ~250 ppb, but for only four flights, the Hg *vs.* CO correlation was statistically significant. The reasons for such a low yield are discussed below.

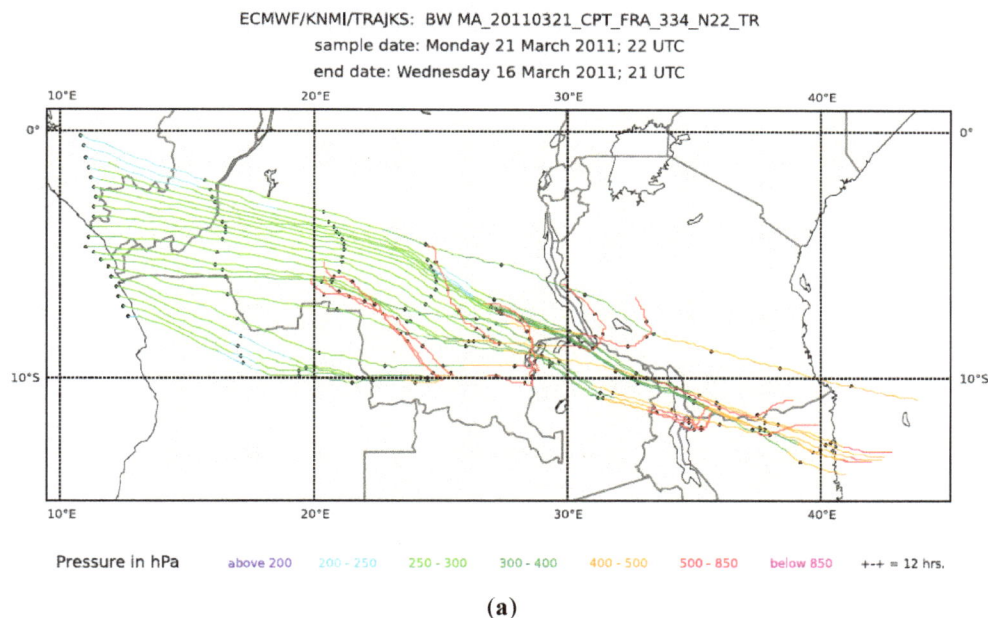

(a)

Figure 6. *Cont.*

257

ECMWF/KNMI/TRAJKS: BW MA_20110321_CPT_FRA_334_N23_TR
sample date: Monday 21 March 2011; 23 UTC
end date: Wednesday 16 March 2011; 22 UTC

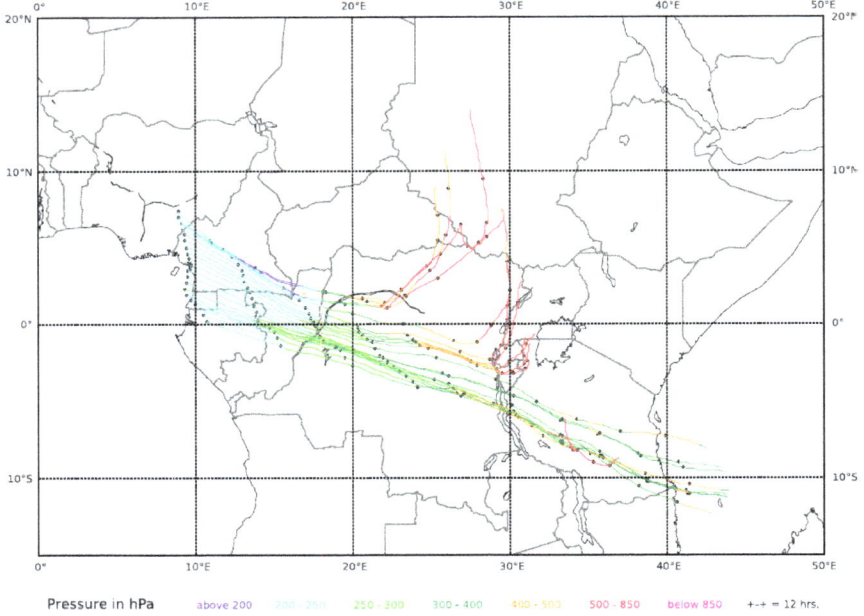

Pressure in hPa   above 200   200 - 250   250 - 300   300 - 400   400 - 500   500 - 850   below 850   +-+ = 12 hrs.

(**b**)

ECMWF/KNMI/TRAJKS: BW MA_20110321_CPT_FRA_334_N24_TR
sample date: Tuesday 22 March 2011; 00 UTC
end date: Wednesday 16 March 2011; 23 UTC

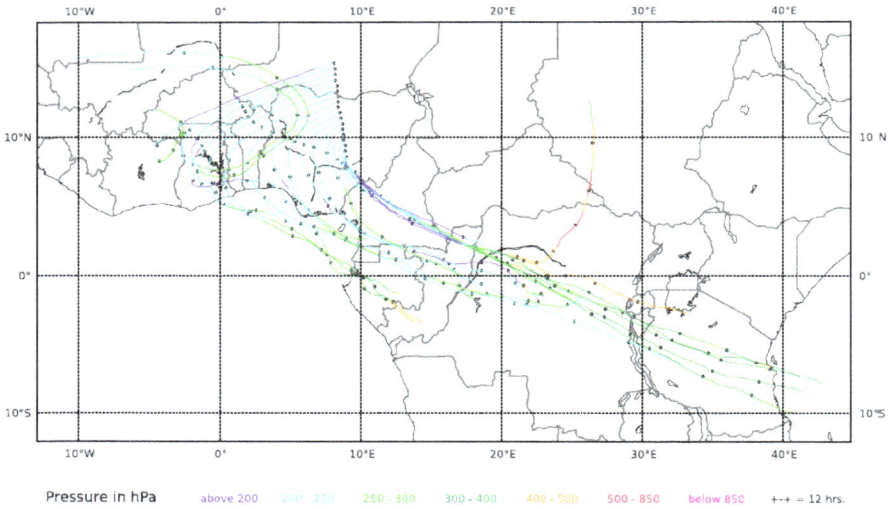

Pressure in hPa   above 200   200 - 250   250 - 300   300 - 400   400 - 500   500 - 850   below 850   +-+ = 12 hrs.

(**c**)

Figure 6. *Cont.*

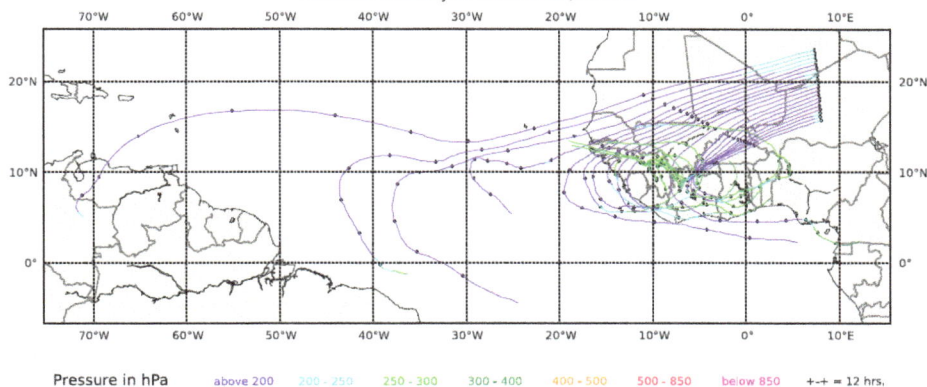

ECMWF/KNMI/TRAJKS: BW MA_20110321_CPT_FRA_334_N01_TR
sample date: Tuesday 22 March 2011; 01 UTC
end date: Thursday 17 March 2011; 00 UTC

(**d**)

Figure 6. Five-day backward trajectories (every 3 min) for Flight 334 from Cape Town to Frankfurt on 21 and 22 March 2011: (**a**) 21–22 UTC; (**b**) 22–23 UTC; (**c**) 23–24 UTC; and (**d**) 0–1 UTC.

Figure 5 and Figure S1 show a typical example of the data obtained during Flight 334 from Cape Town to Frankfurt on 20 and 21 March 2011. The second panel from the top of the data time series in Figure 5 shows somewhat elevated CO mixing ratios of ~100 ppb after ascent, decreasing to a southern hemispheric background of ~75 ppb after 18:30 UTC. CO then increases in the course of the flight up to a maximum of ~240 ppb between 22:45 and 23:10 UTC to subsequently decrease, with another smaller peak, with a maximum of ~175 ppb around 0:20 UTC, to ~125 ppb, before the aircraft crosses the tropopause into the lower stratosphere at ~1:00 UTC. The NOy mixing ratio and particle surface area concentration (the third panel from the top) display a similar pattern, even in the finer structure, whereas CO_2 and CH_4 show only a broad maximum with a somewhat different shape, peaking shortly before 23:00 UTC. The SF_6 mixing ratio increases gradually from ~7.2 ppt after ascent to ~7.3 ppt before entering the stratosphere. Although small, this increase documents that all plume observations are embedded in a broad gradient between southern and northern hemispheric air masses [47]. Mixing ratios of ethane, propane and ethyne, shown in Figure S1, broadly follow the CO pattern, but high mixing ratios of short-lived n-butane and i-butane around 21:15 UTC and at 23:00 UTC indicate the admixture of freshly polluted air. The CH_3Cl mixing ratio (Figure S1, lowermost panel) of ~650 ppt in the tropospheric section of the flight is substantially higher than the background mixing ratio of ~550 ppt, and this implies a large-scale influence of biomass burning. The highest CH_3Cl mixing ratios of almost 700 ppt are found in two samples taken after midnight. They coincide with elevated mixing ratios of CO, ethane, propane, ethyne and NOy (Figure 5, middle panel), but the short-lived butanes have almost disappeared. Such a coincidence is characteristic for aged air from another regional biomass burning plume. Consequently, the CO bulge over equatorial Africa has to be viewed as a composite of several regional plumes. Backward trajectories, shown in Figure 6 for 21–22 UTC (a), 22–23 UTC (b), 23–24 UTC (c) and 0–1 UTC (d), support this view by pointing to different source regions in different sections of the flight. Several of the trajectories exhibit rather fast upward transport from the boundary layer and lower troposphere (purple and red colours). Cloud water content (Figure 5, bottom panel) in several sections of the flight is a sign of convective activity in these areas. The fire map displayed in Figure 7 and the trajectories show that emissions from biomass burning have, to a varying degree, influenced all observations between ~21:00 UTC on 20 March to ~1:00 UTC on 21 March.

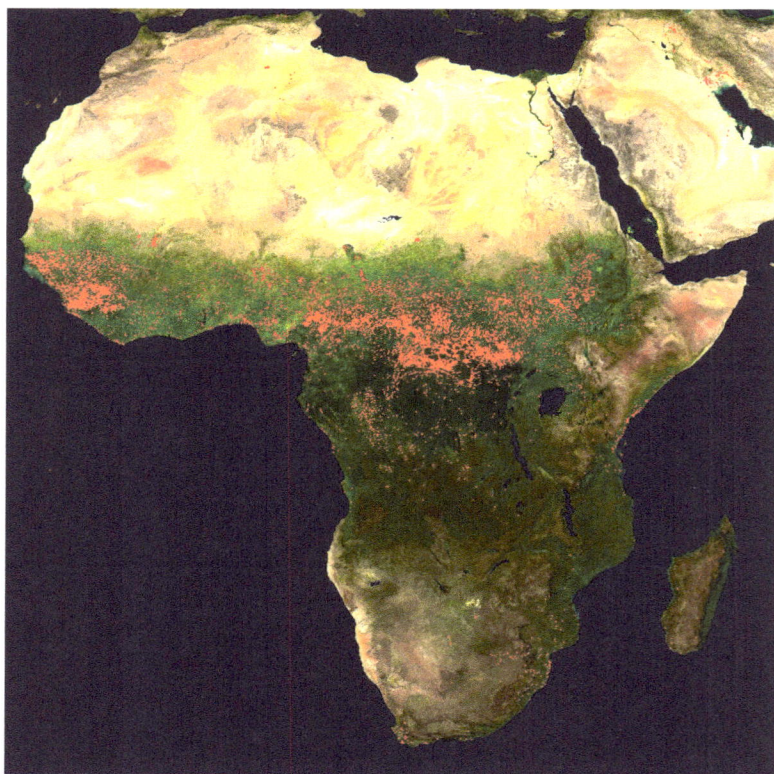

Figure 7. The map of fire counts for the period from 12 to 21 March 2011 (http://rapidfire.sci.gsfc.nasa.gov/firemaps/, accessed on 10 October 2013).

The correlation of Hg *vs.* CO was statistically non-significant for the whole plume starting at 21:16:30 and ending at 01:01:30 UTC, as well as for sections of it, such as between 23:56:30 and 00:56:30 UTC or between 21:21:30 and 23:56:30 UTC. The correlation of Hg *vs.* CO_2 was statistically significant for the whole plume (239.4 ± 94.4 pg·m^{-3}·ppm^{-1} at > 95% level), but statistically non-significant for the sections mentioned above.

In summary, the scarcity of statistically significant Hg *vs.* CO and Hg *vs.* CO_2 correlations in the plumes observed over equatorial Africa is a result of several factors. The plumes are embedded in broad north-south gradients, which violates the assumption of a constant background. Due to their large extent, they consist of a multitude of overlapping smaller plumes from different regions and sources and are thus not homogeneous. In addition, the CO enhancements (ΔCO) against the background are rather small, varying between ~45 ppb during Flights 290 and 291 to Cape Town on 27 and 28 October 2009, to ~165 ppb during Flights 333 and 334 to Cape Town on 20 and 21 March 2011. Assuming that the plumes originate predominantly from biomass burning with a typical Hg/CO emission ratio of 1 pg·m^{-3}·ppb^{-1}, the CO enhancements of this magnitude would produce Hg enhancements of only 0.045 to 0.165 ng·m^{-3}. Such enhancements are difficult to detect with a precision of 0.05 ng·m^{-3} of the mercury measurements, even if the background were constant and the plumes homogeneous. For all of these reasons, the Hg/CO emission ratios derived from these flights will be substantially more uncertain than their statistical uncertainty stated in Table S2.

Three of the four statistically significant Hg *vs.* CO correlations for plumes observed over equatorial Africa yield Hg/CO emission ratios of 2.18–3.36 pg·m^{-3}·ppb^{-1}, which, again, points to the predominant contribution of emissions from biomass burning.

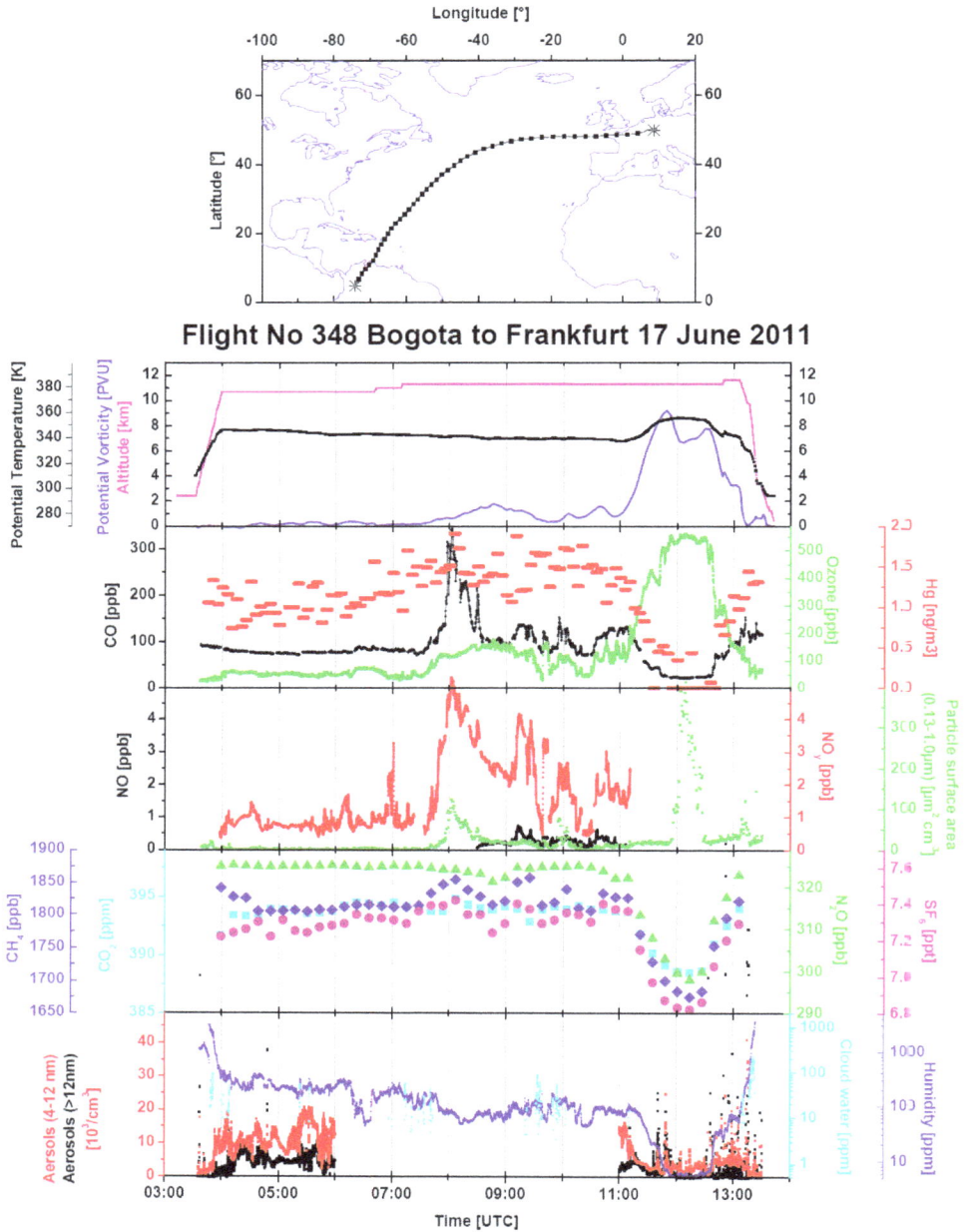

Figure 8. An overview of the data from Flight 348 from Bogotá to Frankfurt on 17 June 2011. The same parameters are displayed as in Figure 5. Additionally, total water content (dark blue) is shown in the bottom panel.

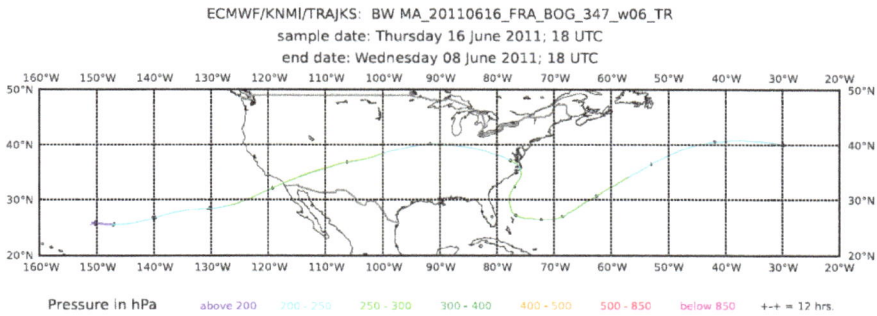

ECMWF/KNMI/TRAJKS: BW MA_20110616_FRA_BOG_347_w06_TR
sample date: Thursday 16 June 2011; 18 UTC
end date: Wednesday 08 June 2011; 18 UTC

Pressure in hPa   above 200   200 - 250   250 - 300   300 - 400   400 - 500   500 - 850   below 850   +-+ = 12 hrs.

(a)

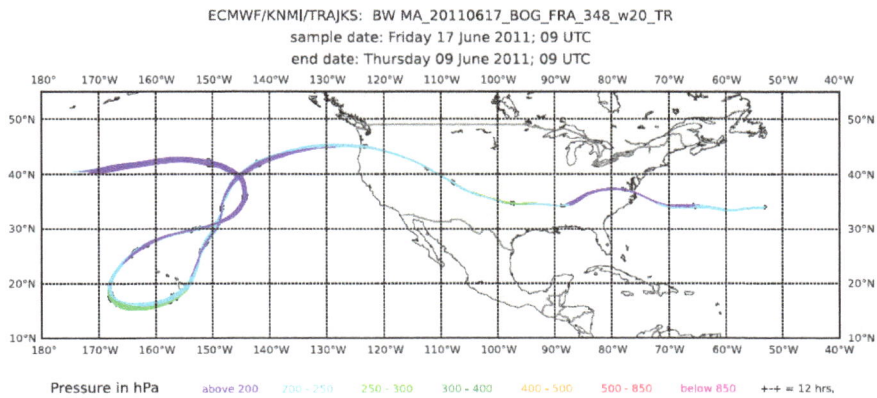

ECMWF/KNMI/TRAJKS: BW MA_20110617_BOG_FRA_348_w20_TR
sample date: Friday 17 June 2011; 09 UTC
end date: Thursday 09 June 2011; 09 UTC

Pressure in hPa   above 200   200 - 250   250 - 300   300 - 400   400 - 500   500 - 850   below 850   +-+ = 12 hrs,

(b)

(c)

Figure 9. (**a**) eight-day backward trajectories for whole air Sample 6 from the CO peak encountered around 17:20 UTC during Flight 347 from Frankfurt to Bogota on 16 June 2011; and (**b**) for whole air Sample 20 from the CO peak encountered around 8:05 UTC during Flight 348 from Bogota to Frankfurt on 17 June 2011. (**c**) The map of the fire counts for 10–19 June 2011 (http://rapidfire.sci.gsfc.nasa.gov/firemaps/, accessed on 10 October 2013).

263

3.4. Plumes Observed during the Flights to and over South America

The plumes encountered over South America during the flights to São Paulo and Santiago de Chile were analysed in detail by Ebinghaus *et al.* [48]. Here, we would only note that based on their chemical signature, backward trajectories and fire maps, these plumes could be unequivocally attributed to biomass burning in the Amazon Basin and its outskirts.

The encounters at 16:47:30 to 17:47:30 during Flight 347 from Frankfurt to Bogotá on 16 June 2011, and at 7:35:30–8:45:30 during the return Flight 348 on 17 June 2011, both above the middle of the Atlantic Ocean at latitudes ranging from 31° to 43°N, are probably due to the same plume. An overview of the data from Flight 348 (Figure 8) shows a sharp CO peak with ~325 ppb at ~8:00 UTC, accompanied by peaks in NOy, aerosol surface area, CH_4 and a small peak of SF_6, the last one originating from anthropogenic emissions. The CO peak coincides also with the highest CH_3Cl, ethane, propane, n-butane, i-butane and ethyne mixing ratios (Figure S2). The low Hg/CO emission ratios of 1.5 ± 0.6 pg·m^{-3}·ppb^{-1} for this flight and 1.3 ± 0.4 pg·m^{-3}·ppb^{-1} for Flight 347 and the peak CH_3Cl mixing ratio indicate that pollutants from biomass burning are by far the most predominant component of these plumes. Backward trajectories in the upper panel of Figure 9 for Samples 6 and 20, taken within the CO peaks observed during Fights 347 and 348, respectively, show a high level transport from U.S. and north-western Mexico. Satellite cloud images point to convective activity over the south-eastern U.S., the Great Plains and north-western Mexico. The map of fire counts for 10–19 June 2011, in the lower panel of Figure 9, shows that numerous fires in the southeast U.S. might be the major source of the observed plumes, with a possible contribution of fires in Southern California and north-western Mexico.

3.5. Hg/CO₂ Emission Ratios

Hg/CO_2 emission ratios are potentially more useful for constraining the mercury emissions, because CO_2 inventories tend to be more accurate than those of CO [21]. Unfortunately, only a few Hg/CO_2 emission ratios have been reported, so far. Table S3 displays the events with significant Hg *vs.* CO_2 correlations. CO_2 data were available only for 46 plume encounters. Among these, for only 17 encounters, the Hg *vs.* CO_2 correlations were statistically significant (significance level $\geq$95%). The yield of statistically significant Hg *vs.* CO_2 correlations is thus somewhat smaller than for Hg *vs.* CO, and the significance of these correlations with mostly only 95% tends also to somewhat smaller values. Eleven of the statistically significant correlations were found for plumes with significant Hg *vs.* CO correlations.

The Hg/CO_2 emission ratios vary over a broad range, from 14.4 to 964 pg·m^{-3}·ppm^{-1}, and those observed during the flights to East Asian destinations vary between 107 and 964 pg·m^{-3}·ppm^{-1}. The Hg/CO_2 emission ratio from the plume observed during Flight 334 to Frankfurt immediately after ascent from Cape Town on 21 March 2011, and during Flight 373 to Chennai after ascent from Frankfurt on 16 January 2012, also fit the range of East Asian plumes. The lowest Hg/CO_2 emission ratios of 14.4 and 21.9 $pg \cdot m^{-3} \cdot ppm^{-1}$ were both derived from plumes encountered during Flights 329 and 334 over equatorial Africa on 24 February 2011 and 21 March 2011, respectively.

The low Hg/CO_2 emission ratios in the plumes of equatorial Africa are comparable to the median of 34.1 $pg \cdot m^{-3} \cdot ppm^{-1}$ (average: 62.7 ± 80.2 $pg \cdot m^{-3} \cdot ppm^{-1}$) of emission ratios observed in the plumes encountered at Cape Point, which, according to their Hg/CO emission ratios, seem to originate predominantly from biomass burning [21]. The Hg/CO_2 emission ratio from the only plume clearly attributed to biomass burning near Cape Point was somewhat higher with 109 ± 27 $pg \cdot m^{-3} \cdot ppm^{-1}$ [49], but this is comparable to 131 ± 53 $pg \cdot m^{-3} \cdot ppm^{-1}$, observed in the plume from biomass burning in the south-eastern U.S. in June 2011 (Flights 347 and 348). Based on the coal mercury content of 0.15–0.45 $\mu g \cdot Hg \cdot g^{-1}$ and a flue cleaning efficiency for mercury of 50%–90%, Brunke *et al.* [21] predicted an Hg/CO_2 emission ratio to be within the range of 2–30 $pg \cdot m^{-3} \cdot ppm^{-1}$. The Hg content in coal consumed in China varies from 0.027 to 0.369 $\mu g \cdot g^{-1}$ [50] and is not much different from that in South Africa. The flue cleaning efficiency for mercury in China is with up to 57% somewhat lower [50], but this difference cannot explain the Hg/CO_2 emission ratios larger than 100 $pg \cdot m^{-3} \cdot ppm^{-1}$ observed over East Asia, Europe and at Cape Point in South Africa [21]. If confirmed by further measurements, high Hg/CO_2 emission ratios would imply a substantial contribution of emissions from other sources than coal burning.

3.6. Hg/CH₄ Emission Ratios

Mercury also frequently correlated with CH_4 in the plumes observed at Cape Point, and the resulting Hg/CH_4 emission ratios helped to constrain the mercury emissions in South Africa [21]. Continuously measured CH_4 data were available only for 26 plume encounters, and of these, only six provided a statistically significant Hg *vs.* CH_4 correlation at a confidence level of at least 95%. The emission ratios listed in Table S4 vary between 4.8 and 41.4 $pg \cdot m^{-3} \cdot ppb^{-1}$. The only Hg/CH_4 emission ratios available for comparison are derived from observations at Cape Point and are centred in the range of up to 6 $pg \cdot m^{-3} \cdot ppb^{-1}$ [21]. The plume encountered during the flight to Cape Town with an emission ratio of 4.8 $pg \cdot m^{-3} \cdot ppb^{-1}$ falls into this range. All other plumes in Table S4 were of mixed or industrial/urban origin, and they have higher Hg/CH_4 emission ratios.

4. Conclusions

Over 100 plumes with elevated mercury concentrations were encountered during the tropospheric sections of the CARIBIC flights from May 2005, until June 2013. In 98 of them, elevated Hg was accompanied by elevated CO mixing ratios. Several Hg plumes without a simultaneous increase in CO were all encountered over the equatorial Atlantic Ocean during the flights to South America and were attributed to the convection of the marine boundary air at the ITCZ. Hg correlated as statistically significant with CO in more than 50% of the observed plumes and with CO_2 in about 30% of the plumes for which CO_2 data were available. Ample ancillary data on the chemical fingerprint of the air within these plumes and backward trajectories provide additional means to identify the origin and the type of the source.

Extensive mercury plumes over equatorial Africa were observed during all flights between Frankfurt and South Africa. These plumes, which extend over thousands of kilometres, are embedded in north-south gradients of mercury, CO and CO_2 and consist of a number of overlapping smaller plumes. Due to the changing background, the inhomogeneity of the plumes and the low precision of the Hg measurements, only a few of the plume encounters provided statistically significant Hg *vs.* CO correlations. Most plumes were observed over East Asia, and relative to the number of flights to East Asian destinations, the yield of plumes with statistically significant Hg *vs.* CO correlations was on par with the African flights. Lower yields of plume occurrence were found for flights to South America and to South Asia. Only two plumes were encountered over North America and one over Europe.

The Hg/CO emission ratios derived from these correlations are consistent with the previously published data compiled by Slemr *et al.* [29] and have a smaller values of ~1 $pg \cdot m^{-3} \cdot ppb^{-1}$ for plumes, which we clearly could attribute to biomass burning using backward trajectories, fire count maps and the presence of chemical tracers for biomass burning, such as CH_3Cl and acetonitrile. Larger values of ~6 $pg \cdot m^{-3} \cdot ppb^{-1}$ and more were found for most of the other plumes. Backward trajectories and the presence of man-made tracers, such as C_2Cl_4 and SF_6 in several of these plumes suggest emissions from urban/industrial sources. Both types of chemical tracers were present in several plumes, with Hg/CO emission ratios between 1 and 6 $pg \cdot m^{-3} \cdot ppb^{-1}$, showing their mixed origin.

Many of the plumes were transported over large distances from the area of their origin to the place of their observation. Backward trajectories point to major source areas in equatorial Africa, East Asia, South America and South Asia. The emissions from equatorial Africa and South America are clearly dominated by biomass burning. The East Asian emissions originate from a large area of East Siberia, Korea, China, the Philippines and the Indochinese Peninsula. They are mostly of urban/industrial origin with a varying contribution from biomass and biofuel burning. The South Asian emissions originate mostly from the Indo-Ganges region of northern India. Like the East Asian ones, they are a mixture of emissions from industrial/urban sources and biomass/biofuel burning. Other mercury source areas were also identified: the Middle East, a region along the northern perimeter of the Black Sea, the Mediterranean Sea and the south-western U.S. The Hg/CO emission ratio and the plume fingerprint for the Middle East region suggests industrial/urban emissions to be dominating. The emissions whose origin area was located toward the northern coast of the Black Sea and toward the Mediterranean had both a substantial contribution from biomass burning. Biomass burning was the major component of the emissions from the south-eastern U.S. during one event. We caution that since only a few plumes were attributed to each, the Middle East, Europe and the U.S., no firm conclusions can be drawn for these areas. We note also that the forest fires at northern mid-latitudes occur in summer when convection processes, which carry them to cruising altitudes, are the most active. Consequently, our observations for these areas are biased in favour of biomass burning and neglect emissions in other seasons, such as, e.g., from residential heating in winter.

Only a few Hg/CO_2 and Hg/CH_4 emission ratios have been reported, so far. The range of the Hg/CO_2 emission ratios from the CARIBIC flights is comparable to the range observed at Cape Point [21]. The Hg/CO_2 emission ratios of 107–964 $pg \cdot m^{-3} \cdot ppm^{-1}$ observed in the plumes over East Asia, however, are substantially higher than 2–30 $pg \cdot m^{-3} \cdot ppm^{-1}$, calculated by Brunke *et al.* [21] for coal burning. If confirmed by further measurements, the higher observed than calculated Hg/CO_2 emission ratios would imply other substantial Hg sources in addition to coal burning.

Acknowledgments: We would like to thank Lufthansa and all members of the CARIBIC team for their continued effort to keep running such a complex project. We thank especially Dieter Scharffe, Claus Koeppel and Stefan Weber for the day-to-day maintenance and operation of the CARIBIC container. Funding from the European Community within the GMOS (Global Mercury Observation System) project and from Fraport AG is thankfully acknowledged. We acknowledge the use of FIRMS data and imagery from the Land Atmosphere Near-real time Capability for EOS (LANCE) system operated by the NASA/GSFC/Earth Science Data and Information System (ESDIS) with funding provided by NASA/HQ.

Author Contributions: All authors are members of the CARIBIC team and contributed to the production of the data on which the paper is based. Franz Slemr calculated the emission ratios, and Peter van Velthoven made the meteorological analyses for each flight and calculated the backward trajectories. All authors discussed the results of the manuscript in all stages of its preparation.

Conflicts of Interest: The authors declare no conflict of interest.

References

1.   Slemr, F.; Schuster, G.; Seiler, W. Distribution, speciation, and budget of atmospheric mercury. *J. Atmos. Chem.* **1985**, *3*, 407–434.

2. Schroeder, W.H.; Munthe, J. Atmospheric mercury—An overview. *Atmos. Environ.* **1998**, *32*, 809–822.

3. Mergler, D.; Anderson, H.A.; Chan, L.H.M.; Mahaffey, K.R.; Murray, M.; Sakamoto, M.; Stern, A.H. Hethyl mercury exposure and health effects in humans: A worldwide concern. *Ambio* **2007**, *36*, 3–11.

4. Scheuhammer, A.M.; Meyer, M.W.; Sandheinrich, M.B.; Murray, M.W. Effects of environmental methylmercury on the health of wild bird, mammals, and fish. *Ambio* **2007**, *36*, 12–18.

5. Nriagu, J.O.; Pacyna, J.M. Quantitative assessment of worldwide contamination of air, water and soils by trace metals. *Nature* **1988**, *333*, 134–139.

6. Nriagu, J.O. A global assessment of natural sources of atmospheric trace metals. *Nature* **1989**, *338*, 47–49.

7. Pirrone, N.; Keeler, G.J.; Nriagu, O. Regional differences in worldwide emissions of mercury to the atmosphere. *Atmos. Environ.* **1996**, *30*, 2981–2987.

8. Pirrone, N.; Allegrini, I.; Keeler, G.J.; Nriagu, J.O.; Rossmann, R.; Robbins, J.A. Historical atmospheric mercury emissions and depositions in North America compared to mercury accumulations in sedimentary records. *Atmos. Environ.* **1998**, *32*, 929–940.

9. Pirrone, N.; Cinnirella, S.; Feng, X.; Finkelman, R.B.; Friedli, H.R.; Leaner, J.; Mason, R.; Mukherjee, A.B.; Stracher, G.; Streets, D.G.; *et al.* Global mercury emissions to the atmosphere from anthropogenic and natural sources. *Atmos. Chem. Phys.* **2010**, *10*, 5951–5964.

10. Pacyna, E.G.; Pacyna, J.M. Global emission of mercury from anthropogenic sources in 1995. *Water Air Soil Pollut.* **2002**, *137*, 149–165.

11. Pacyna, J.M.; Pacyna, E.G.; Steenhuisen, F.; Wilson, S. Mapping 1995 global anthropogenic emissions of mercury. *Atmos. Environ.* **2003**, *37*, S109–S117.

12. Pacyna, E.G.; Pacyna, J.M.; Steenhuisen, F.; Wilson, S. Global anthropogenic mercury emission inventory for 2000. *Atmos. Environ.* **2006**, *40*, 4048–4063.

13. Pacyna, E.G.; Pacyna, J.M.; Sundseth, K.; Munthe, J.; Kindbom, K.; Wilson, S.; Steenhuisen, F.; Maxson, P. Global emission of mercury to the atmosphere from anthropogenic sources in 2005 and projections to 2020. *Atmos. Environ.* **2010**, *44*, 2487–2499.

14. Streets, D.G.; Hao, J.; Wu, Y.; Jiang, J.; Chan, M.; Tian, H.; Feng, X. Anthropogenic mercury emissions in China. *Atmos. Environ.* **2005**, *39*, 7789–7806.

15. Streets, D.G.; Zhang, Q.; Wu, Y. Projections of global mercury emissions in 2050. *Environ. Sci. Technol.* **2009**, *43*, 2983–2988.

16. Lin, C.-J.; Pongprueksa, P.; Lindberg, S.E.; Pehkonen, S.O.; Byun, D.; Jang, C. Scientific uncertainties in atmospheric mercury models I: Model science evaluation. *Atmos. Environ.* **2006**, *40*, 2911–2928.

17. Lindberg, S.; Bullock, R.; Ebinghaus, R.; Engstrom, D.; Feng, X.; Fitzgerald, W.; Pirrone, N.; Prestbo, E.; Seigneur, Ch. A synthesis of progress and uncertainties in attributing the sources of mercury in deposition. *Ambio* **2007**, *36*, 19–32.

18. Slemr, F.; Baumbach, G.; Blank, P.; Corsmeier, U.; Fiedler, F.; Friedrich, R.; Habram, M.; Kalthoff, N.; Klemp., D.; Kühlwein, J.; *et al.* Evaluation of modeled spatially and temporarily highly resolved emission inventories of photosmog precursors for the city of Augsburg: The experiment EVA and its major results. *J. Atmos. Chem* **2002**, *42*, 207–233.

19. Klemp, D.; Mannschreck, K.; Pätz, H.W.; Habram, M.; Matuska, P.; Slemr, F. Determination of anthropogenic emission ratios in the Augsburg area from concentration ratios: results from long-term measurements. *Atmos. Environ.* **2002**, *36*, S61–S80.

20. Jaffe, D.; Prestbo, E.; Swartzendruber, P.; Weiß-Penzias, P.; Kato, S.; Takami, A.; Hatakeyama, S.; Kajii, Y. Export of atmospheric mercury from Asia. *Atmos. Environ.* **2005**, *39*, 3029–3028.

21. Brunke, E.-G.; Ebinghaus, R.; Kock, H.H.; Labuschagne, C.; Slemr, F. Emissions of mercury in southern Africa derived from long-term observations at Cape Point, South Africa. *Atmos. Chem. Phys.* **2012**, *12*, 7465–7474.

22. Slemr, F.; Brunke, E.-G.; Whittlestone, S.; Zahorowski, W.; Ebinghaus, R.; Kock, H.H.; Labuschagne, C. ^{222}Rn-calibrated mercury fluxes from terrestrial surface of southern Africa. *Atmos. Chem. Phys.* **2013**, *13*, 6421–6428.

23. Brenninkmeijer, C.A.M.; Crutzen, P.; Boumard, F.; Dauer, T.; Dix, B.; Ebinghaus, R.; Filippi, D.; Fischer, H.; Franke, H.; Frieß, U.; *et al.* Civil aircraft for the regular investigation of the atmosphere based on an instrumented container: The new CARIBIC system. *Atmos. Chem. Phys.* **2007**, *7*, 1–24.

24. Cooper, O.R.; Moody, J.L.; Parrish, D.D.; Trainer, M.; Ryerson, T.B.; Holloway, J.S.; Hübler, G.; Fehsenfeld, F.C.; Evans, M.J. Trace gas composition of midlatitude cyclones over the western North Atlantic Ocean: A conceptual model. *J. Geophys. Res.* **2002**.

25. Cooper, O.R.; Moody, J.L.; Parrish, D.D.; Trainer, M.; Holloway, J.S.; Hübler, G.; Fehsenfeld, F.C.; Stohl, A. Trace gas composition of of midlatitude cyclones over the western North Atlantic Ocean: A seasonal comparison of O_3 and CO. *J. Geophys. Res.* **2002**, *107*.

26. Dyroff, C.; Zahn, A.; Sanati, S.; Christner, E.; Rauthe-Schöch, A.; Schuck, T.J. Tunable diode laser in-situ CH_4 measurements aboard the CARIBIC passenger aircraft: instrument performance assessment. *Atmos. Meas. Tech. Discuss.* **2013**, *6*, 9225–9261.

27. Schuck, T.J.; Brenninkmeijer, C.A.; Zahn, A. Greenhouse gas analysis of air samples collected onboard CARIBIC passenger aircraft. *Atmos. Meas. Tech.* **2009**, *2*, 449–464.

28. Baker, A.K.; Slemr, F.; Brenninkmeijer, C.A.M. Analysis of non-methane hydrocarbons in air samples collected aboard the CARIBIC passenger aircraft. *Atmos. Meas. Tech.* **2010**, *3*, 311–321.

29. Slemr, F.; Ebinghaus, R.; Brenninkmeijer, C.A.M.; Hermann, M.; Kock, H.H.; Martinsson, B.G.; Schuck, T.; Sprung, D.; van Velthoven, P.; Zahn, A.; *et al.* Gaseous mercury distribution in the upper troposphere and lower stratosphere observed onboard the CARIBIC passenger aircraft. *Atmos. Chem. Phys.* **2009**, *9*, 1957–1969.

30. Talbot, R.; Mao, H.; Scheuer, E.; Dibb, J.; Avery, M.; Browell, E.; Sachse, G.; Vay, S.; Blake, D.; Huey, G.; *et al.* Factors influencing the large-scale distribution of $Hg°$ in the Mexico City area and over the North Pacific. *Atmos. Chem. Phys.* **2008**, *8*, 2103–2114.

31. Neuman, J.A.; Huey, L.G.; Ryerson, T.B.; Fahey, D.W. Study of inlet materials for sampling atmospheric nitric acid. *Environ. Sci. Technol.* **1999**, *33*, 1133–1136.

32. Baron, P.A.; Willeke, K. *Aerosol Measurements: Principles Techniques and Applications*; John Wiley and Sons: New York, NY, USA, 2001; pp. 1–1131.

33. Temme, C.; Einax, J.W.; Ebinghaus, R.; Schroeder, W.H. Measurements of atmospheric mercury species at a coastal site in the Antarctic and over the South Atlantic Ocean during polar summer. *Environ. Sci. Technol.* **2003**, *37*, 22–31.

34. Rutter, A.P.; Schauer, J.J. The effect of temperature on the gas-particle partitioning of reactive mercury in atmospheric aerosols. *Atmos. Environ.* **2007**, *41*, 8647–8657.

35. Amos, H.M.; Jacob, D.J.; Holmes, C.D.; Fisher, J.A.; Wang, Q.; Yantosca, R.M.; Corbitt, E.S.; Galarneau, E.; Rutter, A.P.; Gustin, M.S.; *et al.* Gas-particle partitioning of atmospheric Hg(II) and its effects on global mercury deposition. *Atmos. Chem. Phys.* **2012**, *12*, 591–603.

36. Lyman, S.N.; Jaffe, D.A. Formation and fate of oxidized mercury in the upper troposphere and lower stratosphere. *Nat. Geosci.* **2012**, *5*, 114–117.

37. Sprovieri, F.; Pirrone, N.; Ebinghaus, R.; Kock, H.; Dommergue, A. A review of worldwide atmospheric mercury measurements. *Atmos. Chem. Phys.* **2010**, *10*, 8245–8265.

38. Cantrell, C.A. Technical note: Review of methods for linear least-squares fitting of data and application to atmospheric chemistry problems. *Atmos. Chem. Phys.* **2008**, *8*, 5477–5487.

39. Scheele, M.; Siegmund, P.; van Velthoven, P. Sensitivity of trajectories to data resolution and its dependence on the starting point: In or outside a tropopause fold. *Meteorol. Appl.* **1996**, *3*, 267–273.

40. Heue, K.-P.; Brenninkmeijer, C.A.M.; Wagner, T.; Mies, K.; Dix, B.; Frieß, U.; Martinsson, B.G.; Slemr, F.; van Velthoven, P.F.J. Observations of the 2008 Kasatochi volcanic SO$_2$ plume by CARIBIC aircraft DOAS and the GOME-2 satellite. *Atmos. Chem. Phys.* **2010**, *10*, 4699–4713.

41. Lelieveld, J.; van Aardenne, J.; Fischer, H.; de Reus, M.; Williams, J.; Winkler, P. Increasing ozone over the Atlantic Ocean. *Science* **2004**, *304*, 1483–1487.

42. Kuss, J.; Zülicke, C.; Pohl, C.; Schneider, B. Atlantic mercury emission determined from continuous analysis of the elemental mercury sea-air concentration difference within transects between 50°N and 50°S. *Glob. Biogeochem. Cy.* **2011**.

43. Andreae, M.O.; Merlet, P. Emissions of trace gases and aerosols from biomass burning. *Global Biogeochem. Cy.* **2001**, *15*, 955–966.

44. Mühle, J.; Brenninkmeijer, C.A.M.; Rhee, T.S.; Slemr, F.; Oram, D.E.; Penkett, S.A.; Zahn, A. Biomass burning and fossil fuel signatures in the upper troposphere observed during a CARIBIC flight from Namibia to Germany. *Geophys. Res. Lett.* **2002**.

45. Lai, S.C.; Baker, A.K.; Schuck, T.J.; Slemr, F.; Brenninkmeijer, C.A.M.; van Velthoven, P.; Oram, D.E.; Zahn, A.; Ziereis, H. Characterization and source regions of 51 high-CO events observed during Civil Aircraft for the Regular Investigation of the Atmosphere Based on an Instrument Container (CARIBIC) flights between south China and the Philippines, 2005–2008. *J. Geophys. Res.* **2011**.

46. Baker, A.K.; Traud, S.; Brenninkmeijer, C.A.M.; Hoor, P.; Neumaier, M.; Oram, D.E.; Rauthe-Schöch, A.; Sprung, D.; Schloegl, S.; Slemr, F.; *et al.* Pollution pattern in the upper troposphere over Europe/West Asia and Asia observed by CARIBIC. *Atmos. Environ.* 2014, in press.

47. Maiss, M.; Steele, L.P.; Francey, R.J.; Fraser, P.J.; Langenfelds, R.L.; Trivett, N.B.A.; Levin, I. Sulfur hexafluoride—A powerful new atmospheric tracer. *Atmos. Environ.* **1996**, *30*, 1621–1629.

48. Ebinghaus, R.; Slemr, F.; Brenninkmeijer, C.A.M.; van Velthoven, P.; Zahn, A.; Hermann, M.; O'Sullivan, D.A.; Oram, D.E. Emissions of gaseous mercury from biomass burning in South America in 2005 observed during CARIBIC flights. *Geophys. Res. Lett.* **2007**.

49. Brunke, E.-G.; Labuschagne, C.; Slemr, F. Gaseous mercury emissions from a fire in the Cape Peninsula, South Africa, during January 2000. *Geophys. Res. Lett.* **2001**, *28*, 1483–1486.

50. Tian, H.Z.; Wang, Y.; Xue, Z.G.; Cheng, K.; Qu, Y.P.; Chai, F.H.; Hao, J.M. Trend and characteristics of atmospheric emissions of Hg, As, and Se from coal combustion in China, 1980–2007. *Atmos. Chem. Phys.* **2010**, *10*, 11905–11919.

Benefits of European Climate Policies for Mercury Air Pollution

Peter Rafaj, Janusz Cofala, Jeroen Kuenen, Artur Wyrwa and Janusz Zyśk

Abstract: This paper presents the methodology and results of impact assessment of renewable energy policies on atmospheric emissions of mercury in Europe. The modeling exercise described here involves an interaction of several models. First, a set of energy scenarios has been developed with the REMix (Renewable Energy Mix) model that simulates different levels of penetration of renewable energies in the European power sector. The energy scenarios were input to the GAINS (Greenhouse Gas and Air Pollution Interactions and Synergies) model, which prepared projections of mercury releases to the atmosphere through 2050, based on the current air pollution control policies in each country. Data on mercury emissions from individual sectors were subsequently disaggregated to a fine spatial resolution using various proxy parameters. Finally, the dispersion of mercury in the atmosphere was computed by the chemistry transport model, implemented to the air quality system, Polyphemus. The simulations provided information on changes in concentrations and depositions of various forms of mercury over Europe. Scenarios that simulate a substantial expansion of renewable energies within the power sector indicate extensive co-benefits for mercury abatement, due to the restructuring of the energy system and changes in the fuel mix. The potential for mercury reductions in Europe depends on the rate of fuel switches and renewable technology deployment, but is also influenced by the stringency and timing of the air quality measures The overall scope for co-benefits is therefore higher in regions relying on coal combustion as a major energy source.

Reprinted from *Atmosphere*. Cite as: Rafaj, P.; Cofala, J.; Kuenen, J.; Wyrwa, A.; Zyśk, J. Benefits of European Climate Policies for Mercury Air Pollution. *Atmosphere* **2014**, *5*, 536–556.

1. Introduction

During the last few decades, many studies have investigated the environmental impacts of anthropogenic releases of mercury (Hg). The increasing attention on mercury pollution is mainly driven by the growing evidence of its negative impacts on human health and ecosystems [1]. It is well documented that after its deposition, mercury moves through the water chain and can be transformed by aquatic microorganisms into methylmercury (MeHg), a toxic substance bioaccumulated in fish and seafood. MeHg enters the human body with consumed food and is then transported by blood, passes the blood-brain barrier and causes neurotic dysfunctions [2]. Mercury is also capable of passing the placental barrier and has an immense negative impact on the fetus, decreasing the intelligence quotient of the child. In this way, exposure to mercury might affect the development of specific population groups [3].

Evidence of the negative effects of mercury on human health and the environment has led to intergovernmental negotiations on the preparation of a global legally binding instrument on mercury, which was adopted in 2013 [4]. At the European Union (EU) level, the European Commission addressed the concerns about mercury in 2005 in the "Community Strategy Concerning Mercury" aimed at the reduction of the negative impact of mercury and the risks it poses for the environment and human health [5]. In addition, the EU has supported a number of research projects focused on mercury impacts, as well as on different Hg abatement options.

Combustion of fossil fuels, particularly coal, nowadays represents one of the key sources of anthropogenic emissions of Hg worldwide. In Europe, coal-based power generation contributes the most to the annual emission loads [1]. Therefore, it is very relevant to examine the extent to which the burden from mercury emissions might be reduced due to strategies aiming at a large-scale replacement of coal and other fossil energies with emerging renewable energy sources. This paper presents the methodology and results of the impact assessment of renewable energy policies for the atmospheric emissions of mercury in Europe. The modeling exercise described here was carried out within the EU's Earth Observation for Monitoring and Assessment of the Environmental Impact of Energy Use (EnerGEO) project and involves the interaction of several models tracing the mercury from its sources to its deposition over the European continent.

In this analysis, emissions of mercury are calculated for each country in Europe, including members of the EU and non-EU states, as well as the European part of the Russian Federation. Emission profiles are computed in five-year steps for the next four decades, and the results are highlighted for the years 2005, 2030 and 2050. Mercury emissions through 2050 from fuel combustion and from industrial processes are computed first for the Baseline scenario, which assumes the continuation of current climate and air-quality policies in each country in Europe. The time evolution of mercury releases and its dispersion in the Baseline is then contrasted with a scenario that assumes a rapid penetration of renewable energy sources in the power generation sector.

The paper is structured as follows: the next section describes the modeling framework employed in the scenario analysis. Key features of individual models are provided here together with the explanation of the data flow and the inter-linkages between the four modeling tools involved. Section 3 describes the basic input parameters and data used in the models, as well as the main assumptions behind the mercury scenarios. The results of the impact assessment of renewable energy policies

273

are discussed in Section 4, highlighting the major outcomes of computer simulations. Finally, Section 5 presents conclusions and the implications of the analyses.

2. Modeling Framework

A comprehensive assessment of policies in favor of the substantial deployment of renewables for the emissions of mercury involves a chain of modeling tools that enable one to develop a set of policy-driven emission scenarios and to quantify the spatial distribution of pressures, *i.e.*, the concentration and deposition of mercury over the European territory. The underlying temporal changes in the fuel mix for power generation in individual European countries have been developed by the REMix (Renewable Energy Mix for Sustainable Electricity Supply in Europe) model. REMix simulates different levels of the penetration of renewable energies in a set of energy scenarios, which constitute an input to the GAINS (Greenhouse Gas and Air Pollution Interactions and Synergies) integrated modeling tool. Projections of power generation by source, combined with information on the evolution of emission control measures, are used in GAINS to compute mercury emissions through 2050.

As illustrated in Figure 1, emission scenarios serve as an input to the TNO_MACC-II (Monitoring Atmospheric Composition and Climate) module that spatially disaggregates mercury emissions from individual sectors for selected time periods into the detailed emission source-categories and distributes Hg releases to the atmosphere over the grid with a spatial resolution of 7 by 7 km. Thereafter, spatially distributed emission levels are used to model the atmospheric dispersion of mercury with the use of an extended chemistry transport model implemented on the air quality system, Polyphemus. In this step, future changes in the deposition and concentration of Hg are estimated, while taking into account contributions from natural sources and from mercury sources beyond the European domain. The basic features of the models involved in the assessment are described below.

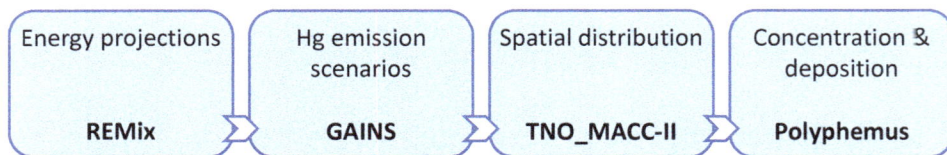

Energy projections	Hg emission scenarios	Spatial distribution	Concentration & deposition
REMix	**GAINS**	**TNO_MACC-II**	**Polyphemus**

Figure 1. Data flow between models used for the mercury assessment.

2.1. REMix

The REMix model analyses electricity generation potentials in Europe and optimizes renewable electricity generation with demand by calculating the cost-effective electric power supply options. REMix is a combination of a high resolution inventory of renewable electricity generation potentials (solar, wind, geothermal energy, biomass, hydro) and electricity demand data with a linear optimization model (50 nodes). It calculates the least cost renewable energy mixes to meet the defined shares of the European electricity demand under given constraints. Resource data with a spatial resolution of ~10 × 10 km and a temporal resolution of 1 h (for solar and wind energy) are input to the inventory [6].

Maps of land use types allowing for sustainable use of resources are generated using a Geographic Information System (GIS). In combination with power plant models, the hourly electricity generation potentials of each source are calculated and provided as input to the linear optimization model. The results of the optimization runs are least-cost electricity generation and transmission structures in Europe under given constraints, provided for each country in Europe and some neighbor countries in the Middle East and North Africa regions. REMix is designed for investigations of the time period between 2005 and 2050 [7].

2.2. GAINS

The GAINS model quantifies emission control potentials and costs for exogenous activity projections considering the physical and economic interactions between pollutants. The energy use in all major economic sectors and 20 fuel types is considered. Besides, the model uses projections of activities in energy-intensive industries and in agriculture. More than 2,000 technologies for air pollution mitigation and at least 500 options to control greenhouse gases (GHG) are included. The model computes emissions of major GHGs, air pollutants, as well as mercury. GAINS analyzes the cost-efficiency of policies and measures to meet air quality and GHG targets, covering the period 1990–2050 [8].

In this work, the implementation of Hg scenarios in GAINS takes advantage of a detailed bottom-up representation of air pollution abatement measures and policies in each country, while being complemented with current legislation on mercury control [9]. Projecting mercury emissions is, however, associated with numerous complexities, since the future Hg levels result from interactions of a range of determinants, measures and policies that simultaneously address multiple pollutants (e.g., particulates or sulfur) and different environmental objectives (e.g., acidification, exposure to fine particles). It is known that alongside measures dedicated to Hg capture, most air pollution control devices (APCD) are able to co-control mercury also to a certain extent [10]. Therefore, the amplification effect of multiple controls is considered in the computation algorithm: the application rate of Hg removing APCDs is derived as an overlap of rates (x) for individual technologies controlling different pollutants. Figure 2 provides a schematic example of the GAINS multi-pollutant framework, whereby an overlap of two technologies—one abating sulfur dioxide (SO_2) and the second abating particulate matter (PM)—is computed as a minimum of application rates if both technologies are used in parallel in a given sector, e.g., a coal-fired power plant. Finally, mercury emissions in GAINS are computed at the level of the three most important forms: elemental (Hg^0), reactive (Hg^{II}) and particulate-bound (Hg_P) mercury. The development of shares of Hg species in future emissions is particularly relevant, due to the differences in their lifetime and atmospheric transport. Speciated emissions for individual countries provide inputs for dispersion modeling tools, such that Hg concentration and deposition levels can be calculated.

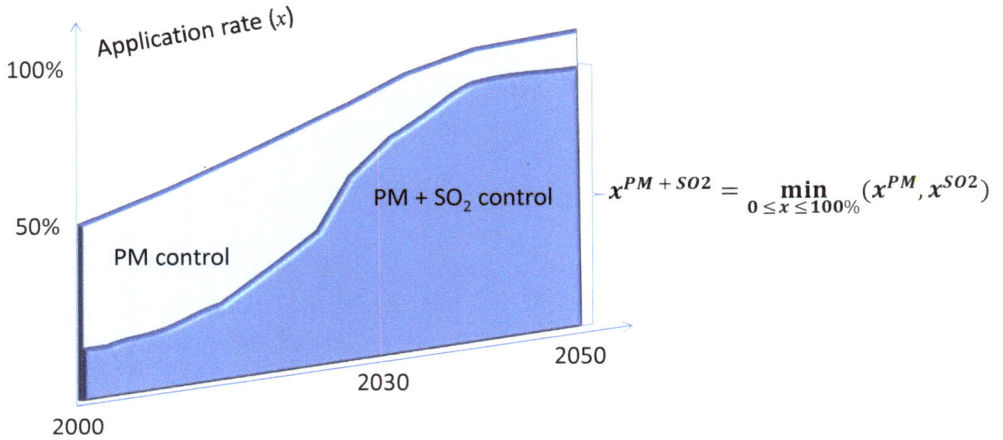

$$x^{PM+SO2} = \min_{0 \leq x \leq 100\%}(x^{PM}, x^{SO2})$$

Figure 2. Illustration of a multi-pollutant technology approach for Hg control in GAINS.

2.3. System for Spatial Distribution of Emissions

For the assessment of concentrations and the deposition of mercury, a correct spatial distribution of emissions over the grid is essential. The gridding procedure, which has been developed within the TNO_MACC-II inventory [11], is used. The spatial distribution system first aggregates the emissions by GAINS activity and sector to a set of 75 emission categories, which are distributed using different proxy parameters, distinguishing area and point sources. For point sources, the system utilizes its own database of point sources, but also The European Pollutant Release and Transfer Register (E-PRTR) is used as a data source for selected countries and pollutants. However, for mercury, this source is not used, because emission estimates from individual plants are not considered to be accurate and consistent enough. Our own database on point sources is set up using, among others, information from commercial databases, such as the Platts database for power plants [12], where capacity is assumed to be proportional to emission strength. The point source database is used as a fractional distribution map for distributing aggregated emissions from the GAINS model, *i.e.*, the point source emissions are first normalized,

276

adding up to 100% for each country, and then, all percentages are multiplied with the emission total for that country for a given pollutant, sector and year. Area sources are distributed using proxy parameters, at a spatial resolution of $1/8° \times 1/16°$ longitude-latitude (approximately 7×7 km). The proxy parameters used include, e.g., population and arable land, but also road, rail and shipping networks. The proxy parameters selected for the distribution of mercury emissions are similar to those selected for the other pollutants. For more information and a complete overview of the spatial distribution system, we refer to Kuenen et al. [11]. A more extensive description of the spatial distribution methodology can also be found in van der Gon et al. [13], which describes an earlier version of the spatially distributed inventory for the year 2005. The calculations are performed using a Structured Query Language (SQL) server platform to reduce computation time. The resulting gridded emissions are aggregated to the 13 source categories to reduce the file size of the output files. The spatial distribution system has been developed for the year 2005. For future years, due to the lack of reliable information, the spatial distribution key is assumed not to change, i.e., reductions by sector are applied evenly to all grid cells within each of the 75 emission categories.

2.4. Polyphemus

Polyphemus is an air quality modeling system, which was used in this study to model atmospheric dispersion of the main mercury forms (Hg^0, Hg^{II}, Hg_p) and species (HgO, HgC1, etc.) over Europe [14]. Its main element is an Eulerian chemistry-transport-model, Polair3D, used for both gaseous and aerosol species. Polair3D tracks multiphase chemistry: (i) gas, (ii) water and (iii) aerosols. Polyphemus is also composed of a library of physical parameterizations, called AtmoData, and a set of programs using AtmoData designed to generate data required by Polair3D, e.g., deposition velocities, vertical diffusion coefficients, emissions, etc. Polyphemus was equipped with a new chemical scheme dedicated to the atmospheric chemistry of mercury, which takes into account the main reactions and processes in the gaseous, aqueous and particulate phases, as presented in Figure 3. This scheme is an upgraded version of the one introduced in Roustan et al. [15]. Its main developments are related to the reactions of mercury with bromine. Additionally, particulate mercury is distributed among 10 different size sections (diameter between 0.01 to 10 µm).

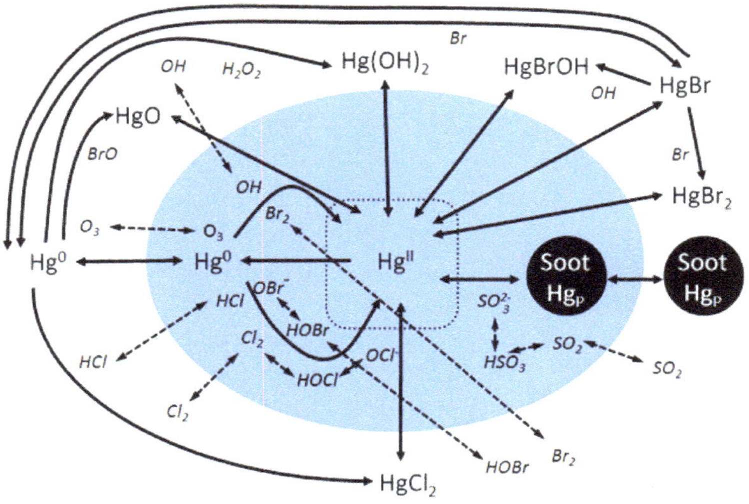

Figure 3. The chemical model for mercury implemented in Polyphemus. Gaseous and aqueous Hg phases are marked by white and blue, respectively. The line arrows show possible transformations of mercury, and the dashed arrows show additional species used in the model that react with mercury.

The dry deposition velocity for gaseous compounds is calculated based on the model and parameters presented in Zhang *et al.* [16,17], whereas for particulate species it is calculated based on [18]. The wet deposition is split between in-cloud and below cloud scavenging with the use of parameterization provided by Binkowski *et al.* [19], Sportisse *et al.* [20], and Seinfeld *et al.* [21]. The validation of the model can be found in Zyśk *et al.* [22,23]. The modeling domain consists of 120×140 cells with a horizontal resolution of $0.5° \times 0.25°$ (along longitude and latitude, respectively). Ten vertical levels are used with the following limits (in meters above the surface): zero; 70; 150; 300; 500; 750; 1,000; 2,000; 3,000; 5,000. The ECMWF (European Centre for Medium-Range Weather Forecasts) data for meteorological parameters and EMEP (European Monitoring and Evaluation Programme) data for natural emissions and reemissions are used. The initial and boundary conditions were set to 0.0012 ppt for HgO, Hg(OH)$_2$, HgCl$_2$, Hg$_P$ and 0.185 ppt for Hg^0.

3. Assumptions for Mercury Scenarios

Two Hg emission scenarios are presented in this paper: (1) the Baseline scenario (BAS), which assumes the development of the European energy system with no measures to control the emissions of GHGs and to deploy renewable electricity sources going beyond the current policies; and (2) the Maximum Renewable Power (MAX) scenario that assumes the decarbonization of the European energy system,

278

including the highest possible electricity (ELE) generation from renewables by 2050. Both scenarios imply a full implementation of recent national legislation on air quality (*i.e.*, policies that were in force or in the final stage of legislative processes as of mid-2012) by 2030, but not strengthening it further between 2030 and 2050.

3.1. Combustion Sources of Mercury

Figure 4 depicts the developments of primary energy demand in the Baseline and the Maximum Renewable Power scenarios. The Baseline assumes an increase of the energy use in Europe by about 7% until 2050, which is a combined effect of a decrease in the EU Member States by 6% and an increase in the rest of Europe by 38%. Even in the Baseline, the share of fossil fuels decreases from 81% in 2005 to 67% by 2050. The share of biomass and other renewable energy increases from 6% to 16%. The Maximum Renewable Power scenario results in a much lower energy demand in 2050 relative to the Baseline. In the EU countries, this demand decreases by more than 30%, and the aggregated reduction for other Europe countries is about 40%. This is a combined effect of a decrease of the demand for final energy, due to a faster implementation of energy efficiency measures and due to the higher share of non-combustion renewable energy sources in power generation.

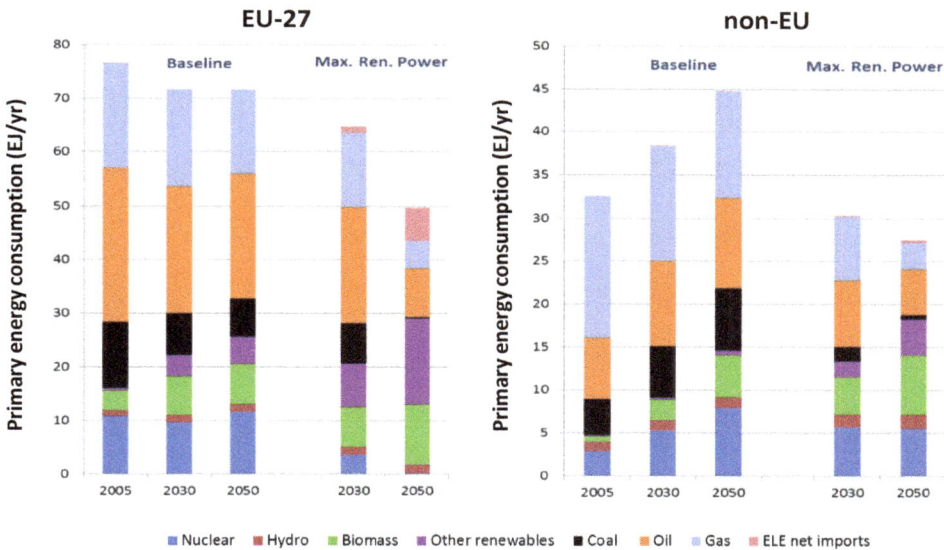

Figure 4. Demand for primary energy in EU-27 and non-EU regions in the Baseline and the Maximum Renewable Power scenarios. Source: REMix and GAINS.

The share of renewables in the primary energy consumption, as projected by the REMix model, increases up to 2050 in the EU countries to 58% and to 46% in the non-EU countries. In addition, more than one fourth of electricity demand is met through imports of solar-based electricity from Africa in the MAX scenario by 2050. For the EU-27, the contribution of solar-power imports to the power generation fuel mix reaches 34%. In this way, the electricity supply for the EU-27 is practically carbon free; only less than 5% originates from coal and gas.

3.2. Non-Combustion Sources of Mercury

Non-combustion Hg sources comprise industrial processes, such as refineries, iron and steel production, non-ferrous metals smelting, the production of non-metallic minerals and chemical products. Projections of activities in industrial processes in the EU-27 are based on inputs from the reference- and low-carbon-scenarios reported by CEC [24]. Estimates for non-EU countries rely on the activities provided by Russ *et al.* [25] or they are derived from trends in macroeconomic drivers (e.g., gross domestic product, value added in individual sectors). The growth in industrial production relative to the present is significantly faster in the non-EU region, as compared to the EU-27. Climate policies, however, result in reduced growth rates for industrial processes in all European states, due to assumptions on higher energy prices, rapid implementation of efficiency measures and overall economic restructuring.

Additional mercury sources of air emissions covered in GAINS include chlorine and gold production, waste incineration and dental-mercury emissions (from cremations). In recent years, the mercury cell capacity in chloralkali plants has been gradually reduced through closures or conversion to non-Hg technologies [26]. This process was driven by environmental regulations and cost reasons. Therefore, we assume that chloralkali facilities using mercury will be phased out in Europe by 2030. Industrial gold mining occurs in a number of European countries. Activity projections for the gold mining sector are derived from the extrapolation of recently observed trends up to 2030 [27], and the production is kept constant thereafter. Estimates of Hg emissions from dental sources are based on the future population growth, annual mortality rates and from the recent trends in the share of cremations [28]. There are few sectors and potential Hg sources in Europe (e.g., intentional mercury use in batteries, lamps or other devices) not modeled explicitly, due to high uncertainties and data scarcity. This study assumes that the activities in Hg-specific sectors listed in this paragraph are not affected by the climate policies.

3.3. Emission Factors and Control Options for Mercury

Uncontrolled emission factors are derived from literature sources on mercury contents in combustible fuels or wastes and from estimates on Hg impurities in raw materials used in production processes. Emissions of mercury in flue gases are estimated in GAINS for each of the major Hg forms by taking into account the removal efficiency and application rate of a wide range of control measures. The model also takes into account the retention of mercury in fly ash. Changes in the mercury speciation in flue gases due to pollution controls are reflected by using the inlet and outlet composition factors. The main parameters applied for the calculation of Hg emissions in Europe are reported in Table 1.

Table 1. Summary of the parameters and characteristics of control options used for Hg calculation in Europe.

Activity Types		Emission Factors[2] (grams Hg/activity)	Speciation-Inlet (%) Hg^0-HG^{II}-Hg_P	Control Measures[3]	Removal Efficiency (%) Hg^0-HG^{II}-Hg_P
Combustion	Hard coal[1]	0.001–0.009 (g/PJ)	55-35-10	CYC	0-0-70
				WSCR	10-40-85
	Lignite[1]	0.007–0.048 (g/PJ)	60-30-10	ESP	15-30-95
	Biomass	0.001 (g/PJ)	50-40-10	ESP + FGD + SCR	35-80-99
	Oil products	0.0001–0.001 (g/PJ)	50-40-10	FF	45-60-99
	Waste	0.6–1 (g/t)	20-60-20	FF + FGD + SCR	60-95-99
	Cremation	2.5 (g/corpse)	80-15-5	FF + FGD + SCR + SI	95-99-99
				GP	70-70-70
Metallurgy	Iron and steel	0.04 (g/t)	80-15-5	CYC	0-0-70
	Non-ferrous metals	0.01–5 (g/t)	80-15-5	ESP	15-30-95
	Industrial gold mining	25 (g/kg)	80-15-5	ESP + FGD + SCR	30-50-99
				FF + FGD + SI	95-99-99
Processes	Oil refineries	0.001–0.02 (g/t)	80-15-5	CYC	0-0-70
	Cement and lime	0.035 (g/t)	80-15-5	ESP	15-30-95
	Other bulk products	0.001 (g/t)	80-15-5	ESP + FGD + SCR	30-50-99
	Chloralkali production	2.5 (g/t Cl)	70-30-0	GP	30-30-0
				SI	95-99-0

Notes: [1] Hg retention in ash 1–17%; [2] ranges reflect regional differences in Hg contents in coal and other fuels (considering calorific values and import/export patterns). [3] Acronyms: CYC, cyclone; ESP, electrostatic precipitator; FF, fabric filter; FGD, flue gas desulfurization; GP, good practices; SCR, selective catalytic reduction; SI, sorbent injection; WSCR, wet scrubber. Further details and data sources provided in Rafaj *et al.* [9].

There are two types of control measures for mercury considered in GAINS. The first set of measures includes "conventional" APCDs, which reduce mercury as a side effect of their operation. As discussed in Section 2.2, the removal efficiency of APCDs for Hg is in most cases reinforced if they are adopted simultaneously. The second set contains technologies directly dedicated to the capture of mercury. The Hg-specific primary abatement measures, such as sorbent injection, are associated with the model with relevant sectors, e.g., waste incineration, crematories or chloralkali production. The implementation rates of air pollution and Hg control strategies are based on recent national legislation, *i.e.*, policies that were in force or in the final stage of the legislative process as of mid-2012 [29].

4. Simulation Results

In 2005, the European emissions of mercury from anthropogenic sources totaled 145 tones (this value lies within a range for respective regions estimated by UNEP [1]), and the future trend up to 2050 is fairly flat. Although the Hg emissions in Europe would have been 35% to 45% higher without the co-control effects of technologies abating other air pollutants, APCDs assumed in the Baseline scenario are insufficient to prevent future emission growth in some of the non-EU countries. Stringent decarbonization policies, as defined in the Maximum Renewable Power scenario, induce overall reductions in Hg emissions of nearly 45 tons in 2050 relative to the Baseline. Cumulatively, the co-benefits from low carbon policies, including a rapid deployment of renewable power sources, are quantified at 1.2 kilotons of avoided mercury emissions in the period 2020–2050 (see the country and sectorial inventory, Tables S1 and Table S2 in the Supplementary Material). Figure 5 (left panel) illustrates the share of remaining emissions by sector in the MAX scenario and indicates the size of Hg avoided due to climate measures. The largest fraction (63%) of emission cuts brought about in the MAX scenario is attributed to the deployment of renewables in the power sector, while the remainder of the reductions originates from changes in industrial activities. Detailed scenario-analysis suggests an increase in Hg emissions over the Baseline from biomass and waste combusting facilities, because of a larger consumption of these fuels for energy purposes.

As is shown in the right panel of Figure 5, emission reductions achieved under the low-carbon policies are very country-specific and depend on a number of factors, e.g., the amount and quality of the coal used for combustion in power and industry sectors in the Baseline, the timing of the adoption of air pollution controls, the characteristics of abatement technologies used in individual countries, *etc.* In general, countries relying on coal as the dominant energy source (for example, Poland or Turkey) are likely to cut their Hg emissions deeper when compared to regions with a more diversified fuel mix.

The emissions of mercury from the power sector already decline in the EU-27 without climate policies (Figure 6). Baseline emissions in 2030 are halved relative to today's levels, and by 2050, they decrease by a factor of four. On the contrary, Hg releases from electricity generation in the non-EU region are twice as high as in the year 2005. The dominant source of mercury from power plants in both regions is the combustion of coal. The contribution of liquid fuels to Hg emissions is negligible. A rapid elimination of coal from the fuel mix, as in the MAX scenario, causes substantial mercury reductions. In EU-27, climate strategies prevent up to 12 tons of Hg from being emitted in 2050. This effect is even more pronounced in the non-EU region, where the switch from coal to renewables causes a Hg decline by 80% below the emissions in 2005. One of the fuels that replaces coal is biomass. Although the mercury content of biomass is small by comparison to coal or lignite,

282

the growth in biomass (and partly waste) combustion results in higher Hg emissions from this source. The corresponding increase in 2050 over the Baseline is 80% for the EU-27 and 15% for the non-EU countries.

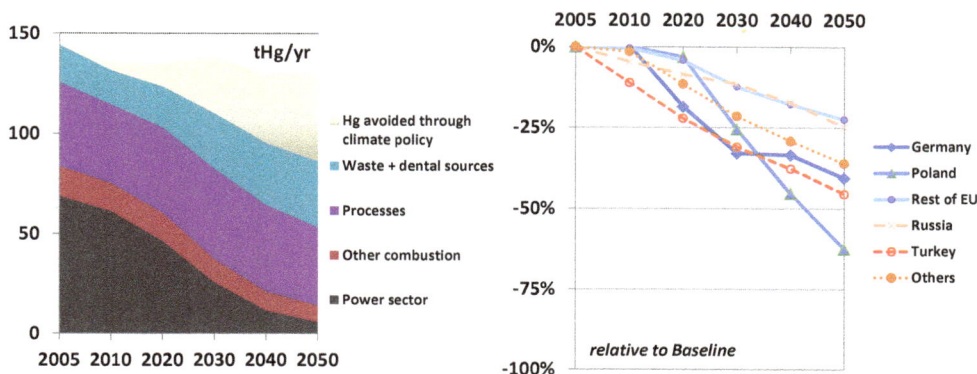

Figure 5. Maximum Renewable Power scenario: mercury emissions in Europe by sector (**left panel**) and reductions for selected regions (**right panel**).

The final step in the scenario assessment of mercury involves the dispersion modeling of spatially-distributed emissions with the use of a chemistry transport model implemented on thePolyphemus platform. First, the results are presented in Figure 7 for the year 2005 in the form of maps showing dry and wet deposition of mercury over the European domain. The majority of the modeled area remains below the deposition levels of $50 \text{ g·km}^{-2}\text{·yr}^{-1}$; nevertheless, there are regions with elevated deposition reaching $80 \text{ g·km}^{-2}\text{·yr}^{-1}$ (Poland, NW Germany and Northern Spain). The observed differences in total mercury deposition in Europe are mainly due to the differences in the deposition of reactive gaseous mercury (Hg^{II}) and mercury bounded in aerosols (Hg_p). Reactive gaseous mercury (Hg^{II}) and mercury bounded in aerosols (Hg_p) are good indicators of coal combustion, because these forms are dispersed in the atmosphere locally, and their deposition strongly depends on local sources. This also explains the spatial feature of the deposition difference, which implies the highestHg abatement potential in areas where coal is used at present.

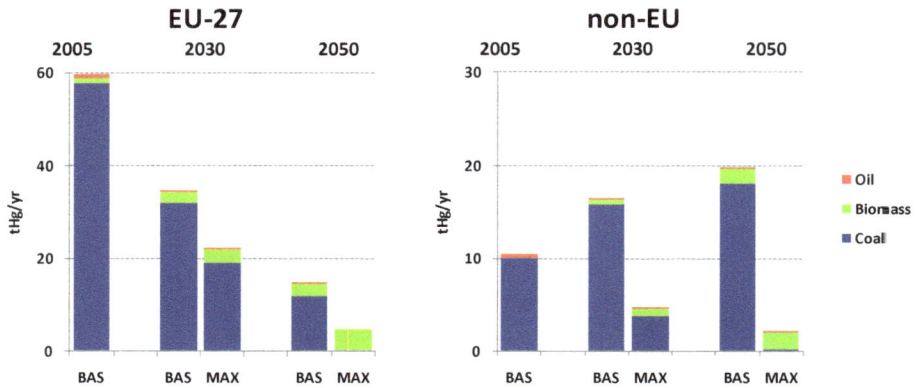

Figure 6. Mercury emissions from the power sector by fuel in the EU-27 and non-EU countries.

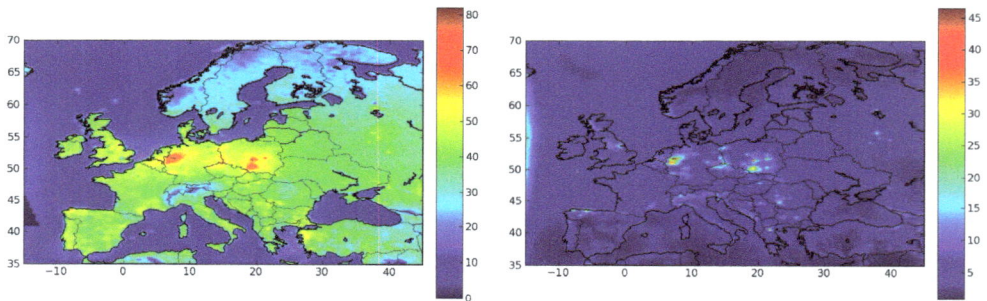

Figure 7. (**Left panel**) Dry deposition of mercury over Europe in 2005. (**Right panel**) Wet deposition of mercury over Europe in 2005. Units: $g \cdot km^{-2} \cdot yr^{-1}$.

As shown in Figure 8 (left panel), there are significant spatial differences expected in overall Hg deposition (wet and dry) in the Baseline between the years 2005 and 2030. In some regions (e.g., The Czech Republic, Germany, Poland), the deposition decreases more than $10 \ g \cdot km^{-2} \cdot yr^{-1}$, whereas an increase is projected in other parts of Europe (Italy, Russia and Turkey). The performed simulations provide insights about future changes in concentrations and depositions of various forms of mercury over Europe as a consequence of policies favoring renewable power sources. The right panel of Figure 8 suggests that fuel mix changes assumed in the MAX scenario in 2030 might induce a reduction in the Hg deposition of $20 \ g \cdot km^{-2} \cdot yr^{-1}$ in some areas (Germany, Poland and Turkey).

Finally, computer simulations allow for the quantification of shares of Hg from natural sources, as well as global and regional anthropogenic sources of Hg deposition over the European territory. Taking Poland as an example, the local anthropogenic sources contributed 30% to the total Hg deposition in 2005. The contribution of anthropogenic emissions originating in other parts of Europe to the total deposition in Poland is estimated at about 5%. The reminder (60%–70%, depending on the period of the year) is attributed to anthropogenic emissions from non-European sources, natural emissions and re-emissions [22].

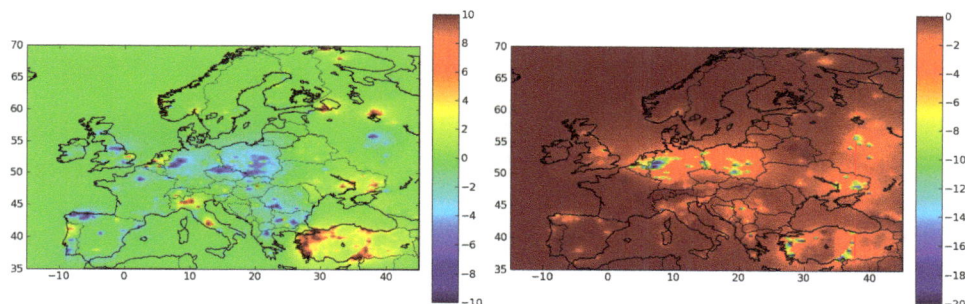

Figure 8. (**Left panel**) The change in the Hg deposition over Europe between 2005 and 2030 for the Baseline scenario. Positive numbers indicate an increase and negative ones a reduction in the deposition levels. (**Right panel**) The difference in the Hg deposition over Europe in 2030 between the Baseline and the Maximum Renewable Power scenarios. Units: $g \cdot km^{-2} \cdot yr^{-1}$.

5. Conclusions

The change in mercury emissions caused by switching from fossil fuels to carbon-free energy sources is an important indicator for the assessment of renewable energy policies. The results of the comprehensive modeling exercise presented in this paper provide an estimate of the effectiveness of the decarbonization of the energy sector and, in particular, of power generation for mercury mitigation. The most important findings and conclusions from this analysis are:

- Current emissions of mercury in Europe are about 145 tons per year and are likely to remain at the same level under the baseline conditions. Coal and lignite combustion for power generation are the largest sources, contributing 60% to the total.
- The operation of air pollution control devices required by the current standards to control air quality (dust, sulfur and nitrogen oxides emissions) reduces the emissions of mercury by about 35%. However, without additional climate measures, pollution controls will have to be significantly tightened in some non-EU countries to reverse the growing trend in Hg emissions by mid-century.
- Decarbonization of the energy system and, in particular, renewable electricity deployment brings extensive co-benefits for mercury abatement. European Hg

285

emissions under the low-carbon policies decrease by one third in 2050 relative to the Baseline, whereby about two thirds of that reduction is achieved in power sector.

- The potential for mercury reduction in each country through implementing renewable electricity generation options depends on the rate of fuel switches and renewable technology deployment, but is also influenced by the stringency and timing of the air quality measures. The overall scope for co-benefits is therefore higher in regions with a high share of coal use and with less stringent policies to control air quality.
- While a rapid decrease in mercury releases from coal-fired power plants under low-carbon strategies is expected, the Hg emissions from biomass and waste burning might slightly increase, due to a larger amount of these fuels utilized in the energy sector.
- Local and regional sources constitute only a fraction of mercury loads deposited over Europe. Due to the long range transport of elemental mercury in the atmosphere, effective strategies for mitigating mercury contamination require global coordinated actions.

Acknowledgments: Research presented here has been carried out within the EnerGEO project (www.energeo-project.eu), funded by the European Community's Seventh Framework Programme.

Author Contributions: All authors designed the modeling experiments, analyzed the data, and contributed extensively to the work presented in this paper. P.R. and J.C. developed the Hg-emission scenarios. J.K. performed the spatial distribution of emissions. A.W. and J.Z. carried out the dispersion modeling of mercury. P.R. prepared the manuscript. All authors discussed the results and implications and commented on the manuscript at all stages.

Conflicts of Interest: The authors declare no conflict of interest.

References and Notes

1. UNEP. *Global Mercury Assessment 2013: Sources, Emissions, Releases and Environmental Transport*; United Nations Environment Programme (UNEP) Chemicals Branch: Geneva, Switzerland, 2013.
2. Munthe, J.; Bodaly, R.A. (Drew); Branfireun, B.A.; Driscoll, C.T.; Gilmour, C.C.; Harris, R.; Horvat, M.; Lucotte, M.; Malm, O. Recovery of Mercury-Contaminated Fisheries. *AMBIO* **2007**, *36*, 33–44.
3. Rice, D.C. Overview of modifiers of methylmercury neurotoxicity: Chemicals, nutrients, and the social environment. *NeuroToxicology* **2008**, *29*, 761–766.
4. UNEP. *Report of the Governing Council, Twenty-Fifth Session (16-20 February 2009), General Assembly, Supplement No. 25*; United Nations Environment Programme: New York, NY, USA, 2009.

5. CEC. *Communication "Community Strategy concerning Mercury"*; Commission of the European Communities: Brussels, Belgium, 2005.

6. Nitsch, J.; Pregger, T.; Scholz, Y.; Naegler, T.; Heide, D.; Tena, D.L.; Trieb, F.; Nienhaus, K.; Gerhardt, N.; Oehsen, A.; *et al. Long-Term Scenarios and Strategies for the Deployment of Renewable Energies in Germany in View of European and Global Developments*; DLR: Stuttgart, Germany, 2012.

7. Scholz, Y. *Renewable Energy Based Electricity Supply at Low Costs—Development of the REMix Model and Application for Europe*; Universität Stuttgart: Stuttgart, Germany, 2012.

8. Amann, M.; Bertok, I.; Borken-Kleefeld, J.; Cofala, J.; Heyes, C.; Höglund-Isaksson, L.; Klimont, Z.; Nguyen, B.; Posch, M.; Rafaj, P.; *et al.* Cost-effective control of air quality and greenhouse gases in Europe: modeling and policy applications. *Environ. Model. Softw.* **2011**, *26*, 1489–1501.

9. Rafaj, P.; Bertok, I.; Cofala, J.; Schöpp, W. Scenarios of global mercury emissions from anthropogenic sources. *Atmos. Environ.* **2013**, *79*, 472–479.

10. Pacyna, J.M.; Sundseth, K.; Pacyna, E.G.; Jozewicz, W.; Munthe, J.; Belhaj, M.; Aström, S. An assessment of costs and benefits associated with mercury emission reductions from major anthropogenic sources. *J. Air Waste Manag. Assoc.* **2010**, *60*, 302–315.

11. Kuenen, J.J.P.; Visschedijk, A.J.H.; Jozwicka, M.; van der Gon, H.D. A multi-year consistent high-resolution European emission inventory for air quality modelling. *Atmos. Chem. Phys.* **2013**, in press.

12. Platts. In *World Electric Power Plants Database*; Global Market Data and Price Assessments; Platts, UDI Products Group: Washington, DC, USA, 2013.

13. Van der Gon, H.D.; Visschedijk, A.; van de Brugh, H.; Dröge, R. *A High Resolution European Emission Data Base for the Year 2005. A Contribution to UBA-Projekt PAREST: Particle Reduction Strategies*; TNO Report TNO-034-UT-2010-01895_RPT-ML; TNO: Utrecht, The Netherlands, 2010.

14. Mallet, V.; Quélo, D.; Sportisse, B.; Ahmed de Biasi, M.; Debry, É.; Korsakissok, I.; Wu, L.; Roustan, Y.; Sartelet, K.; Tombette, M.; *et al.* Technical note: The air quality modeling system Polyphemus. *Atmos. Chem. Phys.* **2007**, *7*, 5479–5487.

15. Roustan, Y.; Bocquet, M.; Musson, G.L.; Sportisse, B. *Modeling Atmospheric Mercury at European Scale with the Chemistry Transport Model Polair 3D*; National Environmental Research Institute: Copenhagen, Denmark, 2005.

16. Zhang, L.; Brook, J.R.; Vet, R. A revised parameterization for gaseous dry deposition inair-quality models. *Atmos. Chem. Phys.* **2003**, *3*, 2067–2082.

17. Zhang, L.; Wright, L.P.; Blanchard, P. A review of current knowledge concerning dry deposition of atmospheric mercury. *Atmos. Environ.* **2009**, *43*, 5853–5864.

18. Zhang, L.; Gong, S.; Padro, J.; Barrie, L. A size-segregated particle dry deposition scheme for an atmospheric aerosol module. *Atmos. Environ.* **2001**, *35*, 549–560.

19. Binkowski, F.S.; Roselle, S.J. Models-3 Community Multiscale Air Quality (CMAQ) model aerosol component 1. Model description. *J. Geophys. Res.: Atmos.* **2003**, *108*, 3:1–3:18.

20. Sportisse, B.; du Bois, L. Numerical and theoretical investigation of a simplified model for the parameterization of below-cloud scavenging by falling raindrops. *Atmos. Environ.* **2002**, *36*, 5719–5727.

21. Seinfeld, J.H.; Pandis, S.N. *Atmospheric Chemistry and Physics: from Air Pollution to Climate Change*; John Wiley & Sons: New York, NY, USA, 2012.

22. Zyśk, J.; Wyrwa, A.; Pluta, M.; Roustan, Y.; Rafaj, P.; Kuenen, J.; Drebszok, K. Modelling of Atmospheric Dispersion of Mercury for Energy Scenarios of the EnerGEO Project. In *Mercury in the environment—identification of hazards to human health*; Falkowska, L., Ed.; Uniwersytet Gdański: Gdańsk, Poland, 2013.

23. Zyśk, J. Modelling of Transport of Mercury in the Atmosphere. In Proceedings of the VII National Scientific Conference "Energy-Environment-Ethics"; Krakow, Poland, 2013.

24. CEC. *A Roadmap for Moving to a Competitive Low Carbon Economy in 2050*; Commission of the Europan Communities: Brussels, Belgium, 2011.

25. Russ, P.; Ciscar, J.-C.; Saveyn, B.; Soria, A.; Szabo, L.; Van Ierland, T.; Van Regemorter, D.; Virdis, R. *Economic Assessment of Post-2012 Global Climate Policies*; Joint Research Centre of the European Community: Seville, Spain, 2009.

26. UNEP. *Global Mercury Cell Production Data*; Global Mercury Partnership Advisory Group; United Nations Environment Programme: New York, NY, USA, 2010.

27. USGS. *2010 Minerals Yearbook—Gold*; Mineral Commodity Summaries; US Geological Survey: Reston, VA, USA, 2012.

28. CremSoc. *International Cremation Statistics 2010*; The Cremation Society of Great Britain: Kent, UK, 2010.

29. Amann, M.; Borken-Kleefeld, J.; Cofala, J.; Heyes, C.; Klimont, Z.; Rafaj, P.; Purohit, P.; Schoepp, W.; Winiwarter, W. *Future Emissions of Air Pollutants in Europe—Current Legistation Baseline and the Scope for Futher Reductions. TSAP Report #1*; International Institute for Applied Systems Analysis: Luxemburg, Austria, 2012.

MDPI AG

St. Alban-Anlage 66

4052 Basel, Switzerland

Tel. +41 61 683 77 34

Fax +41 61 302 89 18

http://www.mdpi.com

Atmosphere Editorial Office

E-mail: atmosphere@mdpi.com

http://www.mdpi.com/journal/atmosphere